Pleas
belov
Bo
a

CANCELLED

13 JUN 1995

13 DEC 1

For con

£10-50

SEED BIOLOGY

PHYSIOLOGICAL ECOLOGY

A Series of Monographs, Texts, and Treatises

EDITED BY

T. T. KOZLOWSKI

University of Wisconsin
Madison, Wisconsin

T. T. KOZLOWSKI. Growth and Development of Trees, Volumes I and II — 1971

DANIEL HILLEL. Soil and Water: Physical Principles and Processes, 1971

J. LEVITT. Responses of Plants to Environmental Stresses, 1972

V. B. YOUNGNER AND C. M. McKELL (Eds.). The Biology and Utilization of Grasses, 1972

T. T. KOZLOWSKI (Ed.). Seed Biology, Volumes I and II — 1972; Volume III — in preparation

YOAV WAISEL. The Biology of Halophytes, 1972

SEED BIOLOGY

Edited by

T. T. KOZLOWSKI

DEPARTMENT OF FORESTRY
UNIVERSITY OF WISCONSIN
MADISON, WISCONSIN

VOLUME I

Importance, Development, and Germination

ACADEMIC PRESS New York and London   1972

ACADEMIC PRESS, INC.
111 Fifth Avenue, New York, New York 10003

United Kingdom Edition published by
ACADEMIC PRESS, INC. (LONDON) LTD.
24/28 Oval Road, London NW1 7DD

LIBRARY OF CONGRESS CATALOG CARD NUMBER: 71-182641

PRINTED IN THE UNITED STATES OF AMERICA

CONTENTS

v

3   Development of Angiosperm Seeds

S. P. BHATNAGAR AND B. M. JOHRI

4   Anatomical Mechanisms of Seed Dispersal

ABRAHAM FAHN AND ELLA WERKER

5   Seed Germination and Morphogenesis

GRAEME P. BERLYN

6   Seed and Seedling Vigor

BRUCE M. POLLOCK AND ERIC E. ROOS

CONTENTS

LIST OF CONTRIBUTORS

Numbers in parentheses indicate the pages on which the authors' contributions begin

S. P. BHATNAGAR (77), Department of Botany, University of Delhi, Delhi, India

GRAEME P. BERLYN (223), School of Forestry, Yale University, New Haven, Connecticut

ABRAHAM FAHN (151), Department of Botany, The Hebrew University of Jerusalem, Israel

C. R. GUNN (1), Botanist New Crops Research Branch, U. S. Department of Agriculture, Agricultural Research Service, Plant Science Research Division, Beltsville, Maryland

B. M. JOHRI (21, 77), Department of Botany, University of Delhi, Delhi, India

T. T. KOZLOWSKI (1), Department of Forestry, University of Wisconsin, Madison, Wisconsin

BRUCE M. POLLOCK (313), U. S. Department of Agriculture, Agricultural Research Service, Crops Research Division, Fort Collins, Colorado

ERIC E. ROOS (313), Department of Agriculture, Agricultural Research Division, National Seed Storage Laboratory, Fort Collins, Colorado

HARDEV SINGH* (21), Department of Botany, University of Delhi, Delhi, India

ELLA WERKER (151), Department of Botany, The Hebrew University of Jerusalem, Jerusalem, Israel

*Present address: Institut for Pflanzenpathologie und Pflanzenschutz, der Universität, Göttingen, Grisebachstr, Germany.

PREFACE

Man's existence and health are directly or indirectly dependent on seeds. This fact has for many years pointed out the urgent need for a comprehensive coverage of information on seed biology. The importance of this work became even greater during the recent years of rapid population increases throughout the world. It was with these thoughts in mind that this three-volume treatise was planned to bring together a large body of important new information on seed biology.

The subject matter is wide ranging. The opening chapter outlines man's dependency on seeds as sources of food and fiber, spices, beverages, edible and industrial oils, vitamins, and drugs. Harmful effects of seeds are also mentioned. Separate chapters follow on seed development, dissemination, germination (including metabolism, environmental control, internal control, dormancy, and seed and seedling vigor), protection from diseases and insects, collection, storage, longevity, deterioration, testing, and certification. These books were planned to be readable and interdisciplinary so as to serve the widest possible audience. They will be useful to various groups of research biologists and teachers, including agronomists, plant anatomists, biochemists, ecologists, entomologists, foresters, horticulturists, plant pathologists, and plant physiologists. The work has many practical overtones and will also be of value to seed producers and users.

These volumes are authoritative, well-documented, and international in scope. They represent the distillate of experience and knowledge of a group of authors of demonstrated competence from universities and government laboratories in England, India, Israel, South Africa, and the United States. I would like to express my deep personal appreciation to each of the authors for his contribution and patience during the production phases. The assistance of Mr. W. J. Davies and Mr. P. E. Marshall in index preparation is also acknowledged.

T. T. KOZLOWSKI

CONTENTS OF OTHER VOLUMES

1

IMPORTANCE AND CHARACTERISTICS OF SEEDS

T. T. Kozlowski and C. R. Gunn

I. Introduction

Seed-producing plants may be grouped into three taxa, Angiospermae, Gymnospermae, and Pteridospermae. The latter group is represented only by fossils from the early Carboniferous to the Cretaceous periods, with its zenith both in numbers and diversity in the late Carboniferous period. Pteridosperms were among the first plants to produce true ovules and seeds (Fig. 1). This type of seed, arising from an integumented female sporangium, is generally considered to be a significant structural ad-

1

vancement in the evolution of plants. The protection afforded by the surrounding integument and the presence of stored nutrients give seed-bearing plants a definite advantage over spore-bearing plants. This ad-

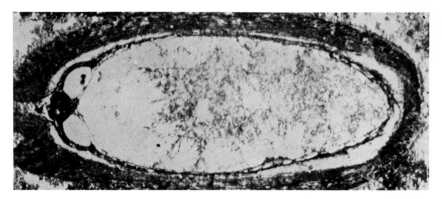

FIG. 1. A fossil pteridosperm seed cut lengthwise showing the micropyle (opening at the apex) and the conspicuous pollen chamber. [Reprinted from Andrews, 1965, (Fig. 88); copyrighted 1947 by Comstock Publishing Company, Inc., used by permission of Cornell University Press.]

vantage was eloquently summarized by Charles Darwin (F. Darwin, 1903) who mused that the sudden appearance in abundance of angiosperms in Cretaceous strata was "an abominable mystery." It still is. The discovery of the pteridosperms merely gave rise to more questions. For additional information on the evolution of the seed habit, the reader is referred to Thomson (1927).

The angiosperms with 220,000 extant species comprise a more successful group, both numerically and in area occupied, than the gymnosperms with 520 extant species. The basic differences between the seeds of the two taxa are that seeds of angiosperms are the product of double fertilization within an ovary, whereas seeds of gymnosperms are the product of a single fertilization and the ovary is absent. Detailed discussions of the formation of gymnosperm and angiosperm seeds are found in Chapters 2 and 3 of this volume.

Another essential feature that permitted seed plants to supersede spore-bearing plants in the Cretaceous period is the partial development of the embryo within the seed while the seed is attached to the mother plant. A noteworthy exception is the seed of *Ginkgo biloba* in which fertilization is long delayed after pollination, sometimes until after the "seed" has fallen to the ground. The fossil pteridosperm seeds may have been similar to ginkgo seeds, because they may not have possessed embryos while attached to the mother plant. No embryos have been found in these fossil seeds. Therefore, it may be inferred that these seeds had small or un-formed embryos when they were shed by the mother plant.

Angiosperms, of which the ancestors date back at least 135 million years, have increased in numbers of individuals and taxa, and have spread in range to become the dominant terrestrial plant group. They, along with the gymnosperms, have dominated spore-bearing plants in recent geo-logical ages because of their capacity to produce seeds. Such structures have a much greater chance to survive the vicissitudes of nature and to produce a new generation than do spores.

Seeds, connecting succeeding generations of spermatophytes, con-stitute a slender thread of life for these plants and are an important source of food for animals and man. We recognize this by using the term seed symbolically:

> From prehistoric times man has understood the role of seeds. Ancient languages, ancient cultures, and our own contain many words and concepts based on this under-standing. The Bible contains several such examples, including the parable of the sower, the use of the word seed to mean off-spring or progeny, and references to good and bad seed. Our language contains both common and technical terms involving seed, although the meanings are often quite unrelated to the subject of plants. The meanings recognize, however, some metaphoric connection in one way or another. Seed is a noun, an adjective, and a verb. Watermen speak of seed oysters, seed pearls, and seed fish. The optician speaks of seeds in glass. The chemist seeds a solution with a crystal to induce crystallization. We speak of the seed of an idea or a plan (Boswell, 1961).

II. Importance of Seeds as Foods

Committed for over a million years to a nomadic life, man settled down about 10,000 years ago when he learned to satisfy his hunger by growing food, especially seed foods. Seeds, the great staple food of the world, feed more people than does any other type of food. The endosperm or cotyle-dons with their rich food reserves for the developing embryo and seedling offer man and other animals a highly nutritious food that can be easily stored. Perhaps the story of Joseph (*Genesis,* 41:25–36) best illustrates the value of the storability of seeds. During the 7 years of plenty he super-vised the storage of seeds to be drawn upon during the predicted 7 years of famine.

The Gramineae contributes more food seeds than any other plant family, viz., rice, wheat, and maize, as well as oats, barley, sorghum, millet, rye, tef, and other edible seeds. Collectively the large-seeded grasses are known as grains or cereals. All great civilizations have been founded on grain crops, primarily because their seeds offered high food value. Approximately 90% of all seeds cultivated is cereal grains (Senti and Maclay, 1961). The three major grains are rice, wheat, and corn. Rice is the staple and often the only food of hundreds of millions of people in China, Japan, India, and Indonesia. The primary production area for rice is the monsoon region of southeastern Asia. The 195 million metric tons of rice produced annually are exceeded by the 292 million metric tons of wheat harvested annually. Produced in nearly all north-temperate countries, wheat has the distinction of being the most widespread seed food crop in the temperate zone. Over 530 million acres are planted to wheat each year. Wheat, the bread-making cereal of the white race, has been cultivated for at least 9000 years. Charred wheat seeds, uncovered at Jarmo in eastern Iraq, are evidence that wheat was one of man's oldest crops (Edlin, 1967).

Earliest samples of maize (2500 BC) from La Perra Cave, Tamaulipas, Mexico, were from pod corn ears which had a weak cob and a husk around each kernel (Mangelsdorf et al., 1967). However, maize ears collected by the Spanish explorers in Mexico had strong ears and no husks around individual kernels. About 90% of the 230 million tons of maize seeds produced each year is used for livestock feed. As a human food, maize is considered inferior to other cereals because of its low protein content and lack of gluten necessary in making loaves. It is used as a flat bread by the people of Mexico and Central and South America. Cornstarch and related products are extensively used in industry. About 1.5 billion lbs of cornstarch are used in the United States annually.

The Leguminosae, the second most important seed food family, provides us with peanuts, soybeans, beans, lentils, peas, chick-peas, horse-beans, and other edible seeds. These seeds are rich (25–40%) either in proteins or in carbohydrates which are essential in a balanced human diet. On a volume basis, the two most important legume seeds are soybeans and peanuts.

Cereal and legume seeds comprise two-thirds to three-fourths of the human diet in parts of Asia and Africa, but only one-third of the diet in the more economically developed countries of Europe, North America, Argentina, Australia, and New Zealand. In these countries a considerable quantity of seeds is fed to livestock and converted into meat and dairy products.

III.   Other Uses of Seeds

Man uses other seeds, such as spices, condiments, and nuts, in his diet as embellishments. One nut, the ubiquitous coconut, is an important element in the diet in some tropical regions. Some of our popular beverages are derived from seeds: coffee and chocolate (cocoa) made from coffee and cacao seeds, beers and ales brewed from barley, and whiskeys and gins fermented from mashes of cereal grains. Seed and seed extracts are also used as medicines. Cotton, a major fiber, is spun from the hairs from cottonseeds.

Another major contribution from seeds is the edible and industrial oils expressed from soybean, peanut, coconut, cotton, palm, sunflower, safflower, rape, flax, sesame, tung, perilla, castorbean, and numerous other seeds. Seed oils supply about one-half of the world's edible oils. Approximately one-half of the seed oils are expressed from soybeans and peanuts.

In the tropics colorful seeds, especially from the Leguminosae and Palmae, are used in making jewelry and other novelties. The noise makers in rattles and maracas are often hard leguminous seeds.

Seeds are sown to produce most of our major nonfood crops including plants that yield fibers, pulp and wood, drugs, and food for livestock. Most of our ornamentals, flowers, trees, and lawn grasses, are started from seeds.

But the role of seeds in human affairs is not all beneficial. For example, seeds disseminate plants which are a burden to man. Weeds plague humans everywhere by reducing their food supply and otherwise affecting their health (e.g., by poisoning livestock and humans). Production of food and fiber plants is characterized by a continuous struggle to control weeds. Losses in crop plants occur primarily as a result of weeds competing with desirable plants for light, water, and minerals. Even with the present level of intensive weed control, it has been estimated that yield of farm, orchard, and garden crops is reduced by about 10% by weeds (Robbins *et al.*, 1952). Plants propagated by seeds also are responsible for some of our most serious social problems. For example, the narcotic drug habits and trades are justifiably viewed with alarm.

IV. Structure of Seeds

A true seed is a fertilized mature ovule that possesses an embryonic plant, stored food material (rarely missing), and a protective coat or coats. The embryo is made up of one or more cotyledons, a plumule (embryonic bud), hypocotyl (stem portion), and a radicle (rudimentary root) (Fig. 2). Application of the term "seed" is seldom restricted to this morphologically accurate definition. Rather, seed is usually used in a functional sense, viz., as a unit of dissemination, a disseminule. In this sense, the term seed embraces dry, one-seeded (rarely two- to several-seeded) fruits as well as true seeds. A fruit is a mature floral ovary which may contain one or more seeds and may include accessory floral parts. Examples of one-seeded fruits which are functional seeds are given in

Table I. These and similar fruits are called seeds in these volumes. At least one seed is always present within the indehiscent fruits listed in Table I; therefore, reproduction is by true seeds.

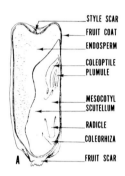

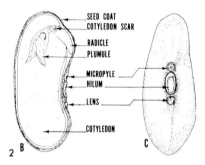

FIG. 2. Diagrams of a seedlike fruit and a true seed. (A) Longitudinal section of a corn caryopsis (seedlike fruit); (B) bean seed, one of the two cotyledons removed; (C) bean seed, showing micropyle, hilum, and lens.

TABLE I
DRY, ONE-SEEDED INDEHISCENT FRUITS REGARDED AS FUNCTIONAL SEEDS

Fruit type	Family
Achene and floral envelope	Polygonaceae
Achene and pappus	Compositae
Carpel	Malvaceae
Caryopsis and floret	Gramineae
Endocarp (with or without exocarp)	Palmae
Loment segment	Leguminosae
Mericarp	Umbelliferae
Nut (and cupule)	Fagaceae
Samara	Ulmaceae
Utricle	Chenopodiaceae

Abuse of the word seed is not restricted to a functional versus a morphological definition. The term seed has been applied to purely vegetative propagation in which no true seed is involved. For example, potato seeds usually are the eyes and adjacent tissue from potato tubers; *Poa bulbosa* seeds are bulbils; and mangrove seeds are seedlings. Use of the term seed in this sense is rejected.

Although seeds normally arise from sexual reproduction, in a number of plants found in various families an asexual method has been substituted. Plants in which embryos arise without fertilization of the ovule are called apomicts. The term "apomixis" is applied to this method of reproduction.

V.   Seed Variability

Although seeds have a number of common characteristics, biologists will also find a number of interesting structural, chemical, and functional differences among seeds of different species.

A.   *External Variation*

Variations in seed size, shape, color, and surface are myriad (Figs. 3, 4, and 5) and are important in seed identification (Chapter 2, Volume III). Some species have little seed variability (stenospermous), whereas others have much variability (euryspermous). Some idea of the extent of variation in seed size may be gained from the fact that a double coconut seed and endocarp may have a fresh weight of 20 lb, whereas 137 million seeds of *Ilysanthes dubia* weigh about 1 lb. Both large and small seeds have been considered primitive (Eames, 1961). Generally, large seeds are associated with perennial, especially woody, plants, Therefore, these seeds and plants are considered primitive (Salisbury, 1942). The problem of illustrating minute seeds (less than one mm in length) has been solved by using scanning electron microscopy (Fig. 5).

Seed shapes have not been adequately studied. As Duke (1969) mused, "which seed shapes are frequent and which are rare?" Common seed shapes are ellipsoid, globose, lenticular, oblong, ovoid, reniform, and sectoroid.

Brown and brown derivatives are by far the most common colors of seeds. Browns and blacks make up over half of the seed colors. Conspicuous colors such as red, green, yellow, and white are infrequent. When they are present, they usually have diagnostic value.

The surfaces of seed coats vary from highly polished to markedly roughened (Chapter 2, Volume III, see also, Figs 4 and 5). Although the surface topography has been used to advantage (Murley, 1951), it is the appendages which have naturally drawn most attention. These appendages include wings, arils, caruncles, spines, tubercles, hairs, and elaiosomes.

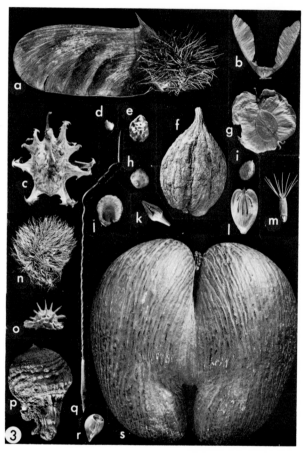

FIG. 3. A selection of seedlike fruits emphasizing variations in size, shape, and fruit-coat topography. (a) *Centrolobium robustum* Martius (0.3 ×); (b) *Acer macrophyllum* Pursh (0.3 ×); (c) *Harpagophytum procumbens* DC (0.3 ×); (d) *Fragaria vesca* L. (1 ×); (e) *Sanguisorba polygama* Nyland (1 ×); (f) *Melocanna baccifera* Skeels (0.3 ×); (g) *Holoptelea integrifolia* Planch (0.7 ×); (h) *Newcastlia cephalantha* Wettstein (1 ×); (i) *Lespedeza crytobotrya* Miq. (0.7 ×); (j) *Althaea rosea* Cav. (1.4 ×); (k) *Gaura odorata* Lag. (0.7 ×); (l) *Heracleum lanatum* Michx. (0.7 ×); (m) *Cnicus benedictus* L. (0.7 ×); (n) *Calligonum setosum* Litwinow (0.7 ×); (o) *Onobrychis* sp. (0.7 ×); (p) *Lithocarpus* sp. (0.3 ×); (q) *Stipa spartea* Trin. (0.7 ×); (r) *Moltkia doerfleri* Wettstein (0.1 ×); (s) *Lodoica maldavica* H. Wendl. (0.16 ×).

B. Internal Variation

An insight into seed anatomy is provided by Martin (1946). His comprehensive study was based on embryo type, size, and placement; food

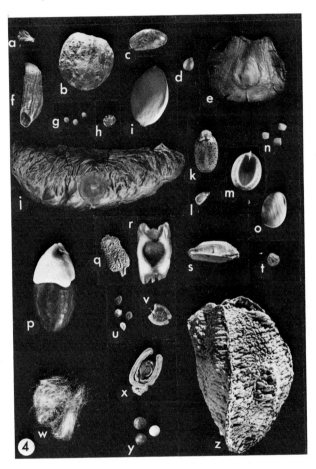

FIG. 4. A selection of true seeds emphasizing variations in shape and seed-coat topography. (a) *Vitis vinifera* Marsh. (0.7 ×); (b) *Aleurites montana* E. H. Wilson (0.7 ×); (c) *Annona squamosa* L. (0.7 ×); (d) *Hybanthus concolor* Spreng. (0.7 ×); (e) *Welwitschia mirabilis* Hook. (0.7 ×); (f) *Pinus taeda* L. (0.7 ×); (g) *Brassica hirta* Moench (0.7 ×); (h) *Vesicaria graeca* Reut. (0.7 ×); (i) *Ginkgo biloba* L. (0.7 ×); (j) *Alsomitra macrocarpa* (Bl.) Roem. (0.3 ×); (k) *Cindoscolus texanus* Small (0.7 ×); (l) *Pyrus prunifolia* Wild. (0.7 ×); (m) *Calycocarpum lyoni* (Pursh) Gray (0.7 ×); (n) *Indigofera hirsuta* L. (1.4 ×); (o) *Gleditsia aquatica* Marsh. (0.7 ×); (p) *Afzelia quanzensis* Welw. (0.7 ×); (q) *Momordica charantia* L. (0.7 ×); (r) *Moringa oleifera* Lam. (0.5 ×); (s) *Citrus grandis* Hassk. (0.7 ×); (t) *Opuntia tortispina* Engelm. & Bigel (0.7 ×); (u) *Cleome integrifolia* Torr. & Gray (0.7 ×); (v) *Passiflora quadrangularis* L. (0.7 ×); (w) *Ipomoea macrorhiza* Michx. (0.7 ×); (x) *Acacia melanoxylum* R. Br. (1 ×); (y) *Vicia sativa* L. (0.7 ×); (z) *Bertholletia excelsa* Lee (0.7 ×).

reserve quantity and quality; and seed size. He organized the seeds according to twelve embryo types shown in Fig. 30, Chapter 2, Volume III.

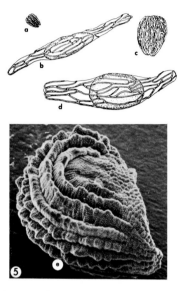

Fig. 5. A selection of minute seeds (less than 1 mm. in length) magnified to show details of the seed coat. a–d, drawings; e, scanning electron microscope photograph. a) *Verbascum blattaria* L. (10×); b) *Dendrobium* sp. (66×); c) *Striga lutea* Lour. (25×); d) *Dendrobium sp. (66×);* e) *Arthrostema ciliatum* R. & P. (55×). b and d reproduced from A. Davis (1946). Copyrighted by American Orchid Society. Used by permission. e photographed by Trevor Whiffin from Whiffin and Rodrigues collection 213, Puebla, Mexico. Used by permission.

Seed coats (Netolitzky, 1926; Harz, 1885; Singh, 1964), cotyledons (Bailey, 1956; Duke, 1969; Lubbock, 1892; Martin, 1946), and endosperms (Brink and Cooper, 1947) have been studied in detail.

C. Chemical Variation

Earle and Jones (1962) compiled the results of chemical analyses for seeds from 113 plant families. The biochemical data are recorded under percent ash, oil, and protein; fraction of alcohol-soluble nitrogen, and trichloroacetic acid-soluble nitrogen; test for alkaloid, starch, and tannin, etc. In a supplement, Jones and Earle (1966) tabulated oil and protein content for seeds of 759 species. Variations in chemical composition of seeds are discussed further in Chapter 2, Volume II.

D. *Physiological Variation*

Considerable variability exists among plant species in sources of carbohydrates and other growth requirements for seed germination and early development of seedlings. Food reserves of angiosperm seeds are stored primarily in the endosperm and, in some species, in the cotyledons. In gymnosperm seeds the reserve foods are stored primarily in the female gametophyte (megagametophyte). Although the endosperm and megagametophyte have similar functions, they vary as to origin and time of formation. The endosperm of angiosperms forms after fertilization and is often triploid. It results from fusion of a diploid-fusion nucleus and a sperm. The megagametophyte of gymnosperms is formed prior to fertilization and is haploid.

The patterns of utilization of endosperm food also vary among species of plants. In some seeds the reserve foods may be digested and absorbed by the embryo before the seed is mature (e.g., beans, peas, sunflowers); in other seeds (e.g., corn, oats, wheat), endosperm foods are not digested until after the mature seeds are planted and imbibe water.

The amount of endosperm present in seeds varies greatly among species of angiosperms and generally is inversely proportional to the size of the embryo. In seeds of some plants (e.g.,*Quercus, Amelanchier*), the endosperm is scant or absent with the embryos filling the entire seed cavity. In such seeds the food reserves are carried in the cotyledons. Other seeds (e.g., *Berberis* and *Ribes*) have very small embryos but extensive endosperm tissue which contains the bulk of the reserve foods of the seed.

Cotyledons may store foods, synthesize them, or do both. In some plants (e.g., *Cornus* and *Acer rubrum*) thin cotyledons which resemble foliage leaves do not store much food but when they emerge from the ground they may turn green and carry on appreciable photosynthesis. In other plants, such as *Pisum* and *Phaseolus,* cotyledons are thick and fleshy and play a major role in storage of food reserves. Germination of seeds having such cotyledons may be epigeous, with the cotyledons emerging from the ground and carrying on limited photosynthesis (e.g., *Fagus, Phaseolus, Robinia*), or it may be hypogeous with the fleshy cotyledons remaining below ground (e.g., *Juglans, Pisum,* and *Quercus*).

There is considerable evidence that in pines, megagametophyte reserves in the seed are important for early phases of seed germination but that normal seedling development also depends greatly on physiological activity of cotyledons after they emerge from the ground (Sasaki and Kozlowski, 1968c, 1969, 1970; Kozlowski and Borger, 1971).

VI. Plant Propagation by Seeds

Biologists have given much attention to seed germination, which is a precarious phase in plant development. Although large quantities of seeds are produced by most plant species and are usually dispersed for considerable distances (Chapter 4, this volume), only a small proportion of these seeds will produce mature plants. The greatest mortality occurs during the germination or seedling stage. High losses of seeds occur because of lack of seed viability, seed dormancy, and removal of, or injury to, seeds by various higher animals, insects, and fungi (Kozlowski, 1971). If a given seed should germinate, the existence of the seedling is tenuous at best as it is subject to attack by a variety of organisms and is extremely sensitive to environmental stresses. For example, germination of *Eucalyptus* seeds under natural conditions was as high as 80–90%; seedling survival was less than 1%, largely as a result of sensitivity of the tender young plants to desiccation, floods, frost, and insects (Jacobs, 1955).

A. *Effects of Environment and Site on Seeds and Germinants*

The specific causes of seed losses vary greatly in different regions. Pregermination losses of *Pseudotsuga menziesii* seeds added up to 46%, largely because of fungus attack (20%), insects and other invertebrates (10%), rodents (8%), and birds (3%). During the first year, 27% of the seeds did not germinate and damping-off fungi accounted for most of these dead seeds (Lawrence and Rediske, 1962). In Oregon only 10% of the *P. menziesii* seed survived from a fall planting until germination the following year. Mice and shrews destroyed about 40% of the seed; birds and chipmunks took 24%, and other factors (nonviability of filled seeds, invertebrates, disease, etc.) accounted for loss of another 25%. Survival of *Tsuga heterophylla* seed was 22%, with mice and shrews destroying 22% of the seed, birds and chipmunks 3%, and other agents 53% (Gashwiler, 1970).

In some areas rodents are a very serious problem to plant development. For example, Abbott (1961) found in Massachusetts that rodents could eat and store all the naturally produced pine seeds except those they failed to detect. Mice and moles consumed more seeds than could be sown economically by direct seeding. By comparison, a study of Cunningham (1960) in Australia showed that up to 80% of the naturally shed seeds of *Eucalyptus regnans* were harvested by insects.

Germination of nondormant seeds often is prevented by unsuitable environmental conditions. This is discussed in considerable detail in Chapter 1, Volume II and will be alluded to here only briefly. Because of wide differences in physical characteristics, temperature, availability of water and mineral nutrients, the establishment of plants varies greatly in different natural seedbeds. Mineral soil often is a good seedbed because of its high infiltration capacity, adequate aeration, and close contact between soil particles and seeds. Litter and duff often are less suitable than mineral soil because they warm more slowly, inhibit root penetration, prevent seeds from contacting the mineral soil, and shade small seedlings. Because of its high water-holding capacity, sphagnum moss often is a suitable seedbed for germination but it may subsequently smother young seedlings. Decayed wood also is an excellent natural seedbed for seeds of forest trees, probably because of its capacity for water retention (Place, 1955).

Requirements for early growth of established seedlings often vary from those for seed germination. Whereas seed germination and seedling survival of *Betula papyrifera* were highest on mineral soils and shaded positions, height growth of seedlings during the first year was greatest on organic seedbeds and in direct light (Marquis *et al.,* 1964). Winget and Kozlowski (1965) also found differences in seed germination and growth of *Betula alleghaniensis* seedlings in various types of naturally occurring seedbeds. Maximum germination rates occurred earlier on moist H-layer humus and on decayed wood (*Tsuga canadensis*) seedbeds than on mineral soil, but total germination was similar for these seedbeds. Rates of height growth and dry weight increment of *Betula* seedlings were greater on seedbeds of humus overlying mineral soil than on sandy loam, silty mineral soil, or decayed wood of *T. canadensis*. Drought resistance of seedlings was greatest on humus seedbeds.

1. APPLIED BIOCIDES

Although it is well known that food plants are the victims of pollution, it is equally true that agriculture is a source of pollution. The growing use of applied chemicals for protection of crops from weeds, pests, and diseases has some deleterious effects on plant regeneration by suppressing seed germination or through toxicity to recently emerged seedlings. For example, a number of insecticides, fungicides, herbicides, and fertilizers decrease the number of germinants. Such widely used chemicals as benzene hexachloride and the fungicide, N-[(trichloromethyl) thio]-4-cyclohexane-1,2-dicarboximide (captan), have been shown to injure roots of young tree seedlings (Simkover and Shenefelt, 1952; Cayford and Waldron, 1967).

Several herbicides check plant establishment by direct suppression of seed germination, toxicity to young seedlings, or both (Kozlowski and Sasaki, 1970). Triazine herbicides, such as 2-chloro-4-(ethylamino)-6-(isopropylamino)-s-triazine (atrazine) and 2-chloro-4,6-bis(ethylamino)-

s-triazine (simazine), did not affect seed germination but were toxic to recently germinated seedlings. Other herbicides at comparable dosages, i.e., *N*-1-naphthylphthalamic acid (naptalam), 2-chloroallyl diethyl-dithiocarbamate (CDEC), *s*-ethyl dipropylthiocarbamate (EPTC), *N,N*-diallyl-2-chloroacetamide (CDAA), and 2,4-dichlorophenoxy-acetic acid (2,4-D) variously inhibited both seed germination and early seedling growth. Both 2,4-D and CDAA greatly suppressed seed germination (Sasaki *et al.,* 1968).

Herbicide toxicity varies greatly with methods of herbicide application because of differences in herbicide losses, leading to variation in herbicide uptake by seedlings. Toxicity of a given herbicide that is applied to the soil often is low, intermediate if it is incorporated in the soil, and greatest if the herbicide is maintained in direct contact, in solution or suspension, with plant tissues. The high absolute toxicity of many herbicides is variously masked in soil cultures because soil-applied herbicides are lost by evaporation, leaching, microbial or chemical decomposition, and irreversible adsorption in the soil (Kozlowski *et al.,* 1967a,b). 2-Chloro-allyl diethyldithiocarbamate applied to the soil surface at 16 or 32 lb/acre had no significant effect on survival or dry weight increment of seedlings (Winget *et al.,* 1963). In contrast, application of CDEC to the soil surface at only 2 lb/acre followed by mixing the herbicide into the soil, significantly reduced survival and dry weight increment of seedlings (Kozlowski and Torrie, 1965). No injury occurred to pine seedlings in the nursery when simazine was applied to the soil surface at 4 or 8 lb/acre or when applied to the foliage. However, when 4 or 8 lb/acre (soil surface basis) of simazine were incorporated into the soil, severe injury resulted and seedlings were eventually killed (Kozlowski and Sasaki, 1970).

In addition to killing seedlings, herbicides cause abnormal developmental changes such as curling, shrivelling, or fusion of cotyledons, and chlorosis, distortion, and growth inhibition of various foliar appendages, as, for example, cotyledons, primary needles, and secondary needles in pines. The primary mechanisms by which herbicides exhibit toxicity are diverse and involve interference with vital processes as well as direct injury to cells and tissues. The toxic action may be exerted by active components of commercial herbicides and sometimes by "inert" ingredients as well as by synergistic effects of both (Sasaki and Kozlowski, 1968a,b).

2. ALLELOPATHY

Seed germination and growth of germinants are inhibited not only by applied chemicals but also by a variety of naturally occurring compounds in plants which are released to the soil (Rovira, 1969; Whittaker, 1970).

Such allelopathic chemicals are released from roots and aerial tissues as well. They may be released to the soil by leaching, volatilization, excretion, exudation and by decay either directly or by activity of microorganisms. Among the naturally occurring compounds that appear to have inhibitory effects on seed germination and growth of neighboring plants are phenolic acids, coumarins and quinones, terpenes, essential oils, alkaloids, and organic cyanides. Allelopathic chemicals are ecologically important because they influence succession, dominance, vegetation dynamics, species diversity, structure of plant communities, and productivity (Whittaker, 1970).

Several investigators have reported inhibitory effects of specific plants on seed germination and growth of adjacent plants. A few examples will be given. Perhaps the best known allelopathic chemical is juglone in *Juglans*. It is washed into the soil from leaves and fruits and inhibits the growth of certain woody and herbaceous plants (Brooks, 1951). The toxic leachates of leaves of *Artemisia absinthium*, (Funke, 1943), *Encelia farinosa* (Gray and Bonner, 1948), and *Ailanthus* (Mergen, 1959) have also been reported. In California the annual vegetation adjacent to naturalized stands of *Eucalyptus camaldulensis* is greatly inhibited. Where *Eucalyptus* litter accumulates annual herbs rarely survive to maturity. Del Moral and Muller (1970) identified several volatile and water-soluble toxins in *Eucalyptus* tissues. The terpenes, cineol and α-pinene, were among products that were identified as being toxic to germinating seeds and seedlings.

The effects of allelopathic chemicals on neighboring plants are modified by several factors such as soil moisture, soil type, and soil microflora. Terpenes produced by *E. camaldulensis* influenced annual grassland flora only after becoming adsorbed to soil particles, whereas phenolic acids influenced growth more directly. Well-drained light soils did not concentrate these toxins, and favorable aeration permitted rapid degradation of the toxins. In dry soils, competition for water combined with allelopathic effects of terpenes and phenolics to produce extensive bare zones. In wet soils, some growth of annual herbs occurred despite presence of the toxins. Readily available soil water apparently lessened the possibility that inhibited plants would be killed by drought. Heavy rains also favored degradation of allelopathic chemicals and washed them deep into the soil and diluted them (Del Moral and Muller, 1970).

Although is is clear that allelopathic chemicals are important under natural conditions, considerable caution should be exercised in interpreting the ecological significance of such toxins on the basis of laboratory experiments. Lerner and Evenari (1961), for example, found in laboratory experiments that leaves of *Eucalyptus rostrata* contained substances that inhibited seed germination. However, tests of soil from beneath *Eucalyptus* trees showed that the allelopathic chemicals did not accumulate in the field to inhibitory levels. Cannon *et al.* (1962) demonstrated in laboratory tests that extracts of *Backhausia* leaves inhibited germination of *Araucaria* seeds. In the field, however, germination of *Araucaria* seeds was higher on litter enriched with *Backhausia* leaves than on natural litter.

B. Seed Dormancy

Embryos of most seeds have a resting stage between development and germination. This resting stage is concurrent with seed and fruit ripening and continues through seed dispersal until external conditions (water, temperature, oxygen, and light) influence the embryo to germinate. Embryos that continue to grow within the seed and fruit (e.g., lack a resting stage) may be categorized as viviparous. Embryos in palm seeds continue to enlarge from inception until the radicle emerges or until the embryo dies. Even though the embryo is constantly enlarging at a slow rate, true germination occurs.

Although not common, vivipary occurs in both monocotyledonous and dicotyledonous plants. In viviparous species, embryos develop into seedlings without a resting stage, while still within the seed and fruit and while still attached to the parent plant. In a few species, such as the red mangrove (*Rhizophora mangle*), there is no static seed stage. Vivipary ensures establishment of red mangroves in an environment where seed germination and subsequent seedling development would be nearly impossible. Some of our crop plants exhibit a viviparous-like condition when the seeds germinate on the parent plant because of external or internal conditions. Some of our common forage legume seeds (e.g., *Melilotus* and *Medicago*), which are excellent examples of species that produce hard (dormant) seeds, may sprout on the plant if external conditions are right as the seeds are maturing. Vivipary is troublesome with some food plants, such as the cereals, in which a short period of seed dormancy is desirable to prevent sprouting and ruin of the grains in storage. Seeds that sprout inside citrus, tomato, and muskmelon fruits are viviparous-like.

Germination of viable seeds while the embryo is in the resting stage may be temporarily (permanently if the seed dies) delayed because of seed dormancy. Seeds of about two-thirds of the species of temperate-zone woody plants show some dormancy (Kramer and Kozlowski, 1960). The causes of seed dormancy have been grouped into five types (Amen, 1963): (*1*) rudimentary embryos, (2) physiologically immature embryos (inactive enzyme systems), (*3*) mechanically resistant seed coats, (*4*) impermeable seed coats, and (*5*) presence of germination inhibitors. A detailed discussion of the nature of seed dormancy is given in Chapter 3, Volume II.

Much research has been conducted on the nature of seed dormancy because of its important implications in plant propagation. Seed dormancy sometimes is disadvantageous and at other times a distinct advantage. Unduly long postponement of seed germination often makes it difficult to establish plant stands. This is the case, for example, with forage legumes such as alfalfa and clover. Seeds of some cultivars of lettuce have such deep-seated dormancy that they do not germinate until after the time of harvest.

Delayed germination is of value to a number of wild and cultivated plants in correlating environmental conditions in a pattern which tends to promote establishment and survival of a species. Seed dormancy enables plants to utilize habitats where unfavorable environmental conditions alternate with favorable ones. This undoubtedly has been important in invasion of plants from aquatic to terrestrial habitats. The prolonged chilling requirement for breaking of dormancy of seeds of temperate-zone plants prevents germination until spring. This tends to insure survival as earlier germination would result in killing of the tender seedlings by cold weather. Temporary dormancy of wild oats has contributed to their spreading so they often become weeds. By comparison, cultivated oats and false wild oats, which germinate in the autumn and are killed by low winter temperatures, do not become weeds. A seed dormancy phase also is helpful in natural reseeding of winter annual legumes which mature seed in the summer. If seeds germinate in the summer the seedlings usually die in the summer heat. However, some seeds germinate in the autumn and the germinants survive.

The degree of seed dormancy varies greatly among species and, within a species, among seeds from the same harvest. Hence, seeds of some wild plants may remain dormant in the soil for many years. The period of germination of such seeds may be spread over several months or sometimes even years. In nature this phenomenon insures establishment of a species even though the early germinants failed to survive severe environmental stresses such as droughts or severe frosts. Another adaptive mechanism involving seed dormancy is the gearing of seed germination in hot and dry regions to the short wet period of the year. In seeds of some desert plants, inhibitors in the seed coats prevent germination. However, when enough rain falls to wet the soil thoroughly the inhibitors are leached out and the seeds germinate (Wareing, 1963).

The following chapters will expand on some of the topics introduced in this chapter and they will discuss important features of seed development, dispersal, germination and metabolism, collection, storage, certification, and protection.

REFERENCES

Abbott, H. (1961). White pine seed consumption by small mammals. *J. Forest.* **59**, 197.

Amen, R. D. (1963). Concept of seed dormancy. *Amer. Sci.* **51**, 408.

Andrews, H. N. (1965). "Ancient Plants and the World They Lived In." Cornell Univ. Press, Ithaca, New York.

Bailey, I. W. (1956). Nodal anatomy and vasculature of seedlings. *J. Arnold Arboretum, Harvard Univ.* **37**, 269.

Beijerinck, W. (1947). "Zadenatlas der Nederlandsche Flora." H. Veenman & Zonen, Wageningen.

Boswell, V. R. (1961). What seeds are and do: An introduction. *Yearb. Agr. (U.S. Dep. Agr.)* p. 1.

Brink, R. A., and Cooper, D. C. (1947). Endosperm in seed development. *Bot. Rev.* **13**, 423 and 479.

Brooks, M. G. (1951). Effect of black walnut trees and their products on other vegetation. *W. Va., Agr. Exp. Sta., Bull.* **347**, 1.

Cannon, J. R., Corbett, N. H., Haydock, K. P., Tracey, J. G., and Webb, L. J. (1962). An investigation of the effect of the dehydroangustione present in the leaf litter of *Backhousia angustifolia* on the germination of *Araucaria cunninghamia* — an experimental approach to the problem of rainforest ecology. *Aust. J. Bot.* **10**, 119.

Cayford, J. H., and Waldron, R. M. (1967). Effects of captan on the germination of white spruce, jack, and red pine seed. *Forest. Chron.* **43**, 381.

Cunningham, T. M. (1960). "The Natural Regeneration of *Eucalyptus regnans*," Bull. 1. School of Forestry, Univ. of Melbourne, Melbourne, Australia.

Darwin, F. (1903). "More Letters of Charles Darwin," Vol. 2, p. 20. Appleton, New York.

Davis, A. (1946). Orchid seed and seed germination. *Am. Orchid Soc. Bull.* **15**, 218.

Del Moral, R., and Muller, C. H. (1970). The allelopathic effects of *Eucalyptus camaldulensis. Amer. Midl. Natur.* **83**, 254.

Duke, J. A. (1969). On tropical tree seedlings. I. Seeds, seedlings, systems, and systematics. *Ann. Mo. Bot. Gard.* **56**, 125.

Eames, A. J. (1961). "Morphology of the Angiosperms." McGraw-Hill, New York.

Earle, F. R., and Jones, Q. (1962). Analyses of seed samples from 113 plant families. *Econ. Bot.* **16**, 221.

Edlin, H. L. (1967). "Man and Plants." Aldus, London.

Funke, G. L. (1943). The influence of *Artemisia absinthium* on neighboring plants. *Blumea* **5**, 281.

Gashwiler, J. S. (1970). Further study of conifer seed survival in a western Oregon clearcut. *Ecology* **51**, 849.

Gray, R., and Bonner, J. (1948). An inhibitor of plant growth from the leaves of *Encelia farinosa. Amer. J. Bot.* **35**, 52.

Harz, C. D. (1855). "Landwirtschaftliche Samenkunde." Parey, Berlin.

Jacobs, M. R. (1955). Growth habits of the eucalypts. *Aust. Forest. Timber Bur.* pp. 1–262.

Jones, Q., and Earle, F. R. (1966). Chemical analyses of seeds. II. Oil and protein content of 759 species. *Econ. Bot.* **20**, 127.

Kozlowski, T. T. (1971). "Growth and Development of Trees," Vol. 1. Academic Press, New York.

Kozlowski, T. T., and Borger, G. A. (1971). Effect of temperature and light intensity early in ontogeny on growth of pine seedlings. *Can. J. Forest Res.* **1,** 57.

Kozlowski, T. T., and Sasaki, S. (1970). Effects of herbicides on seed germination and development of young pine seedlings. *Proc. Int. Symp. Seed Physiol. Woody Plants, Kornik Arboretum, Poland 1968* pp. 19–24.

Kozlowski, T. T., Sasaki, S., and Torrie, J. H. (1967a). Influence of temperature on phytotoxicity of triazine herbicides to pine seedlings. *Amer. J. Bot.* **54,** 790.

Kozlowski, T. T., Sasaki, S., and Torrie, J. H. (1967b). Effects of temperature on phytotoxicity of monuron, picloram, CDEC, EPTC, CDAA, and sesone to young pine seedlings. *Silva Fenn.* **3.**2, 13.

Kozlowski, T. T., and Torrie, J. H. (1965). Effect of soil incorporation of herbicides on seed germination and growth of pine seedlings. *Soil Sci.* **100,** 139.

Kramer, P. J., and Kozlowski, T. T. (1960). "Physiology of Trees." McGraw-Hill, New York.

Lawrence, W. H., and Rediske, J. H. (1962). Fate of sown Douglas-fir seed. *Forest Sci.* **8,** 210.

Lerner, R. H., and Evenari, M. (1961). The nature of the germination inhibitor present in leaves of *Eucalyptus rostrata. Physiol. Plant.* **14,** 221.

Lubbock, J. (1892). "Seedlings." Appleton, New York.

Mangelsdorf, P. C., MacNeish, R. S., and Galinat, W. C. (1967). Prehistoric maize, teosinte, and tripsacum from Tamaulipas, Mexico. *Bot. Mus. Leafl., Harvard Univ.* **22,** 33.

Marquis, D. A., Bjorkbom, J. C., and Yelenosky, G. (1964). Effect of seedbed condition and light exposure on paper birch regeneration. *J. Forest.* **62,** 876.

Martin, A. C. (1946). The comparative internal morphology of seeds. *Amer. Midl. Natur.* **36,** 513.

Mergen, F. (1959). A toxic principle in the leaves of *Ailanthus. Bot. Gaz.* **121,** 32.

Murley, M. (1951). Seeds of Cruciferae of northeastern North America. *Amer. Midl. Natur.* **46,** 1.

Netolitzky, F. (1926). Anatomie der Angiospermen-Samen. *In* "Handbuch der Pflanzen-Anatomie" (K. Linsbauer, ed.), Vol. 10, No. 14. Borntraeger, Berlin.

Place, I. C. M. (1955). The influence of seed-bed conditions on the regeneration of spruce and fir. *Can. Forest. Branch Bull.* **117.**

Robbins, W. W., Crafts, A. S., and Raynor, R. N. (1952). "Weed Control." McGraw-Hill, New York.

Rovira, A. D. (1969). Plant root exudates. *Bot. Rev.* **35,** 17.

Salisbury, E. J. (1942). "Reproductive Capacity of Plants." Bell, London.

Sasaki, S., and Kozlowski, T. T. (1968a). Effects of herbicides on respiration of red pine (*Pinus resinosa* Ait.) seedlings. I. s-triazine and chlorophenoxy acid herbicides. *Advan. Front. Plant Sci.* **22,** 187.

Sasaki, S., and Kozlowski, T. T. (1968b). Effects of herbicides on respiration of red pine seedlings. II. Monuron, diuron, DCPA, dalapon, CDEC, CDAA, EPTC, and NPA. *Bot. Gaz.* **129,** 286.

Sasaki, S., and Kozlowski, T. T. (1968c). The role of cotyledons in early development of pine seedlings. *Can. J. Bot.* **46,** 1173.

Sasaki, S., and Kozlowski, T. T. (1969). Utilization of seed reserves and currently produced photosynthates by embryonic tissues of pine seedlings. *Ann. Bot. (London)* [N.S.] **33,** 473.

Sasaki, S., and Kozlowski, T. T. (1970). Effects of cotyledon and hypocotyl photosynthesis on growth of young pine seedlings. *New Phytol.* **69,** 493.

Sasaki, S., Kozlowski, T. T., and Torrie, J. H. (1968). Effect of pretreatment of pine seeds with herbicides on seed germination and growth of young seedlings. *Can. J. Bot.* **46,** 255.

Senti, F. R., and Maclay. W. D. (1961). Age-old uses of seeds and some new ones. *Yearb. Agr. (U.S. Dep. Agr.)* p. 27.

Simkover, H. G., and Shenefelt, R. D. (1952). Phytotoxicity of some insecticides to coniferous seedlings with particular reference to benzene hexachloride. *J. Econ. Entomol.* **45,** 11.

Singh, B. (1964). Development and structure of angiosperm seed. I. *Bull. Nat. Bot. Gard.* **89,** 1.

Thomson R. B. (1927). Evolution of seed habit in plants. *Trans. Roy. Soc. Can., Sect. 5* [3] **21,** 229.

Wareing, P. F. (1963). The germination of seeds. *Vistas Bot.* **3,** 195.

Whittaker, R. H. (1970). The biochemical ecology of higher plants. *In* "Chemical Ecology" (E. Sondheimer and J. B. Simeone, eds.), pp. 43–70. Academic Press, New York.

Winget, C. H., and Kozlowski, T. T. (1965). Yellow birch germination and seedling growth. *Forest Sci.* **11,** 386.

Winget, C. H., Kozlowski, T. T., and Kuntz, J. E. (1963). Effects of herbicides on red pine nursery stock. *Weeds* **11,** 187.

2

DEVELOPMENT OF GYMNOSPERM SEEDS

Hardev Singh and B. M. Johri

I. Introduction

The gymnosperms attained the zenith of their diversification during the late Palaeozoic–early Mesozoic periods and then declined (Arnold, 1947). Many of the orders became extinct and, presently, they are represented by only seven orders—four of these are monogeneric (Sporne, 1965). Taken as a whole, the group exhibits very diverse seed structures, and it has been possible to follow the development only in the living forms. Therefore, taking development as the theme, stress will be laid on living gymnosperms. In some instances it has been possible to interpret the development of a few fossil seeds on the basis of structures found in the living forms.

It has more or less been agreed upon that following fertilization an ovule becomes a seed (Eames, 1955). Therefore, it might seem logical to deal with only the postfertilization phenomenon in writing on seed development. However, in this chapter, a broad viewpoint has been taken, and all the stages of ovule development leading to seed formation have also been included.

Phylogenetic and morphological aspects of seed structure have not been stressed. The evolution of the ovule has been discussed by P. Maheshwari (1960) and D. L. Smith (1964), and Puri (1970) has dealt with the morphology of the ovule. Furthermore, differences of opinion exist with regard to the terms used for various structures or phenomena encountered in developing ovules. This review, therefore, presents a general account of the development and maturation of the ovule in gymnosperms based on contemporary literature, recent reviews, and books (see Schnarf, 1933, 1937; Chamberlain, 1935; Sterling, 1963; Doyle, 1963; P. Maheshwari and Sanwal, 1963; Favre-Duchartre, 1965; Sporne, 1965; Andrews, 1966; P. Maheshwari and Singh, 1967).

The young gymnospermous ovule comprises a central body called the nucellus which is surrounded by 1 to 3 envelopes, the integuments. The integument leaves the nucellus exposed along a narrow passage called the micropyle. Like the angiospermous ovule, it has a chalaza which is generally very well developed.

The nucellus is usually massive and contains several hypodermal

archesporial cells which later give rise, successively, to primary sporogenous cells, sporogenous cells, megaspore mother cells, megaspores,
and female gametophytes. During this development the ovule becomes
pollinated, a sugary liquid (pollination drop) oozing out of the micropyle
at the time of pollination. The pollen grains settle on the nucellus and
germinate, either immediately or after a resting period, to form pollen
tubes which bore their way through the nucellus. Finally they reach the
female gametophyte. Following pollination the micropyle becomes
occluded and the ovule grows due to the activity of a meristematic tissue
which may be located in the micropylar or the chalazal region. When the
female gametophyte is young, the innermost cells of the nucellus usually
differentiate into densely cytoplasmic cells—the spongy tissue. This
tissue is considered to be the nurse tissue for the female gametophyte.

The female gametopyte is free nuclear, to begin with, and becomes
cellular in later stages of development. On the gametophyte, the archegonia arise either singly or in groups (archegonial complexes). In longisection or transection the gametophyte shows rows of cells radiating
from its center. The egg of gymnosperms is usually a very prominent
cell, several times larger than the other gametophytic cells. Its nucleus is
also fairly large but poorly stainable for chromatin. Following fertilization, the female gametophyte becomes gorged with reserve food material
to form the so-called endosperm.

Gymnospermous seeds usually take a long time (1–3 years) to develop
and may pass through definite growth periods and resting periods usually
during winter. Thus the seeds show a seasonal development. The course
of events leading to the formation of a mature seed may be divided into
three phases: (*a*) prepollination, (*b*) postpollination–prefertilization, and
(*c*) postfertilization.

II. Prepollination Phase

In most conifers, taxads, and *Ginkgo,* the ovule becomes pollinated
when it shows only sporogenous cells or megaspore mother cell; in cycads and *Gnetum,* when it shows a free nuclear gametophyte; and in
Ephedra the archegonia are fully formed at the time of pollination. Thus
the stage of development at which pollination occurs is variable in different gymnosperms. However, except in *Ephedra,* most of them show a
distinct prepollination phase which is characterized by a wide open
micropyle, periclinal and anticlinal divisions in the nucellus, formation
of a pollen chamber in many taxa, cell divisions throughout the ovule, and
undifferentiated xylem and phloem in ovular vascular supply.

A. Initiation

Stages in the initiation of ovules are known only in a few gymnosperms. In cycads the young megasporophylls are embedded in the crown of leaves and are not visible from the outside; in conifers and taxads the young female cone or the ovule is completely enclosed in the winter buds and, therefore, it cannot be recognized externally.

The initiation of ovules has been studied in *Ginkgo* (see Favre-Duchartre, 1956), some conifers (Hagerup, 1933; Singh, 1961; Allen, 1963), *Torreya* (Kemp, 1959), *Taxus* (Pankow, 1962; Loze, 1965), *Ephedra* (Hagerup, 1934; Seeliger, 1954; Pankow, 1962; Lehmann-Baerts, 1967a), *Gnetum* (Lehmann-Baerts, 1967a), and *Welwitschia* (Martens, 1959). There are several reports on the initiation and development of female cones (strobili) in conifers but the development of ovules has not been traced (see Hirmer, 1936; Owens and Smith, 1964). In fact, such studies should be based on both dissected wholemounts of young cones and sections of primordia but, unfortunately, most of the investigations are based on either of the two techniques leading to erroneous interpretations.

In cycads a variable number of ovules initiate laterally on the megasporophyll. The single integument is massive in *Encephalartos* (De Sloover, 1964). The ovules arise in pairs in *Ginkgo,* one each at the tip of a forked peduncle. The latter develop among the leaves on the dwarf shoots (Favre-Duchartre, 1956).

In most conifers the ovule initiates on a humplike structure which arises in the axil of bract scale (Hagerup, 1933; Konar, 1960; Allen, 1963; Owens and Smith, 1964). The humplike structure eventually develops into the commonly called ovuliferous scale (seed–scale complex of Florin, 1954). The latter bears one or more ovules on its upper surface near the base. The micropyle of the ovule in Pinaceae, Araucariaceae and a few podocarps points toward the cone axis; and, in Taxodiaceae, Cephalotaxaceae, and Cupressaceae, it points away from the cone axis (Hagerup, 1933; Singh, 1961; Singh and Oberoi, 1962; Vasil and Sahni, 1964). The ovule in most conifers is orthotropous but in *Podocarpus* and *Dacrydium* it is anatropous (Hagerup, 1933; Konar and Oberoi, 1969). According to Hagerup (1933), the integument arises as one or two lateral outgrowths on the ovular primordium, but these eventually join with each other and form the annular integument. The integument is inserted near the apex of the dome (ovular primordium) in members of Pinaceae, Araucariaceae, *Cryptomeria,* and *Cephalotaxus* (Hagerup, 1933; Singh, 1961; Singh and Chatterji, 1963). But, in most members of the Cupressaceae and in *Sequoia,* the integument is inserted very low on the primordium (Lawson, 1904; Baird, 1953).

In *Cephalotaxus* two ovules arise as lateral protuberances from the secondary axis which arises in the axil of the fertile bract (Singh, 1961). The ovular primordia grow mainly by periclinal divisions which character- ize a leaf primordium (Fig. 1A-D).

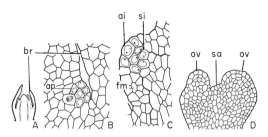

F_IG_. 1. Ovule initiation in *Cephalotaxus drupacea*. (A) Longisection of very young female cone. (B) The portion marked in figure A enlarged to show the initiation of the secondary axis. (C) A later stage in the development of the secondary axis. The tip shows the typical organization characteristic of the shoot apex. (D) Initiation of the two ovules from the secondary axis. (ai) Apical initials; (ap) primordium of secondary axis; (br) bract; (fm) flank meristem; (ov) ovular primordium; (sa) secondary axis; (si) subapical initials. (After Singh, 1961.)

The ovules of taxads are orthotropous and arise terminally on the short secondary shoot (Florin, 1948; Kemp, 1959; Pankow, 1962; Loze, 1965). Kemp, Pankow, and Loze have worked out the ontogenetic transforma- tion of the shoot apex into an ovule in *Taxus* and *Torreya*. The apical initials of the shoot apex give rise to the nucellus, the integument arises from the flank meristem, and the central portion of the ovule is formed from the subapical initials and the pith meristem (Fig. 2A-D).

The process of ovule initiation is very well known in *Ephedra, Gnetum,* and *Welwitschia*. In *Ephedra* the ovule arises by transformation of a short secondary shoot in the axil of the fertile bract (Seeliger, 1954; Pankow, 1962). The ovule develops in the same way as the vegetative axillary bud (Fig. 3A and B), and the two integuments arise acropetally (Fig. 4A) (Lehmann-Baerts, 1967a). Basing his studies on sections, Hagerup (1934) inferred that the integuments are initiated as lateral outgrowths. However, the investigations of Lehmann-Baerts (1967a), based on dis- sected wholemounts, indicate that the integuments are initiated by circu- lar and continuous rings of tissue. The ventral sector of the inner integu- ment shows a precocious (but secondary) growth resulting in an oblique micropyle (Fig. 4B).

The female strobilus of *Gnetum* comprises several acropetally placed collars on the axis. In *Gnetum gnemon* and *Gnetum africanum,* an annular hump of tissue originates on the lower surface of the collars

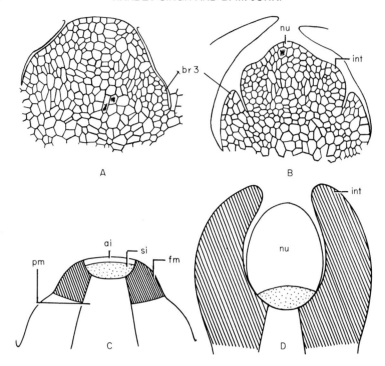

FIG. 2. Ovule initiation in *Taxus baccata*. (A) Longisection of shoot apex of the secondary fertile shoot after the formation of third pair of bracts on it. (B) Initiation of integument and nucellus on the shoot apex. (C,D) Diagrammatic representation of longisection of shoot apex (C) and ovule (D) to show the derivation of parts of ovule from various cyto-histological zones of shoot apex. (ai) Apical initials; (br3) third pair of bracts on the secondary shoot; (fm) flank meristem; (int) integument; (nu) nucellus; (pm) pith meristem; (si) subapical initials. (After Loze, 1965.)

(Fig. 5A-C) (Waterkeyn, 1954; Sanwal, 1962). This seems to be true of *Gnetum ula* also (see P. Maheshwari and Vasil, 1961). The cells of the hump divide periclinally and anticlinally and produce an ovular primordium which differentiates into two regions (Fig. 5D and E). The upper region gives rise to the ovule, and the lower one forms a cushionlike structure (Fig. 5F). Hairs differentiate from the sterile cells between the ovules and on the surface of this cushion. Three envelopes arise acropetally on ovular primordia. According to Hagerup (1934), the outermost envelope initially arises as two outgrowths, and the inner two envelopes as one outgrowth each. Hagerup's study was based on the sections of ovular primordia. Lehmann-Baerts (1967a) has traced the ontogeny of the ovule based on wholemounts, and he finds that the three envelopes arise as annular and continuous rings of tissue. Throughout their growth,

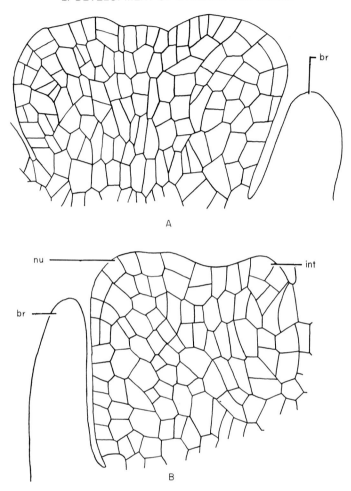

Fig. 3. Ovule initiation in *Ephedra americana*. (A) Longisection of tip of cone axis to show the origin of two shootlike structures (ovular primordia) in the axil of the last pair of bracts. (B) Differentiation of nucellus and integument on an ovular primordium. (br) Last pair of bracts on female cone axis; (int) initials of integument; (nu) nucellus. (After Pankow, 1962.)

the three envelopes remain free from each other (P. Maheshwari and Vasil, 1961).

Welwitschia has also been studied, using sections as well as whole-mounts of ovules (Martens, 1959). In a superbly illustrated article, Martens points out that most of the previous investigators (see Hagerup, 1934) erred in interpreting the structure of the ovule of this enigmatic plant. The ovules initiate in the axil of bracts, and a pair of lateral bracts

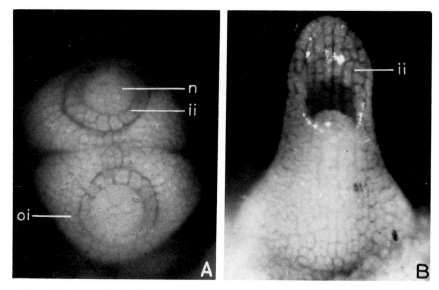

F<small>IG</small>. 4. Ovule initiation in *Ephedra distachya*. (A) Dissected wholemount of the two ovular primorida of a female strobilus to show annular origin of integuments. ×94. (B) Very young ovule; inner integument shows one-sided growth. ×110. (ii) Inner integument; (n) nucellus; (oi) outer integument. (After Lehmann-Baerts, 1967a; courtesy Prof. P. Martens.)

is the first to arise on the ovular primordium, followed by a dorsiventrally compressed wing and the integument. The wing and integument initiate as annular structures, but the outer ring which forms the wing becomes subsequently partitioned into a dorsal and a ventral lobe. Due to inter- calary growth, the integument develops a long micropylar tube which shows variable apical lobing. The structure of the two bracts at the base of the ovule is variable and depends on the position of the ovule in the strobilus. In a well-developed ovule, these bracts fall off very early. The bracts have been referred to by Martens (1959) as *le verticille floral externe*.

Ovules of most living gymnosperms have a conspicuous chalaza which broadly attaches the ovule to the subtending structure. Thus, a funiculus is usually not recognizable in these ovules.

B.   Nucellus

The nucellus grows by periclinal and anticlinal divisions in the primary parietal layer and the epidermis. There are some claims that, in *Ephedra,* the epidermis alone gives rise to the thick nucellus, the parietal tissue being absent. But this does not seem to be supported by the work of Singh and Maheshwari (1962) and Lehmann-Baerts (1967b). When fully

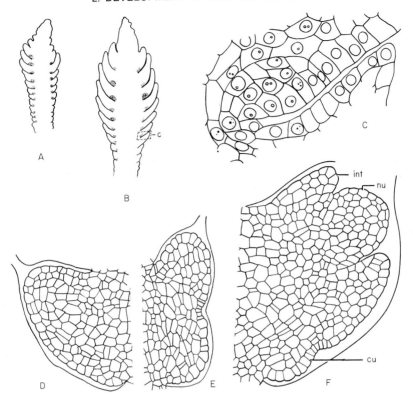

FIG. 5. Ovule initiation in *Gnetum*. (A,B) Longisection of young female strobili; the dotted portion indicates the position of meristematic tissue which later gives rise to ovules. (C) Portion marked C in figure B enlarged to show details of the meristematic tissue. (D–F) Progressive stages in the differentiation of ovule and sterile cushion from the meristematic tissue. (cu) Sterile cushion; (int) integument; (nu) nucellus. *Gnetum africanum* (after Waterkeyn, 1954) in figures A to C, *Gnetum nemon* (after Sanwal, 1962) in figures D to F.

formed the nucellus is usually massive but in some taxads, cupressads, and taxodiacious members it is only about ten cell layers high.

The nucellus is beak-shaped in cycads (De Sloover, 1964) and *Ginkgo* (Favre-Duchartre, 1956), and more-or-less dome-shaped in most other gymnosperms, and remains within or below the micropyle. But in *Agathis*, *Araucaria*, and *Saxegothaea* the nucellus is very massive and protrudes through the micropyle (Favre-Duchartre, 1963). In *Saxegothaea* it is especially prominent and flares out above the micropyle, whereas in *Araucaria* it bends toward one side.

In cycads, *Ginkgo,* and several fossil forms the nucellar epidermis is cutinized. The nucellar tissue generally shows little internal differentiation, but in cycads a central strand of cells becomes demarcated (De

Sloover, 1964; see also Schnarf, 1933). The cells in the strand are elongated along the long axis of the ovule and are poorly cytoplasmic.

The nucellus usually attains its full development before pollination.

C. Pollen Chamber

The pollen chamber is formed by degeneration of cells of the nucellus situated in the micropylar part. Pollen chamber development is a pre-pollination phenomenon and should be distinguished from degeneration of nucellar cells in its micropylar part following landing of the pollen on the nucellus or entry of the pollen tube into the nucellus as seen in cycads (Kershaw, 1912; De Sloover, 1964), *Podocarpus gracillior* (Konar and Oberoi, 1969), *Cryptomeria* (Singh and Chatterjee, 1963), and several other gymnosperms. It is well-developed in fossil ovules, cycads, *Ginkgo,* and *Ephedra.* A poorly developed pollen chamber occurs in *Athrotaxis, Cephalotaxus* (Singh, 1961), *Gnetum* (Waterkeyn, 1960), and taxads and seems to be absent in members of the Pinaceae, Araucariaceae, Podocarpaceae, Cupressaceae, and most members of the Taxodiaceae.

The cells that degenerate to form the pollen chamber are usually not different from the surrounding cells except in cycads and *Ginkgo*. In cycads the cells forming the central strand in the nucellus degenerate (De Sloover, 1964), but in *Ginkgo* these cells become distinct due to their elongation (Fig. 6A). They appear poorly cytoplasmic at the time of formation of the chamber (De Sloover-Colinet, 1963). In *Ginkgo,* formation of the pollen chamber begins by disorganization of subepidermal cells and central cells of the nucellar beak (Fig. 6B). The cutinized nucellar epidermis, which is the last to degenerate, breaks down only in the region of tip (Fig. 6C and D). On the flanks the anticlinal walls of the epidermis become thick and prominent (Favre-Duchartre, 1956; De Sloover-Colinet, 1963). The nucellar cells lining the chamber are densely protoplasmic (Fig. 6D).

In different gymnosperms the shape and structure of the pollen chamber (also called *lagenostome* in fossil members) are quite variable (Favre-Duchartre, 1963); this chamber is especially elaborate in Lagenostomales. In *Salpingostoma dasu* (Gorodon, 1941), it is dome-shaped and prolonged to form a long, tubular structure, the salpynx, and in *Lagenostoma lomaxi* (Oliver and Scott, 1904), the chamber is conical and annular with a core of tissue, the central column. The lagenostome, in Trigonocarpales, is small and often has a short apical beak projecting into the micropyle (Taylor, 1965; see also D. L. Smith, 1964). Among the fossil forms, the simplest pollen chamber is found in Cardiocarpales where it is small and conical (D. L. Smith, 1964). Among the living gymnosperms, the pollen

chamber is narrow, deep and prominent in cycads and *Ginkgo,* and in
Ephedra it extends right down to the top of the female gametophyte
(Singh and Maheshwari, 1962; Lehmann-Baerts, 1967b).

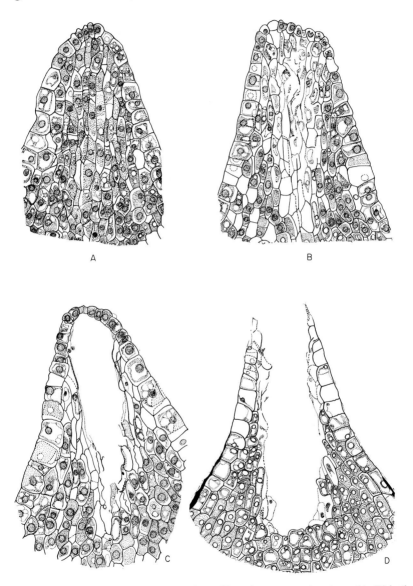

Fig. 6. Longisection of apical portion of nucelli to show successive stages (A–D) in the
development of pollen chamber by degeneration of nucellar cells in *Ginkgo biloba.* The
epidermal cells are the last to degenerate. (After De Sloover-Colinet, 1963.)

D. Megasporogenesis

Hypodermal archesporial cells have been recognized in only a few gymnosperms. Sometimes, the contents of the archesporial cells are not at all distinctive so that the fertile cell becomes recognizable only at the mother cell stage (Konar and Oberoi, 1969).

One, or more, hypodermal archesporial cells have been demonstrated in *Cephalotaxus drupacea* (Singh, 1961), *Cryptomeria* (Singh and Chatterjee, 1963), *Ephedra* (P. Maheshwari and Singh, 1967), *Gnetum* (P. Maheshwari and Vasil, 1961), *Larix decidua* (Strasburger, 1879), *Pinus roxburghii* (Konar, 1960), *Taxus canadensis* (Dupler, 1917), and *Zamia floridana* (F. G. Smith, 1910).

The megaspore mother cell is usually large and elongated with a very prominent wall. In *Ginkgo* this prominence is due to the presence of a special wall layer and the greatly thickened middle lamella (Stewart and Gifford, 1967). The special wall layer is similar in appearance to the middle lamella, except for a tighter arrangement of the fibrillar structure. However, a large part of the prominent cell wall comprises the middle lamella.

In *Encephalartos poggei,* a special callose membrane is present around the cytoplasm of the megaspore mother cell (De Sloover, 1961), and, thus, the wall of the cell is of double nature. This special membrane is connected with the formation of transverse walls between the cells of triad and degenerates during enlargement of the functional megaspore. According to Martens (1966), the presence of callose between the megaspores is a relic of a very old mechanism which tends to make the spores free from each other in a sporangium, like the pollen in anthers.

Studies on the fine structure of the megaspore mother cell of *Ginkgo* have shown that four changes are coincident to the elongation of the cell (Stewart and Gifford, 1967). The endoplasmic reticulum (ER) forms a very intricate system in the micropylar portion of the cell, a large interruption is seen in the nuclear envelope, plastids and mitochondria concentrate in the chalazal part of the cell, and a large vacuole appears in this part. As the cell matures, its micropylar part becomes completely devoid of plastids and mitochondria, and they become densely packed in the chalazal region (Fig. 7). Thus, the chalazal dyad cell receives all these organelles. However, the other organelles, such as dictyosomes, lipid droplets, and multivesicular bodies, do not show such a polar distribution. According to Stewart and Gifford (1967), in several gymnosperms the "kinoplasmic mass," present as a darkly staining region in the chalazal portion of the megaspore mother cell, may be an image produced by the aggregation of organelles that occurs below the nucleus, as in the mature megaspore mother cell of *Ginkgo*. It may be recalled that a concentration

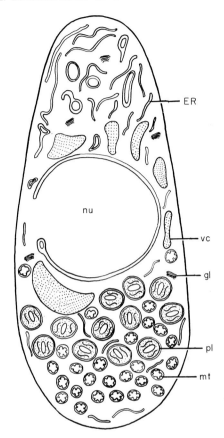

Fig. 7. Diagrammatic representation of fine structure of megaspore mother cell of *Ginkgo biloba*. All the mitochondria and plastids have accumulated in the chalazal portion of cell. (er) Endoplasmic reticulum; (gl) golgi apparatus; (mt) mitochondria; (nu) nucleus; (pl) plastid; (vc) vacuole. (After Stewart and Gifford, 1967.)

of starch grains occurs in the chalazal region of the megaspore mother cell of *Encephalartos* (De Sloover, 1961); the chalazal part gives rise to the functional megaspore.

Various aspects of megasporogenesis, such as the extent of sporogenous tissue and formation of triads, tetrads, and coenomegaspores have been reviewed by P. Maheshwari and Singh (1967).

E. Integument

Cycads, *Ginkgo,* conifers, and taxads show only one integument. During the prepollination stage the integument does not show any internal differentiation and usually comprises parenchymatous tissue with interspersed tanniniferous cells. Vascular bundles and resin or mucilage ducts do not differentiate at this stage.

A ring-shaped collar, comprising thin-walled tissue, is present at the base of the ovule in *Ginkgo* (Favre-Duchartre, 1956), and an epimatium (equivalent to ovuliferous scale) covers a part or the whole of the ovule in podocarps (Konar and Oberoi, 1969). In *Podocarpus* the epimatium is fused with the integument and consists of parenchymatous tissue (several cells contain tannin). An aril initiates as a ring of meristematic tissue at the base of the integument in taxads (Loze, 1965), but its further development is essentially a postpollination phenomenon. An aril is also present in *Phyllocladus*.

Ephedra and *Welwitschia* show bitegmic ovules (Martens, 1959; Singh and Maheshwari, 1962), and the integuments are well developed even before pollination. In *Ephedra* the outer integument is thick and vascularized by three unbranched strands. The inner integument is flimsy and comprises two layers of cells. In *Welwitschia* also the outer integument is massive but it forms the wing.

Gnetum shows three envelopes (integuments; see P. Maheshwari and Vasil, 1961) which are well-developed and vascularized at the time of pollination. The outer integument is several-layered and shows sclereids and laticiferous ducts in the mesophyll. Stomata occur on the outer epidermis. The middle integument comprises six to ten layers interspersed with sclereids; its outer epidermis has stomata. The inner integument is five- or six-layered and forms the micropylar tube. In *Gnetum gnemon* (Sanwal, 1962) the cells of the outer epidermis of the micropylar tube, just above the level of the middle envelope, begin to divide and produce a flange around the tip of the middle envelope. *Gnetum africanum* (Waterkeyn, 1960) has two such flanges.

A unique feature of the prepollination ovules of *G. africanum* and *G. gnemon* is the formation of a structure called an obturator (Waterkeyn, 1960). Just before pollination the micropylar canal is blocked by elongation and divisions in its epidermal and subepidermal layers. The tissue thus formed grows downward and fits into the pollen chamber. In *G. gnemon,* after the formation of pollen tube, the epidermal cells dissolve to give way to the tubes (Fagerlind, 1941). Sanwal (1962) mentioned that the cells blocking the micropyle become thick-walled and develop simple pits in later stages.

III. Pollination Mechanism

Most gymnosperms are wind-pollinated and the pollen grains are received on a drop of nectar (pollination drop) secreted at the micropyle of the ovule. The micropyle is also variously modified, in many plants, to receive the pollen and/or contain the drop; in most conifers the pollen grains seem structurally suitable for wind-pollination (see Doyle, 1945; Dogra, 1964).

A. Structure of Micropyle

The single integument in *Cycas* (De Silva and Tambiah, 1952) and the inner of the two or three integuments in *Ephedra, Gnetum,* and *Welwitschia* (Pearson, 1929; Martens, 1959; P. Maheshwari and Vasil, 1961; Singh and Maheshwari, 1962) are drawn out into a micropylar tube. The latter is short in *Cycas* but very prominent in the other three genera mentioned above. The tube comprises two layers of cells in *Ephedra* (Singh and Maheshwari, 1962), but in the others there are several layers.

Pinaceae show special types of stigmatic micropyles. In *Pinus* and *Picea* the micropyle is wide and extends well beyond the nucellus and over the base of the ovuliferous scale (Doyle and Kane, 1943; McWilliam, 1958). The micropylar tip is produced into two opposite arms which point toward the base of the cone. The arms and the micropylar canal are coated with a sticky film which causes the pollen to adhere to these surfaces. The pollination drop fills the micropylar canal and, sometimes, even fills the space between the arms. After pollination the arms shrivel and dry up.

The micropylar canal in *Larix* and *Pseudotsuga* is short and its tip is asymmetrical with a large stigmatic flap (see Doyle, 1945; Barner and Christiansen, 1960, 1962; Allen, 1963). The stigmatic flap bears hairlike projections and is especially prominent in *Pseudotsuga*. In this genus the pollen grains stick to the stigmatic surface which then collapses; the pollination drop seems to be absent.

In *Abies* and *Cedrus deodara* the micropylar tip is funnel-shaped (Doyle, 1945; Roy Chowdhury, 1961). In the former taxon the funnel is slightly notched in some parts, and in the latter it is deeply scalloped. The flare is stigmatic. The micropylar canal extends over the base of the ovuliferous scale, and the funnel points outward and downward. Following landing of the pollen grains on the flare, the funnel folds over the micropylar opening and pollen grains come to lie on the nucellus. The pollination drop seems to be absent.

Cedrus atlantica and *Cedrus libani* have a micropylar canal which bends over the scale, and the tip of the tube is expanded into a very asymmetrical, thin, and one-sided stigmatic flare. Following pollination, the flare folds over and the nucellus grows up coming in contact with the pollen grains (Doyle, 1945). The pollination drop is absent.

In *Tsuga pattoniana* the integument toward the upper side of the ovuliferous scale forms a lobed stigmatic expansion and, on the other side, shows a broad slit extending down to expose the nucellus (Doyle, 1945). The shape, lobing, and spread of the stigmatic flap are variable. After pollination the flap becomes reflexed and the nucellus grows up to plug the micropylar canal. The pollen grains germinate on the flap *in situ,* and the tubes grow directly and enter the nucellus. The pollination drop is absent.

The tip of the micropyle is simple and astigmatic in *Ginkgo* and in most of the other conifers and taxads.

B. Pollination Drop

Except for a few members (*Abies, Cedrus, Larix, Pseudotsuga,* and *Tsuga*) of Pinaceae, which show a special stigmatic micropyle, and a few more (*Araucaria, Agathis, Tsuga dumosa*) in which the pollen grains do not land on the micropyle, all other gymnosperms investigated so far show a sugary exudation at the micropyle, at the time of pollination. McWilliam (1958) has likened the pollination drop secretion to guttation which also occurs in an atmosphere of high humidity. The fluid serves as a receptor of the wind-borne pollen, as well as a vehicle for transporting it to the nucellus. According to Doyle (1945) the drop serves for flotation of the pollen and as the drop retracts, the pollen is sucked into the micropyle and finally lands on the nucellus. However, McWilliam (1958) has presented evidence that the active absorption of the fluid by the pollen is the prime mover of the grains. In *Cephalotaxus* the exine becomes ruptured and is cast off when the pollen is caught in the pollination drop (Singh, 1961).

Secretion of the drop seems to be a cyclic (24-hour cycle) phenomenon (McWilliam, 1958). It is secreted early in the morning and evaporates or retracts during the day to be secreted again the next morning. The cycle goes on for a few days or until the ovule is pollinated — whichever is earlier. In *Podocarpus* the secretion seems to be a continuous process (Doyle, 1945). Baird (1953; see also McWilliam, 1958) has conclusively proved that the presence of the pollen on the drop causes a fairly rapid, complete, and permanent withdrawal of the fluid.

An impression seems to have been gained in earlier literature that the pollination drop is formed by breaking down of the cells of the nucellar tip; Dogra (1964) mentions this for cycads and van der Pijl (1953) for *Gnetum.* The main evidence against such an impression is that the liquid is exuded even in those plants (e.g., taxads, *Pinus*) in which the nucellar cells do not degenerate before pollination. It is not clear which region of the ovule is responsible for the secretion. McWilliam (1958) stated that, in *Pinus,* the drop is secreted by the cells forming the apex of the nucellus.

Tison (1911) postulated that the pollination drop contains glucose, calcium, and possibly malic acid. The presence of calcium might have been assumed on the basis of usual association of degeneration of nucellar cells with pollination drop formation, and that of malic acid on the idea that it is responsible for the chemotaxis of male gamete or pollen grain. A chemical analysis of the pollination drop of *Pinus nigra* (McWilliam, 1958) does not support this view. Using the methods of chromatography, amino acids and organic acids could not be detected; three sugars (1.25%) were the main organic constituents of the liquid — D-glucose (40 mmoles), D-fructose (40 mM), and sucrose (2.5 mM). Van der Pijl (1953) showed that the slimy pollination drop of *Gnetum gnemon* contains considerable quantities of reducing sugars. However, Ziegler (1959) reported several amino acids, peptides, and malic and citric acids in the pollination drops of *Ephedra* and *Taxus*. Inorganic phosphates and sugars were also reported. Sucrose concentration in the pollination drop of *Ephedra* was especially high (25%). It may be mentioned that even in artificial cultures, the pollen grains of *Ephedra* germinate in a very high concentration of sugar solution (Mehra, 1938).

C. Pollen Grain

The pollen grains are generally produced in large quantities and dispersed by wind. In some conifers the surrounding area becomes clouded by the yellow dust of pollen; the precipitation of yellow dust is spoken of as "sulfur showers." Male cones of some cycads emit a strong odor and several insects visit them when the pollen ripens (see Chamberlain, 1935; Pant and Mehra, 1962). This has lead to the belief that insects might also be the carriers of pollen in cycads. However, confirmation by critical field observations is necessary. Pearson (1929) and van der Pijl (1953) have presented evidence which indicates that *Welwitschia* and *Gnetum gnemon* are entomophilous.

The arrival of the pollen at the micropyle is purely a chance phenomenon (Florence and McWilliam, 1956; McWilliam, 1959). However, in a freshly pollinated ovule there is a greater concentration of pollen around the rim of the micropyle than on any other surface of the cone (Sarvas, 1955). This seems to be due to the sticky area of the micropyle–pollination drop rather than any bioelectrical phenomenon (McWilliam, 1959) although the pine pollen carries a negative electrostatic charge in the range of 4–7 mV.

When caught in the pollination drop, the pollen grains usually float on the latter's surface (Baird, 1953). In cycads, *Ginkgo, Pinus,* and *Picea* the pollen grains show a germinal furrow which closes in dry weather but becomes wide open in high humidity. The pollen tube comes out through this furrow. In several other plants such as *Ephedra,* members of Cupressaceae, and *Cephalotaxus* the exine is cast off before the pollen germinates.

In most gymnosperms the pollen caught at the micropyle reaches the nucellus where it germinates. Examples of pollen grains germinating a short distance from the nucellus, or inside the micropyle, are *Gnetum* (P. Maheswari and Vasil, 1961), *Tsuga, Pseudotsuga, Abies* (Doyle, 1945), and Caytoniales. In a few gymnosperms, such as *Agathis, Araucaria, Saxegothaea,* and *Tsuga dumosa* the pollen lands away from the micropyle and germinates *in situ* (Dogra, 1964). The pollen tubes grow toward the micropyle.

A feature of pollen grains in many conifers, which seems to be associated with the pollination mechanism is their winged nature. Two-winged pollen grains are found even in the extinct families of conifers (Doyle, 1945). Among the living conifers two-winged pollen grains are common in members of Pinaceae and *Podocarpus. Pherosphaera* and *Dacrydium* show three-winged pollen (Sterling, 1963). Doyle (1945) put forward the view that the wings help to orient the pollen grain in the pollination drop; the orientation is particularly necessary where the ovules are inverted, e.g., *Pinus* and *Podocarpus.* McWilliam (1958) did not support this view since he could not find any preferred orientation of pollen on the nucellus.

IV. Male Gametophyte

The terminology used by various authors for different cells of the male gametophyte of gymnosperms is quite varied. The male gametophyte has been reviewed by Sterling (1963) who, surprisingly, suggested that the microspore be called the "embryonal cell"—a term the usage of which is already well established in connection with the development of the proembryo in gymnosperms.

Except in *Gnetum* and *Welwitschia,* the development of the male gametophyte in gymnosperms follows a fairly uniform pattern. Prothallial cells are present in cycads (one cell), *Ginkgo* (one ephemeral cell and one persistent cell; De Sloover-Colinet, 1963), Pinaceae (two cells), Podocarpaceae (two cells that divide to form several secondary prothallial cells), Araucariaceae (six to forty secondary prothallial cells derived from two cells; Hodcent, 1963), and *Ephedra* (one cell + one nucleus or cell). The prothallial cells are absent in members of Cupressaceae, Taxodiaceae, *Cephalotaxus,* and Taxales (Sterling, 1963). After cutting off of prothallial cells, the nucleus of the central cell divides, giving rise to the tube cell and antheridial cell. The latter divides to form a sterile stalk cell and fertile body or generative cell (Fig. 8A-D) which eventually forms two equal or unequal male gametes. Multiple male gametes are known in *Microcycas* and a few species of *Cupressus* (Sterling, 1963).

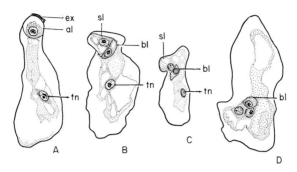

Fig. 8. Male gametophyte of *Cephalotaxus drupacea*. (A) Young pollen tube showing antheridial cell and tube nucleus. The exine of pollen grain is seen above the antheridial cell. (B) Antheridial cell has divided. (C) The wall around stalk cell is not distinct. (D) Stalk and body cells have reached near the tube nucleus. The stalk nucleus has dense cytoplasm around it. (al) Antheridial cell; (bl) body cell; (ex) exine; (sl) stalk cell; (tn) tube nucleus. (After Singh, 1961.)

In both *Gnetum* and *Welwitschia* the pollen grains are three-celled at the time they are shed. A controversy exists regarding the nature of the three cells in both of these genera. Of the three cells, two represent a tube cell and body cell, and the third is interpreted either as a prothallial cell or a stalk cell. The precise sequence by which the three cells arise in *Welwitschia* is not known. However, similar information on *Gnetum* has been very well illustrated and documented by several investigators (P. Maheshwari and Vasil, 1961).

A detailed study of the ontogenetic sequence indicated that a stalk cell is absent in *Gnetum,* but a prothallial cell is present (Negi and Madhulata, 1957; Waterkeyn, 1959; see also Battaglia, 1958). However, while reviewing this feature, Sterling (1963) seemed to have ignored the ontogenetic evidence and, purely on the basis of his belief that "prothallial cells are eliminated in evolution before" the stalk cell (he calls this a sterile cell), insists that the cell in dispute is the stalk cell and not the prothallial cell. In a preliminary study, Waterkeyn (1954) labeled the third cell *pied* (= stalk), but a more exhaustive study (Waterkeyn, 1959) led him to call it the prothallial cell.

The structure of male gametes is quite variable. In cycads and *Ginkgo* the gametes are top-shaped with a conspicuous spiral of cilia. Another interesting feature of the male gametophyte of these plants is the haustorial nature of the pollen tube which grows into the flanks of the nucellus. Thus, the grain end of the pollen tube grows toward the archegonia and contains male gametes. In members of Cupressaceae, Taxodiaceae, and Araucariaceae (Eames, 1913; Burlingame, 1913), mitosis of the body cell is followed by a cell plate to form two male cells, but in Pinaceae, *Cephalotaxus,* and *Ephedra* only male nuclei are formed. In Podocarpaceae and Taxales some plants show male nuclei, whereas others have male cells (see Konar and Oberoi, 1969).

In *Podocarpus gracilior* the nucleus of the mature body cell is eccentrically placed. A wall is not formed following division of the nucleus, and the resulting male nuclei are more-or-less similar in size and appearance. However, the more centrally placed nucleus enlarges quickly, and the other (now smaller) nucleus is wholly or partly extruded from the cytoplasm of the functional male cell. The larger nucleus comes to lie in the center of the cell (Konar and Oberoi, 1969).

There is no unanimity of opinion concerning the structure of male gametes of *Taxus* (Favre-Duchartre, 1960). For this plant, the description given by Favre-Duchartre conforms to that of *Podocarpus gracilior* except that an evanescent wall is formed between the nuclei in *Taxus.*

The ultrastructure of male gametes has been studied in a few gymnosperms. The gametes are usually devoid of a cell wall. In *Zamia integrifolia* the nucleus comprises the major bulk of the gamete (Norstog, 1967, 1968). The numerous flagella are connected through very long basal bodies to the spiral band. The latter comprises an outer electron-dense layer in which the basal bodies are embedded, a middle granular layer of varying thickness, and an inner fibrous layer. The latter contains microtubules which lie parallel to one another but are arranged obliquely. Mitochondria are uniformly distributed in the cytoplasm.

Light microscope observations on the cytoplasm of male gametes of *Biota* revealed that it is differentiated into three zones (Singh and Oberoi, 1962). The middle zone is most prominent, dense, and contains organelles such as mitochondria, amyloplasts, vesicle groups, and ribosomes (Chesnoy, 1969). The outermost zone shows only a few Golgi vesicles.

At the time of shedding, the pollen grain contains one or several nuclei. A one-celled shedding stage is uncommon, and has been reported in *Taxus* (Dupler, 1917), *Cryptomeria* (Singh and Chatterjee, 1963), and *Cupressus* (Konar and Banerjee; 1963). More commonly, the mature pollen comprises a variable number of prothallial cells, a tube nucleus, and an antheridial cell.

In most gymnosperms the pollen tube is prominent, short, and unbranched, and as it makes its way through the nucellus, the surrounding cells become crushed and disorganized (Willemse and Linskins, 1969). In *Ephedra* the pollen tube is narrow and very short. Since the pollen chamber is rather deep, the pollen grains land more or less on the female gametophyte. The pollen tube shows an exceptional structure and behavior in cycads and *Ginkgo*. In a few podocarps, as the pollen tube comes in contact with the female gametophyte, it expands and gives out numerous small branches which grow downward all around (see Konar and Oberoi, 1969). Usually, the pollen tubes reach the female gametophyte when it shows mature archegonia but, in *Taxus,* several podocarps, and members of Taxodiaceae and Cupressaceae the pollen tube enters when the female gametophyte is only at the free nuclear stage or has just become cellular (P. Maheshwari and Singh, 1967). Development of the archegonia is usually hastened by early arrival of the pollen tube.

V. Postpollination–Prefertilization Phase

There is usually a long time lapse between pollination and fertilization, except for *Ephedra* in which fertilization occurs about 10 to 15 hours after pollination. During this period, the ovule shows several distinct changes.

A. Closing of Micropyle

Concerning *Gnetum,* Rodin and Kapil (1969) state, "Soon after pollination, or perhaps before pollination in some species, cells lining the micropylar tube enlarge and fill the tube with 'closing tissue' (Berridge, 1911). This tissue is not known among any other living gymnosperms but was discussed by Berridge in relation to the Bennettitales." It may be mentioned that Berridge examined only a few ovules of *Gnetum* which were past the stage of fertilization and, therefore, she did not mention the presence of plugging tissue in prepollination stages. Further, although prepollination plugging of the micropyle seems to be rare (known to occur in two species of *Gnetum;* see Section II,E), the postpollination plugging of the micropyle in gymnosperms occurs almost universally. Rodin and Kapil (1969) seem to be in error in writing that the "closing tissue" is not known among any other living gymnosperm.

The occlusion of the micropyle usually takes place by elongation, followed by cell division in the epidermal and/or subepidermal cells lining the micropyle (Fig. 9A–C) (Looby and Doyle, 1942; Singh, 1961; P. Maheshwari and Vasil, 1961). In *Pinus* the epidermal cells do not undergo any change, but the subepidermal cells elongate radially causing an annular bulge in the micropyle. As the swelling proceeds, the canal becomes closed (McWilliam, 1958). In *Cedrus* the plugging tissue does not seem to be formed around the entire micropylar canal (Roy Chowdhury, 1961).

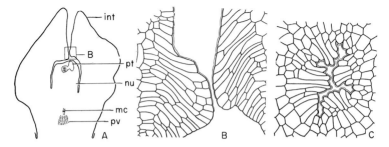

FIG. 9. Postpollination closing of micropyle in *Cephalotaxus drupacea*. (A) Longisection of ovule after pollination. (B) Portion marked B in figure A enlarged to show the elongation and divisions in cells that close micropyle. (C) Transection of closed micropyle. (int) Integument; (mc) megaspore mother cell; (nu) nucellus; (pv) pavement tissue; (pt) pollen tube. (After Singh, 1961.)

In *Ephedra* the cells of the inner integument lining the micropylar canal do not undergo any change, but the outer integument shows small papillate outgrowths directed toward the inside (Singh and Maheshwari, 1962; Lehmann-Baerts, 1967b). As the ovule matures, the papillae elongate and become thick-walled (Fig. 10A and B). They help in closing the space between the two integuments and may also seal off the micropyle by pressing inward on the micropylar tube.

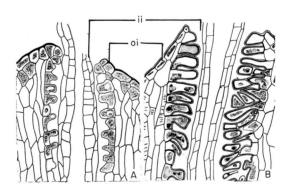

FIG. 10. Postpollination closing of micropyle in *Ephedra gerardiana*. (A) Longisection of micropylar canal showing numerous papillate projections from outer integument. (B) The papillae have enlarged and become thick-walled. (ii) Inner integument; (oi) outer integument. (After Singh and Maheshwari, 1962.)

B. Enlargement of Ovules

Enlargement of the ovule is the most conspicuous change in the appearance of a pollinated ovule. However, its anatomical aspects have not been studied in detail in any gymnosperm. Frequently, the result of this enlargement is erroneously described as the integument and nucellus either being free from or united with each other up to the base/three-fourth length/tip of ovule. Both Schnarf (1933) and Sporne (1965) have drawn attention to this phenomenon.

In some living gymnosperms such as cycads and *Cephalotaxus,* the enlargement results from activity of a meristem which is located in the massive chalazal region of the ovule. Some enlargement may occur in the micropylar region also, but this activity is limited compared to the massive growth of the chalaza (Fig. 11A–F). As a result of this unequal en-

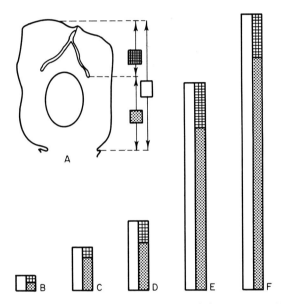

FIG. 11. Postpollination enlargement of ovule in *Encephalartos poggei.* (A) Longisection of ovule at free-nuclear stage. (B–F) Histograms to show the contribution of chalazal and micropylar portions to total length of the ovule at prepollination stages (B,C), and successively older postpollination stages (D–F). Greater portion of mature seed (F) is formed by chalazal portion. (Modified from data and diagrams of De-Sloover, 1964.)

largement of the two regions, in the lower massive portion of the ovule, the integument and nucellus cannot be demarcated (Fig. 12A–D). Therefore, they are described as fused even though no ontogenetic fusion takes place. In such ovules, a greater part of the seed coat arises from activity of the chalaza. Judging from the photographs of longisections of ovules of *Macrozamia* (see Figs. 26–29 of Baird, 1939) the unequal enlargement of different parts of the ovule seems to begin before pollination.

In ovules of some fossils such as *Pachytesta* and *Trigonocarpus* the nucellus has been described to be free from the integument right up to

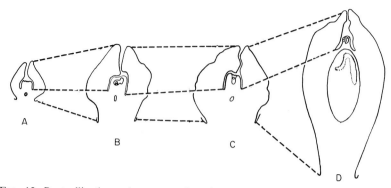

FIG. 12. Postpollination enlargement of ovule of *Cephalotaxus drupacea*. (A) Longi-section of prepollination ovule. (B–D) Postpollination ovules in successive stages of development. Enlargement of chalazal portion (below the level of insertion of integument) of the ovule is evident in figure D. (Modified from Singh, 1961.)

the base of ovule (Taylor, 1965). This appearance might as well be due to the greater enlargement of the micropylar region, as compared to the chalazal region of the ovule. In such ovules the seed coat arises chiefly from the integument. In some living gymnosperms also, e.g., *Callitris* (Baird, 1953), *Sequoia* (Looby and Doyle, 1942), and *Athrotaxis* (Brennan and Doyle, 1956), the nucellus seems to be free from the integument right up to the base of the ovule. This condition appears to be brought about by an initial low insertion of the integument on the ovular primordium, and the presence of a distinct meristem in the nucellus extending from below the sporogenous tissue to the insertion of the integument (Fig. 13A–D). The latter seems to be keeping pace with growth of the nucellus (Fig. 14A–E).

On the basis of a study of nearly mature seeds of four species of *Gnetum,* Rodin and Kapil (1969) have classified the seed coats in this genus into three categories: (*a*) species with three (it is implied that the three layers are derived, one each, from the three integuments) free or nearly free layers of testa, represented by *Gnetum gnemon;* (*b*) species in which the outer two layers of seed coat are fused except the apical one-sixth portion, represented by *Gnetum ula;* and (*c*) species in which the outer two layers are completely fused, represented by *Gnetum montanum* and *Gnetum neglectum.* It may be mentioned that the three integuments initiate independently (see P. Maheshwari and Vasil, 1961; Lehmann-Baerts, 1967a), and the fusion of integuments mentioned by Rodin and Kapil (1969) does not seem to be real; such an appearance may be due to unequal growth of various parts of the ovule. In *G. ula* there may be only one meristem which is responsible for the growth of outer and middle envelopes so that they appear fused in later stages. Nothing is known

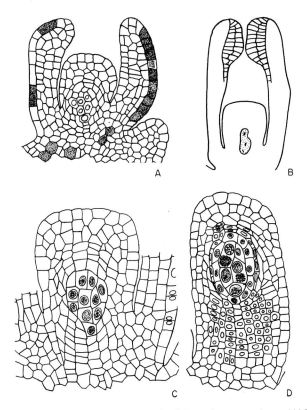

FIG. 13. Postpollination enlargement of ovule of *Sequoia sempervirens*. (A) Longisection of prepollination ovule. (B) Outline diagram of a postpollination ovule; note closed micropyle. (C) Nucellus enlarged to show meristematic cells in chalazal portion. (D) Later stage; nucellus has grown due to activity of meristematic cells. (After Looby and Doyle, 1942.)

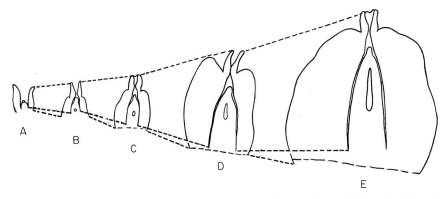

FIG. 14. Postpollination enlargement of ovule of *Callitris robusta*. (A) Longisection of ovule at pollination. (B–E) Successive stages of development to show greater enlargement of micropylar portion of ovule as compared to chalazal portion. (Modified from Baird, 1953.)

about the initiation of integument in *G. montanum* and *G. neglectum* (where the outer two integuments have been reported to be completely fused). At any rate, there is an urgent need to study the mechanics of growth of the ovule and the location and functioning of the meristems connected with postpollination growth.

In the young ovules of *Ephedra* the two integuments are inserted at nearly the same level (P. Maheshwari, 1935; Singh and Maheshwari, 1962; Pankow, 1962; Lehmann-Baerts, 1967b), but during later stages, the inner integument seems to be inserted at a much higher level (Fig. 15A and B). This may be due to the activity of a meristem which brings

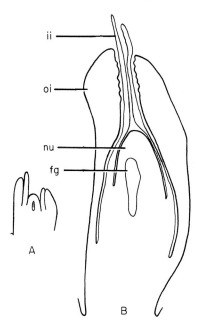

Fig. 15. Enlargement of ovule of *Ephedra foliata*. (fg) Female gametophyte; (ii) inner integument; (nu) nucellus; (oi) outer integument. (A) Longisection of young ovule; the two integuments are inserted at the same level. (B) Older ovule; the outer integument is inserted at a lower level as compared to inner one. (After P. Maheshwari, 1935.)

about the growth of the region which lies below the nucellus and inner integument, whereas the growth of the outer integument may be caused by an independent meristem.

C. Nutritive Tissue

The usual form of nutritive tissue is one or more layers of densely cytoplasmic cells which differentiate around the sporogenous cells, become progressively conspicuous during subsequent stages, and disorganize when the female gametophyte becomes cellular. This tissue is usually spoken of as "spongy tissue" or tapetum. The spongy tissue is very well-developed in cycads, *Ginkgo,* members of Pinaceae, Taxodiaceae, Cupressaceae, Araucariaceae, and Podocarpaceae. It is poorly developed in *Taxus canadensis* (Dupler, 1917) and is absent in *Athrotaxis* (Brennan and Doyle, 1956), *Austrotaxus* (Saxton, 1934), *Callitris* (Baird, 1953), *Cephalotaxus* (Singh, 1961), *Ephedra* (Lehmann-Baerts, 1967b), *Gnetum* (P. Maheshwari and Vasil, 1961), *Taxus cuspidata* (Sterling, 1948), *Torreya* (Coulter and Land, 1905) and *Welwitschia* (Pearson, 1929).

In *Cephalotaxus drupacea* the nucellar cells, in the vicinity of the sporogenous tissue at the chalazal end, undergo periclinal divisions to form a pavement tissue (Singh, 1961). This tissue comprises small cells arranged in radiating rows, and converging toward the megaspore mother cell. In later stages the cells show compound starch grains and become crushed during enlargement of the free-nuclear gametophyte. A pavement tissue is present in *Gnetum* also and is very prominent due to the dense cytoplasm of cells (P. Maheshwari and Vasil, 1961). *Callitris* and *Actinostrobus* show a pavement tissue similar to *Gnetum* and *Cephalotaxus* (Looby and Doyle, 1940; Baird, 1953; Saxton, 1913).

The spongy tissue is derived from the sporogenous cells which do not function as megaspore mother cells or from nucellar cells that surround the sporogenous tissue. And, accordingly, the spongy tissue has been designated as primary or secondary tapetum (see Thomson, 1905). However, this nomenclature has not been followed widely; Chamberlain (1935) did not mention it. The tapetum usually forms a uniformly thick layer around the female gametophyte but, in *Sequoiadendron* (Looby and Doyle, 1942), it is rather thick at the chalazal end of the gametophyte and becomes thinner higher up where it is represented by an occasional degenerated cell at the micropylar end. In *Fitzroya* the spongy tissue is discontinuous (Doyle and Saxton, 1933).

The cells of the spongy tissue are usually uninucleate and densely cytoplasmic. Occasionally, they may contain starch grains as in *Pinus* (Ferguson, 1904). In *Ginkgo* the cell walls of the spongy tissue, in direct contact with the gametophyte, enlarge and their inner walls often partly dissolve (Lee, 1955). The nuclei of these cells enlarge and become nearly twice the size of other nuclei. Occasionally, these nuclei become polyploid.

When the female gametophyte is in the free-nuclear stage in *Zamia floridana,* the tapetal cells, especially those near the gametophyte, contain numerous fine lipoidal bodies (Pettitt, 1966). The cell wall of the spongy tissue is acetolysis-resistant and perforated. The walls of the innermost tapetal cells are somewhat different in appearance and represented by a thin sheet of resistant material of which the inner surface is covered by sudanophilic droplets or merely an aggregation of droplets. The droplets coalesce to form clusters that arrange themselves against the tapetal cell wall. Pettitt made similar observations on *Cycas revoluta, Encephalartos,* and *Pinus excelsa.* He presented evidence to suggest that these droplets are responsible for the deposition of material on the outer surface of megaspore membrane. Favre-Duchartre (1956) made similar observations for *Ginkgo.*

The spongy tissue eventually degenerates and becomes compressed between the female gametophyte and outer tissues of the ovule. According to Pettitt (1966), this compressed layer of acetolysis-proof material has been interpreted as the megaspore membrane in plants such as cycads and *Ginkgo.* In *Cedrus deodara* the single-layered spongy tissue is persistent, and its remnants can be distinguished as a distinct layer even in the mature seed (Roy Chowdhury, 1961).

D. Female Gametophyte

Except for *Gnetum* and *Welwitschia,* the female gametophyte of gymnosperms follows a fairly uniform pattern of development (see P. Maheshwari and Singh, 1967). The megaspore undergoes several free nuclear divisions, a large vacuole appears in the center of the young gametophyte, and the nuclei come to lie in a thin layer of cytoplasm along the periphery of the sac. During earlier stages the nuclear divisions are generally synchronous. The number of free nuclei is almost constant for a species, perhaps depending on the size of the gametophyte. For example, twelve mitotic waves occur in *Cephalotaxus drupacea* (Favre-Duchartre, 1957). At the end of the last mitosis, secondary spindles develop so that every nucleus becomes connected by spindle fibers to six adjacent nuclei. The spindles lie parallel to the surface of the gametophyte. Anticlinal walls are laid down centripetally, and the gametophyte takes on the appearance of a honeycomb. Each honeycomb cavity is an "alveolus," sometimes called a primary prothallial cell. The alveoli do not have walls near the central vacuole of the gametophyte and are, thus, open at their inner ends. In cross section they appear hexagonal. Each honeycomb cavity grows toward the center of the gametophyte, with the nucleus at the open end and the spindles still connecting the adjacent nuclei. The persisting spindles seem to guide the laying down of the wall material. As the alveoli extend inward, the circumference of the vacuole decreases and some alveoli become closed. The closure of an alveolus is initiated by the lagging behind of its nucleus. As the nucleus lags behind, the spindles connecting it with the neighboring nuclei dip down (along with the nucleus) and the middle regions of the spindles together with the membranes which have developed on them approach each other. The approaching spindles coalesce, and, thus, the nuclei previously separated by the closing alveolus become connected by spindle fibers. The membranes on this new spindle continue inward from the point of closure. Ultimately, the nucleus passes back slightly into the closed alveolus and is ready to initiate division which will result in the formation of cellular tissue. Thus, different alveoli become closed at variable distances from the center of the gametophyte.

Some of these undivided and closed alveoli, situated at the micropylar end, function as archegonial initials. Other alveoli undergo a series of periclinal divisions so that the gametophyte appears to be formed of rows of radiating cells. However, this arrangement may be lost in older gametophytes owing to the laying down of irregular walls. A few minor variations of this mode of development are known in *Actinostrobus, Athrotaxis, Callitris,* and *Sequoia* (P. Maheshwari and Singh, 1967).

The archegonial initial divides periclinally, giving rise to a large central cell and a much smaller neck initial which undergoes a few divisions to form a short neck. The central cell enlarges considerably, with its nucleus lying just below the neck. Its cytoplasm becomes highly vacuolate. This phase is referred to as the foam stage of the archegonium. Eventually, the cytoplasm becomes dense and numerous "proteid vacuoles" come to lie in it. The nucleus of the central cell usually divides forming an ephemeral ventral canal cell or nucleus and the egg nucleus. The latter descends a little into the archegonium and enlarges considerably. The cells surrounding the archegonium divide actively, acquire dense cytoplasm and prominent nuclei, and form the jacket. Generally, this jacket is one-layered but, sometimes, two or three layers are present.

The archegonia occur singly in cycads, *Ginkgo, Ephedra,* members of Pinaceae, Podocarpaceae, Taxales, Cephalotaxaceae, and Araucariaceae, but they are grouped into one or more complexes which show a common jacket layer in Cupressaceae and Taxodiaceae. The archegonia are usually placed at the micropylar end of the gametophyte but, in several taxa, the archegonial complexes are placed laterally.

In both *Gnetum* and *Welwitschia* the free-nuclear gametophyte shows several small vacuoles and the nuclei are arranged all over the cytoplasm. Laying down of the walls takes place by free cell formation. In *Gnetum,* a few cells or free nuclei lying in the micropylar region of the gametophyte become prominent as the pollen tube arrives and function as egg nuclei (P. Maheshwari and Vasil, 1961). In *Welwitschia* the cells at the micropylar end of the gametophyte contain two or three nuclei, and several of these cells grow up through the nucellus in the form of slender "embryo sac tubes" (Pearson, 1929). The cytoplasm and the nuclei also migrate in the same direction. An embryo sac tube continues to elongate until a pollen tube is encountered. The nucleus which lies foremost in the embryo sac tube is apparently the one to be fertilized. It is stated that fusion takes place in the body of the pollen tube.

A peculiar feature of the gametophyte of *Ginkgo* and several other fossil plants is the presence of a columnlike structure at the micropylar end, called the *tent pole*. It is formed by active cell divisions in the portion of the gametophyte situated between the archegonia.

In cycads the portion of the gametophyte around the archegonia grows up in an annular fashion, leaving a depression in the middle, the archegonial chamber. A similar depression is also present in those conifers (e.g., *Biota;* Singh and Oberoi, 1962) that show a micropylar archegonial complex. In most conifers the mature archegonia are somewhat deep-seated owing to the upward growth of gametophytic cells situated around the neck of archegonia. A similar deep-seating of the archegonia in *Ephedra* is due to a very large number (thirty to forty) of neck cells.

The gametophyte is usually oval in outline, but in *Ephedra* and *Gnetum* the gametophyte has the outline of an inverted flask. In *Ephedra* the pointed and drawn-out chalazal portion of the gametophyte penetrates deep into the chalaza and functions as a haustorium.

The mature egg cell of most gymnosperms is large and prominent and contains a dense and nearly nonvacuolate cytoplasm which is gorged with numerous densely staining bodies. Its nucleus is usually large and filled with nucleoplasm, but the chromatin is inconspicuous.

The cytoplasm of the central cell (showing the foam stage) of *Pinus nigra* is electron transparent, and the various organelles, such as mitochondria, plastids, and dictyosomes, are concentrated near the periphery of the cell or around the nucleus (Camefort, 1962, 1965a). The numerous vacuoles present at this stage are bound by a single incomplete membrane and may even contain cytoplasmic organelles. Eventually, the enclosing membrane disappears and the contents become as dense as the general cytoplasm. The vacuoles apparently do not seem to give rise to any inclusions of the mature egg cell.

The organelles in the micropylar portion of cytoplasm of the central cell of *Juniperus communis* are arranged in a special manner and have been described as *asteroids* (Chesnoy, 1967). A small vacuole lies immediately below the nucleus of the central cell, and the asteroids are located below the vacuole. The asteroids comprise mitochondria and leucoplasts arranged in a radial manner around a mass of ribosomes and microtubules. It has been suggested that this area of organelle assembly might also be the center of organelle multiplication. *Biota orientalis* also shows a similar organization of organelles in its central cell (Chesnoy, 1969). Peculiar annulated lamellae are found in the central cell of *Sciadopitys verticillata* (Gianordoli, 1969). Similar structures have only been described to be occurring in animal cells.

Electron-microscopic observations on the egg cytoplasm of *Ginkgo, Larix, Pinus,* and *Pseudotsuga* have shown that the classically described proteid vacuoles are only islets of cytoplasm that have become isolated by a double membrane derived from the ER (Camefort, 1965a,b; Thomas and Chesnoy, 1969). They arise by the deformation of plastids in the young central cell. Microbodies and vesicular bodies partly covered by single membranes are common (Fig. 16). These cytoplasmic nodules usually remain connected with the general cytoplasm by short peduncles.

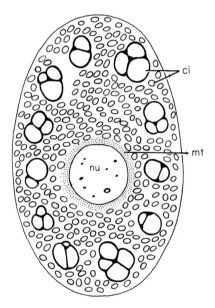

FIG. 16. Schematic representation of fine structure of *Larix* egg. (ci) Cytoplasmic inclusions; (mt) mitochondria; (nu) nucleus. (After Camefort, 1968.)

Plastids covered by a membrane of ER are also present in the cytoplasm. Cytoplasmic inclusions in the egg of cupressacious plants are only of one type and are similar to the microbodies described above (Chesnoy, 1967, 1969).

The egg cytoplasm contains a large number of mitochondria. The latter are scattered in the mature egg of *Biota* and *Pinus,* but in *Larix* and *Pseudotsuga* (Camefort, 1967; Chesnoy, 1969; Thomas and Chesnoy,

1969) the mitochondria accumulate around the egg nucleus to form the perinuclear zone which is very rich in mitochondrial deoxyribonucleic acid (DNA) and even stains with the Feulgen nuclear reaction.

Several papers have appeared on the structure of the egg nucleus. Shimamura (1956) showed that the nucleus is Feulgen-negative in *Pinus thunbergii*. In *Ginkgo* (Lee, 1955) and *Zamia* (Bryan and Evans, 1956) the large egg nucleus was shown to contain a small globule of Feulgen-positive chromatin. Favre-Duchartre (1956), however, reported the occurrence of several fine chromatic filaments in the egg nucleus of *Ginkgo*. A similar organization of the chromatin has also been reported in *Pinus* on the basis of electron-microscopic studies (Camefort, 1959, 1964). The chromatic filaments are formed by the assemblage of fibrils which are about 100 Å in diameter. Camefort has reported numerous (around 100) nucleoli in the large egg nucleus of *Pinus;* the nucleoli comprise only fibrils which are around 50 Å in diameter. In most other organisms the nucleoli comprise fibrils and granules. In *Pinus* nucleolus shows a few zones in its internal structure, but these zones are formed merely by the compact or sparse arrangement of the fibrils.

E. Megaspore Membrane

The megaspore membrane is very prominent in many fossil seeds, cycads and *Ginkgo,* but is recognizable in almost all gymnosperms. It is sudanophilic and, therefore, can be broadly classified as a lipid (Pettitt, 1966).

The membrane covers the entire gametophyte, but in cycads it has been shown to become ruptured in the region of the archegonial chamber (Chamberlain, 1935). The thickness of the membrane does not seem to be uniform around the gametophyte, and the variations in this regard have already been pointed out by P. Maheshwari and Singh (1967). The presence of the membrane is mentioned in most of the papers dealing with structure and development of the ovule, but there are only few publications dealing with the detailed structure of the membrane during its ontogeny.

The first comprehensive investigation on the structure and chemical composition of the megaspore membrane was reported by Thomson (1905), and this work has been referred to in various publications on the morphology of gymnosperms. Although his techniques were primitive, his observations seem accurate. Broadly, he regarded the membrane as composed of an outer suberized "exosporium" and an inner cellulose-pectinacious "endosporium," the latter comprising two homogeneous layers. He compared the membrane with the exine–intine of pollen grains and considered the two structures to be fundamentally alike in their morphology and chemical composition. Influenced perhaps by this likeness, the pollen-centered terms exine and intine have come to be applied for exosporium and endosporium, respectively. More recently, Erdtman's (1952, 1957) terminology of pollen wall stratification has been used for the megaspore membrane.

Von Lürzer (1956) investigated the structure of the membrane in several conifers by acetolysing the gametophyte. The endospore dissolves away, and the exospore persists after such treatment. The latter can be differentiated into an outer sculptured sexine and an inner smooth nexine. The sexine is piloid, the pila being free from each other in the Pinaceae, but their swollen terminal portions are united in the Cupressaceae.

By using refined histochemical methods, Favre-Duchartre (1956) described stratification and development of the megaspore membrane in *Ginkgo biloba*. At the early free-nuclear stage the gametophyte is surrounded by a pectocellulose layer (the intine) and later another membrane (the exine), which is sudanophilic and comprises perpendicular rodlike elements, differentiates outside the intine. The exine grows in thickness, and this has been ascribed to the deposition of particles originating in the surrounding spongy tissue. As the development proceeds, the gametophyte becomes covered by yet another homogeneous sudanophilic layer which lies between the gametophyte and the megaspore membrane and has been termed the cuticle.

The megaspore membrane of *Encephalartos poggei* has been studied in great detail; the work begun by Martens (1957) has been extended by De Sloover (1963). This is based on unacetolysed as well as acetolysed material, stained with various histochemical dyes and studied with both light and electron microscopes. It has been shown that each of the two main layers can be further differentiated into two layers. Moreover, another membrane (which is continuous with the radial walls of the gametophyte) differentiates in later stages, and this finally comprises a cuticle and cellulose layer. The membrane covering the gametophyte can thus be differentiated into six layers at maturity. During the free-nuclear stages of the gametophyte, the megaspore membrane comprises two layers—an outer layer of rodlike structures (pila of Erdtman, 1952) standing vertically to the surface and an inner homogeneous layer. As the gametophyte becomes cellular, the rodlike structures increase in size and a new layer (basiosexine) develops just below the radial layer. The inner homogeneous layer differentiates into two zones—a sporopollenine zone (nexine) and a pectic zone (medine). During further growth the base (endosexine) of each rodlike structure fragments repeatedly, the middle portion narrows somewhat, and the top (ectosexine) becomes slightly swollen. Eventually, the megaspore membrane is crushed and loses its elaborate differentiation. The rodlike structures become flattened and they coalesce with each other irregularly. De Sloover (as also mentioned by Favre-Duchartre, 1956) found the deposition on exine of granules that originated in the spongy tissue, but he did not consider this process as exine formation. He presumes that exine is a distinct layer before the granules begin to deposit.

Thus, the concept of the presence of exine and intine as components of the megaspore membrane is fairly well-established. However, Pettitt (1966), on the basis of his study of the megaspore membrane as well as the pollen coat of several gymnosperms, has presented an entirely new interpretation. He considers that the structure which has been habitually labeled the megaspore membrane represents the remains of the spongy tissue. The two structures have similar chemical composition. Pettitt believes that the true megaspore membrane is represented by a thin acetolysis-proof membrane closely associated with the peripheral cells of the embryo sac. This membrane is the one labeled "cuticle" by Favre-Duchartre and De Sloover. The membrane looks like a cellular reticulum in surface view, and Pettitt has shown that this is because the membrane has strong (but stubby) anticlinal walls. This interpretation is certainly a novel one and has good arguments in its favor but requires confirmation.

VI. Fertilization

The mechanism and cytological aspects of fertilization have been described for many gymnosperms which show archegonia and the account is more or less similar for these plants. Besides several old publications on this subject, some additional ones are by McWilliam and Mergen (1958) for *Pinus,* Singh (1961) for *Cephalotaxus,* and Roy Chowdhury (1961) for *Cedrus* (see also Linskens, 1969).

The broad pollen tube forces its way through the nucellus of which the cells that lie in the way become crushed. The tube finally ruptures the neck cells of the archegonium and delivers its contents (usually two male gametes, a stalk nucleus, and a tube nucleus) into the egg cytoplasm. In cycads the grain end of the pollen tube releases the male gametes into the archegonial chamber (Chamberlain, 1935). It is not clear as to how the large motile gamete finds its way to the egg which is covered by closely adhering neck cells. Once inside the egg cytoplasm, the nonfunctional nuclei usually do not move and persist *in situ* (or occasionally exhibit a freak type of division) for some time. In *Ephedra,* one of the male gametes may fuse with the ventral canal nucleus, and the diploid nucleus may undergo a few irregular divisions (Khan, 1943). The fusion has been termed "double fertilization" although in no way does it resemble double fertilization as it occurs in angiosperms.

The functional male gamete (usually carrying a portion of male cytoplasm) approaches the egg nucleus and, after making contact, sinks into it and also becomes lenticular. Finally the two nuclei form a continuous surface, the nuclear membranes disappear at the point of contact, and the chromatin of each nucleus condenses, forming a prophase configuration. The two groups of chromatin approach each other, and numerous fibrils appear in the fusion nucleus. While the two chromatin groups merge, the fibrils become arranged in a bipolar spindle. Meanwhile, the chromatin threads shorten and thicken and become arranged in a metaphase plate which is somewhat oblique to the long axis of the archegonium. The chromosomes divide longitudinally and pass to the opposite poles. Some aspects of this behavior have been confirmed by a study using the electron microscope (Camefort, 1968).

An electron-microscopic study of the behavior of cytoplasm of the male and female gametes during fertilization and subsequent development, in members of Pinaceae, has led to the concept of the formation of a new cytoplasm (neocytoplasm) which arises from the nucleoplasm of the female gamete, perinuclear zone of the egg nucleus, and a part of the cytoplasm associated with the male gamete. The mitochondria are contributed mainly by the female cytoplasm, and the plastids only by the male cytoplasm (Camefort, 1969). The neocytoplasm alone takes part in formation of the embryonal cytoplasm, while the remaining cytoplasm of the zygote degenerates. In *Biota* the neocytoplasm seems to be derived mainly from the male gamete (Chesnoy, 1969). It is evident that the varied sources of neocytoplasm in gymnosperms are likely to have implications in cytoplasmic inheritance.

VII. Embryogeny

One of the important differences between early embryogeny of angiosperms and gymnosperms is that the latter show a free-nuclear phase, whereas in angiosperms division of the zygote is followed by wall formation (Johansen, 1950). Exceptions are now well-established in both groups. Among the gymnosperms in *Sequoia, Gnetum,* and probably also *Welwitschia,* division of the zygote is followed by a cell wall (Buchholz, 1939; P. Maheshwari and Vasil, 1961); and, among the angiosperms, the young embryo is said to be free-nuclear in *Paeonia* (Wunderlich, 1966). Gymnosperms seem to represent a heterogeneous group and the development of embryo roughly falls into four types: (*a*) cycad and Ginkgo; (*b*) conifer; (*c*) Ephedra and (*d*) Gnetum and Welwitschia.

In embryo development all the stages before the elongation of suspensor are spoken of as proembryo.

A. Cycad and Ginkgo Type

The zygote nucleus divides *in situ* followed by several free-nuclear
divisions. The nuclei become evenly distributed throughout the entire
proembryo. Sometimes, evanescent walls develop during the free-nuclear
period (Favre-Duchartre, 1956; Chamberlain, 1910). In later stages the
free nuclei are more-or-less evenly distributed in the cytoplasm in *Ginkgo,*
but in cycads the free nuclei concentrate at the base of the proembryo
and the upper portion contains only a few nuclei in the thin cytoplasm.
Subsequently, only the nuclei at the base divide whereas the upper nuclei
show signs of degeneration (Bryan, 1952). At the time of wall formation
there are 256 free nuclei in *Ginkgo;* 512 in *Cycas circinalis* (L. N. Rao,
1963), and 512 or 1024 in *Dioon.* The newly formed cells fill the entire
proembryo in *Ginkgo,* but in cycads the cells form only in the lower
portion of the proembryo. Following wall formation, the cells at the base
divide and function as embryonal cells, whereas the upper cells elongate
to form a massive suspensor. The distal embryonal cells keep on adding
to the elongating suspensor so that embryonal cells are pushed deep into
the central portion of female gametophyte. In *Zamia* and *Cycas* (and
possibly in other cycads also) the embryonal cells of the first layer elongate somewhat to form a conspicuous cap on the meristematic embryonal
cells (Bryan, 1952; K. Maheshwari, 1960). The cap cells persist for some
time but, eventually, degenerate and do not contribute to the tissues of
mature embryo. More than one young embryo may be present in some
seeds and this usually results from the formation of more than one zygote
(simple or archegonial polyembryony).

B. Conifer Type

In embryo development of conifers and taxads (see Doyle, 1963), the
first mitosis is usually intranuclear and the resulting nuclei are formed
within the nucleoplasm of the zygote. The nuclear membrane of the
zygotic nucleus disappears at the end of mitosis (Camefort, 1968). The
two nuclei move toward the base of the archegonium where further synchronous mitoses take place (Fig. 17A). All the free nuclei lie in the
chalazal portion of the zygote (Fig. 17B) which contains dense cytoplasm (termed "neocytoplasm" by Camefort, 1969). The number of free
nuclei, before walls are initiated, is variable: four in *Athrotaxis;* eight in
Pinus; sixteen in *Cephalotaxus;* thirty-two in *Podocarpus andinus;* and
sixty-four in *Agathis* (Roy Chowdhury, 1962). The cell walls are laid

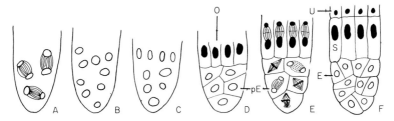

FIG. 17. Proembryo development in conifers. (A–D) Lower portion of archegonia showing synchronous division (A) of free nuclei, their subsequent arrangement and wall formation (D) to give two tiers of cells. (E) Proembryo showing internal division of cells. (F) Three-tiered mature proembryo. (E) Embryonal tier; (O) open tier; (pE) primary embryonal tier; (S) suspensor tier; (U) upper tier. (After Doyle, 1963.)

down on the telophase spindle of the last mitoses and also on the secondary spindles which may arise after the last mitoses (Dogra, 1967). Following wall formation, there arises a lower group of variously arranged (arranged in one layer in members of Pinaceae) primary embryonal cells and an upper group of cells arranged in a single layer, the open tier (primary upper tier of Dogra, 1967) (Fig. 17B–D). The cells of the open tier have no walls on the upper side and are, therefore, continuous with the general cytoplasm of the proembryo. All the cells now divide (the internal division, Fig. 17E) and this results in doubling (in podocarps the cells become binucleate) the number of cells of the lower tier, now called the *embryonal tier;* and the cells of the open tier divide transversely to form an upper tier and a lower suspensor tier (Fig. 17F). The cells of the upper tier eventually degenerate, the suspensor elongates, and the embryonal cells divide to form a mass. The distal cells of the mass elongate giving rise to a massive suspensor — the secondary suspensor — to which are later added more cells of the embryonal mass.

   In most members of Pinaceae the proembryo shows four tiers of four cells each. The lower two tiers comprise the embryonal tier; next is the suspensor tier (habitually called the *rosette tier*); and uppermost is the upper tier comprising cells open at the top. The cells of the suspensor tier do not elongate but divide to form lobes of an evanescent mass of cells which has been termed *rosette embryos.* Since these masses never form an embryo, the usage of this term appears objectionable (Doyle, 1963; Dogra, 1967). The function of the suspensor is taken up by the distal embryonal tier.

   Minor variations of this basal plan of development occur in *Actinostrobus, Athrotaxis, Callitris, Cupressus sempervirens, Fitzroya, Sequoia, Torreya,* and *Widdringtonia* (Doyle, 1963). However, these variations have been resolved to interpret the embryogenies as conforming to the basal plan.

Embryogeny of the araucarians seems to differ radically from the rest of the conifers and taxads. The free nuclei remain in the middle of the proembryo and wall formation results in a central group of cells surrounded by a jacket of peripheral cells (Eames, 1913; Burlingame, 1915). The proximal cells of the jacket develop into cap cells, and the distal cells into suspensor. The central cells are the embryonal cells. The cap cells elongate somewhat and look prominent but eventually degenerate.

Most conifers show polyembryony — both simple and cleavage. The latter usually results from the different rates of elongation of the components of suspensor or secondary suspensor, so that a suspensor cell or a group of them bearing one or a few embryonal cells become separated from the embryo system and develop independently.

C. *Ephedra Type*

Although the embryogeny of only a few species of *Ephedra* has been worked out, the various accounts are more or less similar (Lehmann-Baerts, 1967b). The zygote nucleus divides *in situ* and the two nuclei move apart, sometimes occupying the two poles of the zygote. Two more mitoses follow to produce eight nuclei (Fig. 18A and B). Each nucleus is surrounded by a densely staining cytoplasmic sheath with radiating strands (Fig. 18B). At the eight-nucleate stage an independent cell wall develops around each nucleus resulting in eight units. The latter develop further and put out a tubular outgrowth. The nucleus may divide before the formation of the outgrowth (Fig. 18C) or it moves into the tube and divides. A transverse wall is then laid down giving rise to an embryonal cell and the suspensor cell which elongates (Fig. 18D). The embryonal cell comes to lie deeper in the gametophyte; it divides and the distal cells elongate to form secondary suspensor (Fig. 18E–H). *Ephedra* shows both simple and cleavage polyembryony.

D. *Gnetum and Welwitschia Type*

Several zygotes are formed in the female gametophyte. A prominent cell wall surrounds the zygote. There is minor variability in the early behavior of the zygote in different species of *Gnetum* (P. Maheshwari and Vasil, 1961). In *Gnetum gnemon* the zygote generally puts out a branched tube and the nucleus migrates into one of the branches. The tubes are very tortuous and grow in all directions in the endosperm. Later, the tube becomes transversely septate. Finally, a terminal cell differentiates at the tip of the tube and, by divisions in all planes, forms a globular embryo. In *Gnetum africanum,* a row of cells is produced by successive division of the zygote, and each of these cells forms a tube. In *Gnetum*

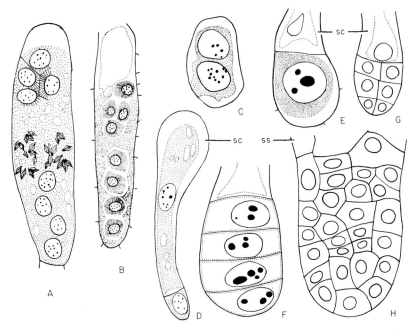

FIG. 18. Embryo development in *Ephedra distachya*. (A) Eight-nucleate proembryo; the four nuclei in the upper region are becoming separated from each other by cell walls. (B) Later stage showing eight embryonal units. (C,D) Two-nucleate and two-celled embryonal unit. (E–H) Successive stages in the enlargement and segmentation of embryonal cell, and formation of secondary suspensor. (sc) Suspensor cell; (ss) secondary suspensor. (After Lehmann-Baerts, 1967b.)

ula the zygote divides to form two cells which develop into tubular structures. The latter divide further, followed by the elongation of daughter cells to produce a bunch of elongated and uninucleate tubes which grow in all directions in the female gametophyte (Fig. 19A–D). The tube has a prominent cell wall and the nucleus usually lies at the tip. A peculiar cell is formed near the tip (Fig. 19E–H); it divides in all planes producing a globular mass of cells (Fig. 19I–L). In *Welwitschia* the development of the proembryo is not known in detail, but it seems similar to *Gnetum* (Pearson, 1929; Martens and Waterkeyn, 1969). In both these genera simple as well as cleavage polyembryony are regular features.

Following development of the globular mass, the cells lying in its distal part elongate resulting in a massive secondary suspensor, and the proximal cell divides to form the embryonal mass. The distal cells of the mass keep on adding to the secondary suspensor so that the embryonal mass is pushed deep into the female gametophyte.

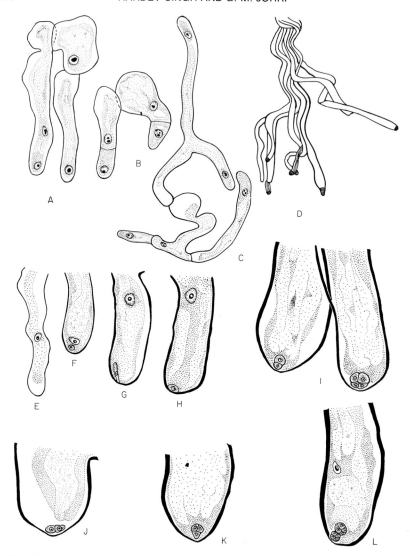

FIG. 19. Embryo development in *Gnetum ula*. (A–C) Early behavior of zygote derivatives to form tubular structures. (D) A bunch of long and tortuous tubular structures. (E–H)
Successive stages in formation of "peculiar" cell at the tip of primary suspensor tubes which
show a thick wall. (I–L) Division in peculiar cell and its derivatives. (After Vasil, 1959.)

Unlike the variations in development of the proembryo, later stages
in embryo development seem fairly uniform. A variable number of
cotyledons (two in cycads, *Ginkgo, Ephedra, Gnetum, Welwitschia* and
many conifers; and several in members of Pinaceae) arises laterally at the
plumular pole, and the radicular pole differentiates by activity of a group

of meristematic cells. A thick caplike tissue — the coleorhiza — is present in cycads. Plumular leaves are present in *Ginkgo,* cycads, and *Cedrus.*

In *Gnetum* and *Welwitschia* the hypocotyl produces a vascularized lateral protuberance — the feeder. In a mature seed the feeder is more prominent than the embryo proper (P. Maheshwari and Vasil, 1961; Martens and Waterkeyn, 1964).

VIII.   Maturation of Seed

With development of the embryo the seed coat begins to harden, and the reserve food materials start accumulating in the female gametophyte which may now be called *endosperm.* As the embryo is maturing the seed becomes dehydrated; the nucellus becomes greatly compressed and is represented by a thin caplike perisperm over the female gametophyte. A conspicuous red and fleshy aril develops around the seed of taxads and some podocarps.

A.   Development of the Seed Coat

As indicated earlier (Section VB) the seed coat may develop mainly from the tissue derived from the chalazal portion of the ovule (cycads, members of Pinaceae, *Cephalotaxus*) or both chalaza and integument may contribute (*Gnetum, Ephedra,* cupressads), or it may be derived mainly from the integument (*Trigonocarpus,* Pentoxyleae). In podocarps the ovuliferous scale (epimatium) forms the outer portions of the seed coat. All three integuments seem to contribute to the seed coat in *Gnetum,* but in *Ephedra* and *Welwitschia* only the outer integument is involved. Whatever the mode of origin, the initial changes are an increase in the number of cell layers in the particular region, followed by differentiation of the mucliage canal (in cycads) or resin ducts (in conifers), differentiation of xylem and phloem in the provascular strands (in cycads, *Cephalotaxus,* taxads, *Ephedra, Gnetum, Welwitschia,* and several fossil forms), deposition of tannin in several cells distributed all over the young seed coat and, finally, differentiation of three layers in the seed coat — the outer parenchymatous sarcotesta, the middle sclerenchymatous sclerotesta, and the innermost thin-walled endotesta (Fig. 20A). In members of Pinaceae a few layers of cells of the ovuliferous scale, in the vicinity of the ovule, give rise to a wing; and in Araucariaceae the wing seems to arise from the entire bract scale. In *Welwitschia* also the seeds are winged but the wing arises from the outer integument. Rudimentary wings derived from the integument are present in a few members of Cupressaceae and Taxodiaceae.

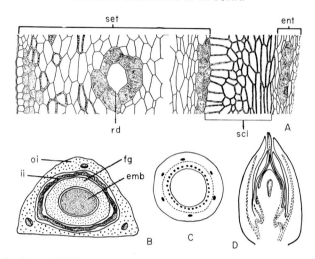

Fig. 20. Seed coat. (A) Transection of part of nearly mature seed coat of *Cephalotaxus drupacea* to show the three layers — sarcotesta, sclerotesta, and endotesta. (B) Transection of seed of *Ephedra foliata;* a ring of three vascular bundles is seen in the outer integument. (C) Transection of seed coat of *Bowenia spectabilis* to show two rings of vascular traces. (D) Longisection of ovule of *Gnetum gnemon;* vascular traces are seen in the three integuments. (emb) Embryo; (ent) endotesta; (fg) female gametophyte; (ii) inner integument; (oi) outer integument; (rd) resin duct; (scl) sclerotesta; (set) sarcotesta. (A — after Singh, 1961; B — after Khan, 1943; C — after Kershaw, 1912; D — after Sanwal, 1962.)

Except in conifers, most of the gymnospermous ovules are vascularized. *Cephalotaxus* and *Cedrus* have vascularized ovules, and thus seem exceptional among the conifers (Singh, 1961; Roy Chowdhury, 1961). The simplest pattern of vasculature is seen in *Ginkgo,* taxads, and *Ephedra* where two to four unbranched (sparingly branched in *Torreya* and *Austrotaxus*) vascular bundles, arranged in a ring, traverse the entire length of the seed coat (Florin, 1948; Eames, 1952). In *Ephedra* only the outer integument is vascularized (Fig. 20B), and the three bundles entering the ovule contribute a few tracheids to the base of the nucellus (Singh and Maheshwari, 1962). In *Cephalotaxus,* two inverted vascular strands lying at opposite radii are present (Singh, 1961). The traces give out a horizontal branch each, in the middle region, toward the inner part of the ovule. In cycads and *Welwitschia* there is a double system of vascular bundles (Fig. 20C); only the outer integument is vascularized in *Welwitschia*. The outer system comprising several traces traverses the sarcotesta, and the bundles may fuse and branch to form a network. The inner system supplies the endotesta and consists of several traces (Kershaw, 1912).

In *Gnetum,* all three layers of the seed coat contain varying amounts of vascular tissue (Fig. 20D). A single ring of bundles enters the base of the seed, and each bundle bifurcates giving rise to an outer series which supplies the sarcotesta. The inner series again bifurcates giving rise to bundles which supply sclerotesta and endotesta (see P. Maheshwari and Vasil, 1961).

The sarcotesta is quite variable and is greatly reduced or may be absent as in taxads and *Ephedra*. This region usually shows vascular tissue, tannin cells, and mucilage/resin ducts. The sarcotesta has a stomatiferous epidermis and is mainly parenchymatous sprinkled with thick-walled sclereids in *Gnetum* and several conifers such as *Cephalotaxus, Cryptomeria,* and *Podocarpus* (Singh, 1961; A. R. Rao and Malviya, 1963; Rodin and Kapil, 1969). Laticifers also occur in *Gnetum*. The sclerotesta is the stony layer which may be smooth or ribbed, the latter condition is especially prominent in fossil forms such as *Pachytesta* (Taylor, 1965). The sclerotesta is usually prominent in the micropylar region. The endotesta is generally composed of a few layers of parenchyma and forms a papery covering around the endosperm.

A thick and fleshy aril is present in taxads and *Phyllocladus* (Schnarf, 1937). It is comprised of thin-walled cells which seem rich in cell sap. The epidermis of the aril shows numerous stomata. In *Torreya* the aril contains cells with aromatic contents. The aril is usually nonvascularized. In *Taxus,* a stub of vasculature is present at the base of the aril, and in *Torreya* the vasculature of the seed coat traverses the aril.

A unique feature of the ovule of *Ephedra* is the presence of a hypostase. This is a curved plate of thin-walled cells lying near the chalazal end of the female gametophyte. The central vasculature of the seed ends just below the hypostase.

B. Endosperm

As young embryos start developing in the seed, the central portion of the female gametophyte breaks down to form the "corrosion cavity." The young embryos are pushed into this cavity by elongation of the suspensors, and, thus, it serves to nourish the developing embryos.

In *Ephedra,* during postfertilization stages, the micropylar portion of the female gametophyte grows upward. The growth extends into the deep pollen chamber and may plug a part of it (Lehmann-Baerts, 1967b).

The most conspicuous change in the female gametophyte during seed maturation is deposition of reserve foods such as fat, starch, and protein. These seem to be utilized at the time of seed germination. Thus, the female gametophyte in gymnosperms serves the dual function of bearing gametes and nourishing the embryo.

The main cytological changes associated with the maturing endosperm are nuclear divisions which are not followed by wall formation. Up to seven free nuclei have been observed in the maturing gametophytic cells of *Cephalotaxus* (Fujita, 1961). The nuclei eventually fuse to form large polyploid masses.

IX. Development in Relation to Time

Unlike the angiosperms, where the time lag between ovule initiation and seed maturation may be a month or so, in gymnosperms the same process usually takes a much longer time (see also Schnarf, 1933). As many gymnosperms are temperate-zone plants, the ovules pass through one or two winter rests, with activity renewed in spring. In tropical gymnosperms, such as cycads and *Gnetum,* ovule development is a continuous process but, even in these, it takes about a year from the initiation to maturation of the seed. Ovule development on a plant or a stand is usually synchronous. Therefore, in a particular situation, on a tree or a stand of trees, the dates of pollination or fertilization in different ovules are more or less the same; usually, over the years, these dates are about the same. In most gymnosperms, pollination takes place in spring, but in *Cedrus deodara* it takes place in autumn (Roy Chowdhury, 1961).

Roughly, the gymnosperms show a 1-, 2-, or 3-year type of life cycle. Some species of *Podocarpus* show a 1-year type of life cycle, whereas the others have a 2-year type (Konar and Oberoi, 1969). A 1-year life cycle is exhibited by cycads, *Ginkgo, Ephedra, Gnetum,* and most cupressads and taxodiaceous members. In these the ovules are usually initiated in autumn, and they pass the winter in the sporogenous tissue stage (Fig. 21). They develop further during the spring when megasporogenesis is completed and pollination occurs. Cycad ovules usually show a free-nuclear gametophyte at the time of pollination. Fertilization takes place during the next 3–4 months, and the seeds mature and are shed by autumn. The ovules of most fossil gymnosperms were shed before fertilization, whereas in *Ginkgo* the ovules may be shed before or soon after fertilization (Eames, 1955; Andrews, 1966), and further development of the seed takes place on the ground. In *Gnetum* the embryo is still immature at the time of seed shedding (P. Maheshwari and Vasil, 1961).

Cephalotaxus, Torreya, and a few other gymnosperms show a 2-year type of life cycle (Singh, 1961). The ovule passes through two winter rests and the time lapse between pollination and fertilization is usually 1 year (Fig. 22). Following fertilization, the seeds mature within 3 to 6 months and are shed from the plant (Fig. 22).

Pinus is the best known example of a 3-year type of life cycle. The cones initiate in autumn or early winter, and pollination occurs in spring

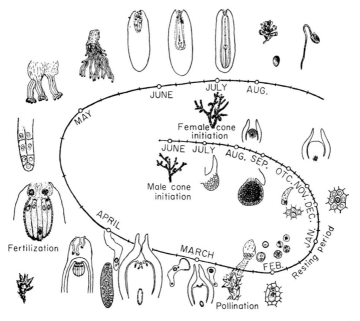

FIG. 21. Diagrammatic representation of the life cycle of *Biota orientalis* in relation to time. The sexual cycle is completed in 1 year. (After Singh and Oberoi, 1962.)

(Konar, 1960). Fertilization takes place 1 year later, and the female cones mature and shed seeds during the next summer (Fig. 23). Thus, the ovule passes through three winters. During the first winter the ovules show sporogenous cells, a free-nuclear female gametophyte in the second, and nearly mature embryo in the third (Fig. 23).

X. Conclusions

Gymnosperms comprise an important group of plants which yield timber, resins, essential oils, drugs, and edible nuts and are used in paper manufacture and in the starch industry (Thieret, 1958; P. Maheshwari and Singh, 1960). Some gymnosperms such as *Welwitschia* have a rather restricted distribution and need to be conserved and propagated (Kers, 1967). They provide a link with the past and are very interesting academically. Except for cycads and *Ginkgo,* these plants are usually propagated by seeds and, therefore, a study of the development of seeds in gymnosperms becomes especially significant. Procedures to forecast the crop of seeds in a year have been worked out for a few commercially important plants (Allen, 1941; Mathews, 1963). Breeding with a view to combine favorable characteristics is also well-advanced.

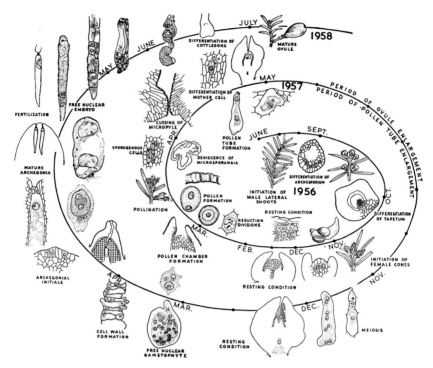

FIG. 22. Chronology of reproductive phases of *Cephalotaxus drupacea*. The seeds require 2 years from initiation to maturity. (After Singh, 1961.)

Many gaps exist in our present knowledge of seed development. More information is needed on origin of the nucellus in *Ephedra,* extension of the pollen chamber in cycads, role of the pollination drop, postpollination growth of ovules, alveolation in female gametophytes, structure of the megaspore membrane, cytochemistry of the egg and other allied structures, and fertilization and embryo development in *Welwitschia.*

The nucellus usually grows by periclinal and anticlinal divisions in the primary parietal layer and epidermis. There are some claims that in *Ephedra* the epidermis alone gives rise to the thick nucellus, the parietal tissue being absent. Working on *Ephedra distachya,* Lehmann-Baerts (1967b) showed that the nucellus is of dual origin. He believes that this is also the case in *Ephedra foliata* and *Ephedra gerardiana.* However, he suggests that in other species, such as *Ephedra intermedia* and *Ephedra helvetica,* the nucellus may be only epidermal in origin. For these reasons, the different species should be investigated exhaustively.

It has been reported that in *Bowenia* and *Encephalartos* the pollen chamber lies in the upper beak-shaped portion of the nucellus (Kershaw,

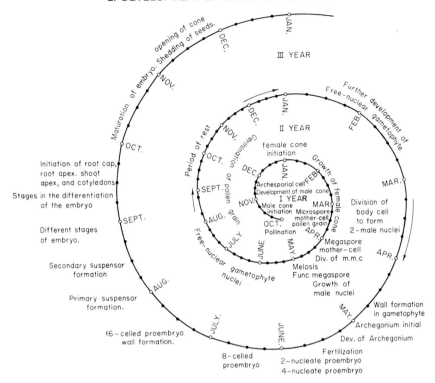

FIG. 23. Development of seed in relation to the months or years in *Pinus roxburghii*, showing a 3-year type of life cycle. (Div. of m.m.c.) reduction divisions in megaspore mother cell (func. megaspore) functional megaspore. (After Konar, 1960.)

1912; De Sloover, 1964). After the pollen grains land in the pollen chamber, a group of cells lying at its base start degenerating and, thus, extend the lower limit of the chamber. Since formation of the pollen chamber is a prepollination phenomenon, the postpollination extension is called the lower pollen chamber or intermediary chamber. The pollen grains come to lie and germinate in the lower cavity. In most other gymnosperms the pollen grains germinate in the pollen chamber. When investigated critically, a similar chamber may be found in other cycads, as well as in *Ginkgo*. These plants seem specialized since they have a pollen chamber for receiving pollen and an intermediary chamber for its germination.

There are conflicting views and paucity of experimental data on the role of the pollination drop (see Section III, B). Investigations on the problem encounter several difficulties. The chief one is the production of the fluid in microquantities at the micropyle for only a few days in the year. A small field laboratory may have to be set up and special methods devised for on-the-spot observations and for handling the microquantities of fluid. Once the system is better known, a simulant may be used to confirm some of the results. The pollination drop seems to occur only in gymnosperms.

One of the morphologically fascinating aspects of seed development is postpollination growth of the ovule. It brings about a reorientation of the existing tissues in some plants. Unfortunately, this is one of the most neglected and misinterpreted phases of development.

Wall formation in the female gametophyte of most gymnosperms takes place by alveolation. The alveoli are hexagonal in cross section and grow centripetally as long cells, their nuclei being located at the inner open end. Many alevoli close at the inner end, long before reaching the center of the gametophyte. The exact mechanism of closure of the few alevoli which grow up to the center of the gametophyte, has been demonstrated for *Athrotaxis* only.

The megaspore membrane has been traditionally interpreted as homologous to the pollen coat and even with a similar stratification. However, Pettitt (1966) believes that only a thin cuticle represents the megaspore membrane, and in earlier studies the compressed remnants of spongy tissue seem to have been mistaken for the membrane.

The egg nucleus is usually large and full of nucleoplasm. The chromatin (as studied by electron microscopy) has been reported to occur as thin and very long threads (Camefort, 1964). In view of the frequent occurrence of fibrils as inclusions in nuclei (Wergin *et al.* 1970), evidence for regarding the threads present in the gymnospermous egg nuclei as chromatin, needs to be strengthened, possibly by treating the nuclei with deoxyribonuclease.

Following fertilization, neocytoplasm arises in the zygote and this alone forms the cytoplasm of the embryo. The leftover cytoplasm degenerates to form the plug. Embryo development conforms to four types: (a) cycad and Ginkgo; (b) conifer; (c) Ephedra, and (d) Gnetum and Welwitschia. The araucarians need to be investigated more vigorously, as their mode of embryo development does not seem to conform to any of the above types. *Welwitschia* seems to show a peculiar mode of fertilization which, along with embryogeny of the plant, has not been exhaustively studied. The mature embryo in most gymnosperms is large and shows two or more cotyledons.

The seed coat may develop from the chalaza, or integument, or be contributed by both chalaza and the integument. It usually comprises three layers—outer parenchymatous sarcotesta, middle stony sclerotesta, and innermost parenchymatous endotesta. The seeds are winged in several plants. The seed coat is vascularized by a ring of two to four un-branched bundles, or there may be a double system of vascular bundles. In *Gnetum* all the three layers of the seed coat contain varying amounts of vascular tissue. The female gametophyte matures into endosperm in the seed.

Gymnosperms usually take 1–3 years to complete their life cycle. But the exact data, for most of the plants, are not available in respect of the seasonal progress of the reproductive cycle.

Contributions to fill these gaps would be very rewarding, and better understanding of seed development in this group is urgently called for in view not only of the academic interest but also the economic importance of gymnosperms.

Acknowledgments

We are grateful to Dr. (Mrs.) Vani Hardev for reviewing the manuscript and help in various ways; to Mrs. Krishna Kumar, Miss Vatsala, and Mr. Megh Pillay for translating some French literature; to Mr. Krishan Lal and Mr. R. S. Saini for help in preparing the illustrations; and to Mr. D. C. Jain for typing the manuscript.

References

Allen, G. S. (1941). A basis for forecasting seed crop of some coniferous trees. *J. Forest.* **39**, 1014.

Allen, G. S. (1963). Origin and development of the ovule in douglas-fir. *Forest Sci.* **9**, 386.

Andrews, H. N. (1966). "Studies in Palaeobotany." John Wiley, New York.

Arnold, C. A. (1947). "An Introduction to Palaeobotany." McGraw-Hill, New York.

Baird, A. M. (1939). A contribution to the life history of *Macrozamia reidlei. J. Proc. Roy. Soc. West. Aust.* **25**, 153.

Baird, A. M. (1953). The life history of *Callitris. Phytomorphology* **3**, 258.

Barner, H., and Christiansen, H. (1960). The formation of pollen, the pollination mech-anism, and the determination of most favourable time for controlled pollination in *Larix. Silvae Genet.* **9**, 1.

Barner, H., and Christiansen, H. (1962). The formation of pollen, the pollination mech-anism and determination of most favourable time of controlled pollination in *Pseudo-tsuga menziesii. Silvae Genet.* **11**, 89.

Battaglia, E. (1958). L'evoluzione del gametofito maschile dalle gimnosperme alle angio-sperme sulla base recenti observazioni in *Gnetum. Caryologia* **11**, 217.

Berridge, E. M. (1911). Some points of resemblance between gnetalean and bennettitalean seeds. *New Phytol.* **10**, 140.

Brennan, M., and Doyle, J. (1956). The gametophytes and embryogeny of *Athrotaxis. Sci. Proc. Roy. Dublin Soc.* [N.S.] **27**, 193.

Bryan, G. S. (1952). The cellular embryo of *Zamia* and its cap cells. *Amer. J. Bot.* **39**, 433.

Bryan, G. S., and Evans, R. I. (1956). Chromatin behaviour in the development and maturation of the egg nucleus of *Zamia umbrosa. Amer. J. Bot.* **43**, 640.

Buchholz, J. T. (1939). The embryogeny of *Sequoia sempervirens* with a comparison of the sequoias. *Amer. J. Bot.* **26**, 248.

Burlingame, L. L. (1913). The morphology of *Araucaria braziliensis*. *Bot. Gaz.* **55**, 97.

Burlingame. L. L. (1915). The morphology of *Araucaria braziliensis*. III. Fertilization, the embryo and the seed. *Bot. Gaz.* **59**, 1.

Camefort, H. (1959). Structures des filaments de chromatine du noyau de l'oosphère chez le *Pinus laricio* Poir. (var. *austriaca*). Étude en microscopie électronique. *C. R. Acad. Sci.* **249**, 1790.

Camefort, H. (1962). L'organisation du cytoplasme dans l'oosphère et la cellule centrale du *Pinus laricio* Poir. (var. *austriaca*). *Ann. Sci. Natur.: Bot. Biol. Veg.* [12] **3**, 269.

Camefort, H. (1964). Observations sur la structure des chromosomes et des nucléoles de l'oosphère des pins. *C. R. Acad. Sci.* **259**, 4335.

Camefort, H. (1965a). Une interprétation nouvelle de l'organisation du protoplasme de l'oosphére des pins. *In* "Travaux biol. veg. dédiés à Lucien Plantefol," pp. 407–436. Masson et Cie, Paris

Camefort, H. (1965b). L'organisation du protoplasme dans la gamète femelle, ou oosphère, du *Ginkgo biloba* L. *J. Microsc. (Paris)* **4**, 531.

Camefort, H. (1967). Observations sur les mitochondries et les plastes de la cellule centrale et de l'oosphère du *Larix decidua* Mill. (*Larix europea* D.C.). *C. R. Acad. Sci.* **265**, 1293.

Camefort, H. (1968). Cytologie de la fécondation et de la proembryogénèse chez quelques gymnospermes. *Bull. Soc. Bot. Fr.* **115**, 137.

Camefort, H. (1969). Fécondation et proembryogénèse chez les Abiétacées (notion de néocytoplasme) *Rev. Cytol. Biol. Veg.* **32**, 253.

Chamberlain, C. J. (1910). Fertilization and embryogeny in *Dioon edule*. *Bot. Gaz.* **50**, 415–429.

Chamberlain, C. J. (1935). "Gymnosperms Structure and Evolution." Chicago Univ. Press, Chicago, Illinois.

Chesnoy, L. (1967). Nature et évolution des formations dites "astéroïdes" de la cellule centrale de l'archegone du *Juniperus communis* L. Etude en microscopie photonique et électronique. *C. R. Acad. Sci.* **264**, 1016.

Chesnoy, L. (1969). Sur la participation du gamète mâle à la constitution du cytoplasme de l'émbryon chez le *Biota orientalis* Endl. *Rev. Cytol. Biol. Veg.* **32**, 273.

Coulter, J. M., and Land, W. J. G. (1905). Gametophyte and embryo of *Torreya taxifolia*. *Bot. Gaz.* **39**, 161.

De Silva, B. L. T., and Tambiah, M. S. (1952). A contribution to the life history of *Cycas rumphii* Miq. *Ceylon J. Sci., Sect. A* **12**, 1.

De Sloover, J.-L. (1961). Etude sur les Cycadales. I. Méiose et mégasporogenèse chez *Encephalartos poggei* Asch. *Cellule* **62**, 105.

De Sloover, J.-L. (1963). Etude sur les Cycadales. II. La paroi mégasporale chez *Encephalartos poggei* Asch. *Cellule* **63**, 333.

De Sloover, J.-L. (1964). Etude sur les Cycadales. III. Nucelle gametophyte femelle et embryon chez *Encephalartos poggei* Asch. *Cellule* **64**, 151.

De Sloover-Colinet, A. (1963). Chamber pollinique et gamétophyte mâle chez *Ginkgo biloba*. *Cellule* **64**, 129.

Dogra, P. D. (1964). Pollination mechanisms in gymnosperms. *In* "Recent Advances in Palynology" (P. K. K. Nair's, ed), pp. 142–175. Nat. Bot. Garden, Lucknow.

Dogra, P. D. (1967). Seed sterility and disturbances in embryogeny in conifers with particular reference to seed testing and tree breeding in Pinaceae. *Stud. Forest. Suec.* **45**, 1–97.

Doyle, J. (1945). Development lines in pollination mechanism in the Coniferales. *Sci. Proc. Roy. Dublin Soc.* **24**, 43.

Doyle, J. (1963). Proembryogeny in *Pinus* in relation to that in other conifers – a survey. *Proc. Roy. Irish Acad., Sect. B* **62**, 181.

Doyle, J., and Kane, A. (1943). Pollination in *Tsuga pattotiana* and in species of *Abies* and *Picea*. *Sci. Proc. Roy. Dublin Soc.* **23**, 57.

Doyle, J., and Saxton, W. T. (1933). Contributions to the life history of *Fitzroya*. *Proc. Roy. Irish Acad., Sect. B* **41**, 191.

Dupler, A. W. (1917). The gametophytes of *Taxus canadensis*. *Bot. Gaz.* **64**, 115.

Eames, A. J. (1913). The morphology of *Agathis australis*. *Ann. Bot. (London)* **27**, 1.

Eames, A. J. (1952). Relationships of the Ephedrales. *Phytomorphology* **2**, 79.

Eames, A. J. (1955). The seed and *Ginkgo*. *J. Arnold Arboretum, Harvard Univ.* **36**, 165.

Erdtman, G. (1952). "Pollen Morphology and Plant Taxonomy: Angiosperms." Almqvist and Wiksells, Stockholm.

Edrtman, G. (1957). "Pollen and Spore Morphology/Plant Taxonomy: Gymnospermae, Pteridophyta and Bryophyta (illustrations)." Almqvist and Wiksells. Stockholm and New York.

Fagerlind, F. (1941). Bau und Entwicklung der *Gnetum*-Gametophyten. *Kgl. Sv. Vetenskapsakad., Handl.* **19**, 1.

Favre-Duchartre, M. (1956). Contribution à l'étude de la reproduction chez *Ginkgo biloba*. *Rev. Cytol. Bio. Veg.* **17**, 1.

Favre-Duchartre, M. (1957). Contribution à l'étude de la reproduction chez *Cephalotaxus drupacea*. *Rev. Cytol. Biol. Veg.* **18**, 305.

Favre-Duchartre, M. (1960). Contribution à l'étude des spermatozoides de *Taxus baccata*. *Rev. Cytol. Biol. Veg.* **21**, 329.

Favre-Duchartre, M. (1963). A propos des nucelles des plantes a ovules. *Ann. Acad. Reims E.R.S.* **1**, 1.

Favre-Duchartre, M. (1965). A propos des gamétophytes femelles et des archégones des plantes ovulées. *Ann. Sci. Natur.: Bot. Biol. Veg.* [12] **6**, 147.

Ferguson, M. (1904). Contributions to the life history of *Pinus* with special reference to sporogenesis, the development of gametophytes and fertilization. *Proc. Wash. Acad. Sci.* **6**, 1.

Florence, R. G., and McWilliam, J. R. (1956). The influence of spacing on seed production. *Z. Forstgenet. Forstpflanzenzuecht.* **5**, 97.

Florin, R. (1948). On the morphology and relationships of the Taxaceae. *Bot. Gaz.* **110**,31.

Florin, R. (1954). The female reproductive structures of conifers and taxads. *Bot. Rev.* **29**, 367.

Fujita, T. (1961). Multinucleate endosperm cells in *Cephalotaxus drupacea* Siebold et Zuccarini, *J. Jap. Bot.* **36**, 29.

Gianordoli, M. (1969). Observations sur l'organisation du cytoplasme et sur la présence de lamelles annelées dans la cellule centrale du *Sciadopitys verticillata* (Sieb. Zucc.). *Rev. Cytol. Biol. Veg.* **32**, 183.

Gorodon, W. T. (1941). On *Salpingostoma dasu*, a new Carboniferous seed from East Lothian. *Trans. Roy. Soc. Edinburgh* **60**, 427.

Hagerup, O. (1933). Zur Organogenie und Phylogenie der Konifern-Zapfen. *Kgl. Dans. Vidensk. Selsk., Mat.-Fys. Skr.* **10**, 1–82.

Hagerup, O. (1934). Zur Abstammung einiger Angiospermen durch Gnetales und Coniferae. *Kgl. Dan. Vidensk. Selsk., Mat.-Fys. Skr.* **11**, 1.

Hirmer, M. (1936). Die Blüten der Coniferen. I. Entwicklungsgeschichte und vergleichende Morphologie des weiblichen Blüten Zapfen der Coniferen. *Bibl. Bot.* **114**, 1.

Hodcent, E. (1963). Etude cytologique de la formation des grains de pollen à partir de leurs cellules mères chez *Araucaria columnaris* (Forster). *C.R. Acad. Sci.* **257,** 489.

Johansen, D. A. (1950). "Plant Embryology." Chronica Botanica, Waltham, Massachusetts.

Kemp, M. (1959). Morphological and ontogenetic studies on *Torreya californica.* II. Development of the megasporangiate shoot prior to pollination. *Amer. J. Bot.* **46,** 249.

Kers, L. E. (1967). The distribution of *Welwitschia mirabilis* Hook. f. Svensk. *Bot. Tidsskr.* **61,** 97.

Kershaw, E. M. (1912). Structure and development of the ovule of *Bowenia spectabilis. Ann. Bot. (London)* **26,** 625.

Khan, R. (1943). Contributions to the morphology of *Ephedra foliata* Boiss. II. Fertilization and Embryogeny. *Proc. Nat. Acad. Sci., India* **13,** 357.

Konar, R. N. (1960). The morphology and embryology of *Pinus roxburghii* Sar. with a comparison with *Pinus wallichiana* Jack. *Phytomorphology* **10,** 305.

Konar, R. N., and Banerjee, S. K. (1963). The morphology and embryology of *Cupressus funebris* Endl. *Phytomorphology* **13,** 321.

Konar, R. N., and Oberoi, Y. P. (1969). Studies on the morphology and embryology of *Podocarpus gracilior* Pilger. *Beitr. Biol. Pflanz.* **45,** 329.

Lawson, A. A. (1904). The gametophytes, archegonia, fertilization and the embryo of *Sequoia sempervirens. Ann. Bot. (London).* **18,** 1.

Lee, C. L. (1955). Fertilization in *Ginkgo biloba. Bot. Gaz.* **117,** 79.

Lehmann-Baerts, M. (1967a). Etude sur les Gnétales. VIII. Ontogenése ovulaire chez *Gnetum africanum* et *Ephedra distachya. Cellule* **66,** 313.

Lehmann-Baerts, M. (1967b). Etude sur les Gnétales. XII. Ovule, gamétophyte femelle et embryogenèse chez *Ephedra distachya* L. *Cellule* **67,** 53.

Linskens, H. F. (1969). Fertilization mechanisms in higher plants. *In* "Fertilization: Comparative Morphology, Biochemistry and Immunology" (C. B. Metz and A. Monroy, eds.), Vol. 2, pp. 189–253. Academic Press, New York.

Looby, W. J., and Doyle, J. (1940). New observations on the life history of *Callitris. Sci. Proc. Roy. Dublin Soc.* **22,** 241.

Looby, W. J., and Doyle, J. (1942). Formation of gynospore, female gametophyte and archegonia in *Sequoia. Sci. Proc. Roy. Dublin Soc.* **23,** 35.

Loze, J.-C. (1965). Etude de l'ontogenèse de appareil reproducteur femelle de l'if *Taxus baccata. Rev. Cytol. Biol. Veg.* **28,** 211.

McWilliam, J. R. (1958). The role of the micropyle in the pollination of *Pinus. Bot. Gaz.* **120,** 109.

McWilliam, J. R. (1959). Bioelectrical phenomena in relation to pollination in *Pinus. Silvae. Genet.* **8,** 59.

McWilliam, J. R., and Mergen, F. (1958). Cytology of fertilization in *Pinus. Bot. Gaz.* **119,** 246.

Maheshwari, K. (1960). Morphology and embryology of *Cycas circinalis.* M.Sc. Thesis, University of Delhi.

Maheshwari, P. (1935). Contributions to the morphology of *Ephedra foliata* Boiss. I. The development of the male and female gametophytes. *Proc. Indian Acad. Sci., Sect. B* **1,** 586.

Maheshwari, P. (1960). "Evolution of the Ovule," 7th Sir A.C. Seward Memorial Lecture. Birbal Sahni Institute of Palaeobotany, Lucknow.

Maheshwari, P., and Sanwal, M. (1963). The archegonium in gymnosperms: A review. *Mem. Indian Bot. Soc.* **4,** 103–119.

Maheshwari, P., and Singh, H. (1960). Economic importance of conifers. *J. Univ. Gauhati* **11,** 1.

Maheshwari, P., and Singh, H. (1967). The female gametophyte of gymnosperms. *Biol. Rev.* **42**, 88.

Maheshwari, P., and Vasil, V. (1961). "Gnetum." Counc. Sci. Industr. Res. New Delhi.

Martens, P. (1957). Bec nucellaire, micropyle et paroi megasporale chez *Encephalartos poggei* Asch. *Cellule* **58**, 152.

Martens, P. (1959). Etude sur les Gnétales. III. Structure et ontogenèse du cône et de la fleur femelles de *Welwitschia mirabilis*. *Cellule* **60**, 171–286.

Martens, P. (1966). Du mégasporange cryptogamique à l'ovule gymnospermique. *In* "Trends in Plant Morphogenesis" (E. G. Cutter, ed.), pp. 155–169. Longmans, London.

Martens, P., and Waterkeyn, L. (1964). Etudes sur les Gnétales. VII. Recherches sur *Welwitschia mirabilis*. IV. Germination et plantules: Structure fonctionnement et productions du méristème caulinaire apical. *Cellule* **65**, 1.

Martens, P., and Waterkeyn, L. (1969). Sur l'embryogénèse de *Welwitschia mirabilis*. *Rev. Cytol. Biol. Veg.* **32**, 331.

Mathews, J. D. (1963). Factors affecting the production of seeds by forest trees. *Forest. Abstr.* **24**, 1.

Mehra, P. N. (1938). The germination of pollen grains in artificial cultures in *Ephedra foliata* Boiss. and *Ephedra gerardiana* Wall. *Proc. Indian Acad. Sci., Sect. B* **8**, 218.

Negi, V., and Madhulata (1957). Male gametophyte and megasporogenesis in *Gnetum. Phytomorphology* **7**, 230.

Norstog, K. J. (1967). Fine structure of the spermatozoids of *Zamia* with special reference to the flagellar apparatus. *Amer. J. Bot.* **54**, 831.

Norstog, K. J. (1968). Fine structure of the spermatozoids of *Zamia:* Observations on the microtubule system and related structures. *Phytomorphology* **18**, 350.

Oliver, F. W., and Scott, D. H. (1904). On the structure of the Palaeozoic seed *Lagenostoma lomaxi*, with a statement of the evidence upon which it is referred to *Lyginodendron*. *Phil. Trans. Roy. Soc. London, Ser. B* **197**, 193.

Owens, J. N., and Smith, F. H. (1964). The initiation and early development of the seed cone of Douglas fir. *Can. J. Bot.* **42**, 1031.

Pankow, H. (1962). Histogenetische Studien an den Blüten einiger Phanerogamen. *Bot. Stud.* **13**, 1.

Pant, D. D., and Mehra, B. (1962). "Cycas." Central Book Depot, Allahabad.

Pearson, H. N. W. (1929). "Gnetales." Cambridge Univ. Press, London and New York.

Pettitt, J. M. (1966). A new interpretation of the structure of the megaspore membrane in some gymnospermous ovules. *J. Linn. Soc. London, Bot.* **59**, 253.

Puri, V. (1970). "The Angiosperm Ovule," Presidential Address (Section of Botany), 57th Session, pp. 1–35. Indian Sci. Congr., Khargpur.

Rao, A. R., and Malviya, M. (1963). Sclereids in *Cryptomeria japonica* D. Don. *Proc. Nat. Inst. Sci. India, Part B* **29**, 551.

Rao, L. N. (1963). Life-history of *Cycas circinalis* L. II. Fertilization embryogeny and germination of the seed. *J. Indian Bot. Soc.* **42**, 319.

Rodin, R. J., and Kapil, R. N. (1969). Comparative anatomy of the seed coats of *Gnetum* and their probable evolution. *Amer. J. Bot.* **56**, 420.

Roy Chowdhury, C. (1961). The morphology and embryology of *Cedrus deodara* Loud. *Phytomorphology* **11**, 283.

Roy Chowdhury, C. (1962). The embryogeny of conifers: A review. *Phytomorphology* **12**, 313.

Sanwal, M. (1962). Morphology and embryology of *Gnetum gnemon* L. *Phytomorphology* **12**, 243.

Sarvas, R. (1955). Investigations into the flowering and seed quality of forest trees. *Comm. Inst. For. Fenn.* **45**, 1.

Saxton, W. T. (1913). Contributions to the life history of *Actinostrobus pyrimidalis* Miq. *Ann. Bot. (London)* **27**, 321.

Saxton, W. T. (1934). Notes on conifers. VIII. The morphology of *Austrotaxus spicata* Compton. *Ann. Bot. (London)* **48**, 411.

Schnarf, K. (1933). "Embryologie der Gymnospermen." Gebrüder Borntraeger, Berlin.

Schnarf, K. (1937). "Anatomie der Gymnospermen-Samen." Gebrüder Borntraeger, Berlin

Seeliger, I. (1954). Studien am Sporossvegetationskegel von *Ephedra fragilis* var. *campylopoda* (C. A. Mey) Stapf. *Flora (Jena)* **141**, 114.

Shimamura, T. (1956). Cytochemical studies on the fertilization and proembryo of *Pinus thunbergii*. *Bot. Mag. (Tokyo)* **69**, 524.

Singh, H. (1961). The life history and systematic position of *Cephalotaxus drupacea* Sieb. et Zucc. *Phytomorphology* **11**, 153.

Singh, H., and Chatterjee, J. (1963). A contribution to the life history of *Cryptomeria japonica* D. Don. *Phytomorphology* **13**, 429.

Singh, H., and Maheshwari, K. (1962). A contribution to the embryology of *Ephedra gerardiana*. *Phytomorphology* **12**, 361.

Singh, H., and Oberoi, Y. P. (1962). A contribution to the life history of *Biota orientalis*. *Phytomorphology* **12**, 373.

Smith, D. L. (1964). The evolution of the ovule. *Biol. Rev.* **39**, 137.

Smith, F. G. (1910). Development of the ovulate strobilus and young ovule of *Zamia floridana*. *Bot. Gaz.* **50**, 128.

Sporne, K. R. (1965). "The Morphology of Gymnosperms." Hutchinson, London.

Sterling, C. (1948). Gametophyte development in *Taxus cuspidata*. *Bull. Torrey Bot. Club* **75**, 147.

Sterling, C. (1963). Structure of the male gametophyte in gymnosperms. *Biol. Rev.* **38**, 167.

Stewart, K. D., and Gifford, E. M., Jr. (1967). Ultrastructure of the developing megaspore mother cell of *Ginkgo biloba*. *Amer. J. Bot.* **54**, 375.

Strasburger, E. (1879). "Die Angiospermen und die Gymnospermen." Gustav Fischer, Jena.

Taylor, T. N. (1965). Paleozoic seed studies: A monograph on the American species of *Pachytesta*. *Palaeontographica* **117B**, 1–46.

Thieret, J. W. (1958). Economic botany of the cycads. *Econ. Bot.* **12**, 3.

Thomas, M-J., and Chesnoy, L. (1969). Observations relatives aux mitochondries Feulgen positives de la zone périnucléaire de l'oosphère du *Pseudotsuga menziesii* (Mirb.) Franco. *Rev. Cytol. Biol. Veg.* **32**, 165.

Thomson, R. B. (1905). The megaspore membrane of the gymnosperms. *Univ. Toronto Stud., Biol. Ser.* **4**, 85.

Tison, A. (1911). Remarques sur les gouttelettes des ovules de coniferes. *Mem. Soc. Linn. Normandie* **24**, 51.

van der Pijl, L. (1953). On the flower biology of some plants from Java. *Ann. Bogor.* **1**, 77.

Vasil, V. (1959). Morphology and embryology of *Gnetum ula* Brougn. *Phytomorphology* **9**, 167.

Vasil, V., and Sahni, R. K. (1964). Morphology and embryology of *Taxodium mucronatum* Tenore, *Phytomorphology* **14**, 369.

von Lürzer, E. (1956). Megasporenmembranen bein einigen Cupressaceen. *Grana Palynol.* **1**, 70.

Waterkeyn, L. (1954). Etude sur les Gnétales. I. Le strobile femelle, l'ovule et de graine de *Gnetum africanum* Welw. *Cellule* **56**, 103.

Waterkeyn, L. (1959). Etude sur les Gnétales. II. La strobile mâle, la microsporogenèse et la gamétophyte mâle de *Gnetum africanum* Welw. *Cellule* **60**, 7.

Waterkeyn, L. (1960). Etude sur les Gnétales. IV. Le tube micropylaire et la chambre pollinique de *Gnetum africanum* Welw. *Cellule* **61**, 81.

Wergin, W. P., Gruber, P. J., and Newcomb, E. H. (1970). Fine structural investigation of nuclear inclusions in plants. *J. Ultrastruct. Res.* **30**, 533.

Willemse, W. T. M., and Linskens, H. F. (1969). Development du microgamétophyte chez le *Pinus sylvestris* entre la méiose et la fécondation. *Rev. Cytol. Biol. Veg.* **32**, 121.

Wunderlich, R. (1966). Zur Deutung der eigenartigen Embryoentwicklung von *Paeonia*. *Oesterr. Bot. Z.* **113**, 395.

Ziegler, H. (1959). Über die zusammensetzung des "Bestäubungstropfens" und den Mechanismus seiner Sekretion. *Planta* **52**, 587.

3

DEVELOPMENT OF ANGIOSPERM SEEDS

S. P. Bhatnagar and B. M. Johri

I. Introduction

The end product of the flowering process is a fruit with one or many seeds. There can be no second opinion as to the great importance of seeds. Seeds themselves, or the products obtained therefrom, have probably been used by man since before recorded history. Orville L. Freeman (1961) very aptly remarked, "Seeds are the germ of life, a beginning and an end, the fruit of yesterday's harvest and the promise of tomorrow's."

Because of their immense importance, it becomes imperative to look into the processes that lead to the formation of seeds. The morphological and histological changes which occur in the ovule form the main theme of this chapter.

As early as 3000 years ago, the Arabs were conscious of the role of pollen in "date" culture. However, this was almost forgotten until the seventeenth and eighteenth centuries when Camerarius and Koelreuter established the role of pollen in the formation of seeds. Amici (1824) discovered, for the first time, germinating pollen grains on the stigma of *Portulaca*. Six years later, he (Amici, 1830) traced the pollen tube to the mouth of the ovule. Amici (1847) and Hofmeister (1849) demonstrated the presence of a preexisting germinal vesicle (egg) which, under the stimulation of the pollen tube, gave rise to the embryo.

Strasburger (1884) observed *syngamy,* which is the fusion of one of the male gametes with the egg nucleus. For several years the fate of the second male gamete remained enigmatic. In 1898, Nawaschin showed that the second male gamete fuses with the polar nuclei, and this phenomenon is referred to as *triple fusion.* Syngamy and triple fusion together constitute double fertilization—a process unique to angiosperms.

In addition to having more-or-less showy appendages, the flower usually has less prominent or even hidden sex organs (Fig. 1). The sex organs comprise stamens and pistil and are normally essential for seed set. The stamens constitute the male or the pollen-producing parts; within the pollen grains are formed the sperms or male gametes. Usually the pistil, which is the female or seed-forming organ, consists of the ovary, style, and stigma. The ovary contains ovules, the progenitors of seeds.

The seed can be defined as a fertilized and ripened ovule. Popularly, some single-seeded fruits, such as the caryopsis of Gramineae and cyp-

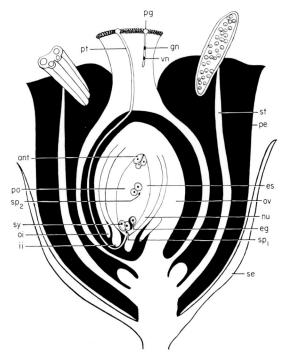

FIG. 1. Diagrammatic sectional view of a flower to show various parts; note the germination of pollen grains, path of pollen tube, and its entry into embryo sac. (ant) Antipodal cells; (eg) egg; (es) embryo sac; (gn) generative nucleus; (ii) inner integument; (nu) nucellus; (oi) outer integument; (ov) ovule; (pe) petal; (pg) pollen grain; (po) polar nuclei; (pt) pollen tube; (se) sepal; (sp_1, sp_2) sperm cells; (st) stamen; (sy) synergid; (vn) vegetative nucleus. (After Hill *et al.,* 1967.)

sella of Compositae are also considered "seeds." Seeds, in turn, are enclosed in the fruit which is the ripened ovary. There are certain fruits which, even in nature, do not contain seeds. Such fruits are formed without fertilization and are referred to as parthenocarpic.

The first step in the formation of seed is the opening of the flower bud. This signifies sexual maturity.

II. Ovule

The ovule or megasporangium is the forerunner of seed. A normal ovule (Fig. 1) has a stalk called funiculus by which it is attached to the placenta. The ovular envelopes, the integuments, enclose a massive (or scanty) nucellus.

Generally five types of mature ovules have been recognized. These are orthotropous, anatropous, amphitropous, campylotropous, and hemianatropous. In the orthotropous or atropous ovule the micropyle and the funiculus are in the same line. Such ovules occur in Polygonaceae, Urticaceae, Cistaceae, and Piperaceae. When the micropyle and hilum are near each other, as a result of unilateral growth, the ovule is called anatropous. This type is most prevalent in the angiosperms, especially in Sympetalae. In the campylotropous ovule the curvature is much less than in the anatropous type, and this is common in Resedaceae and Leguminosae. When the curvature affects also the nucellus and embryo sac, so that the latter becomes horseshoe-shaped, as in some Alismaceae, Butomaceae, and Centrospermae, the ovule is called amphitropous. In the hemianatropous or hemitropous type the funiculus is at right angles to the nucellus and integuments, e.g., in *Ranunculus*. A peculiar type of ovule found in some members of Plumbaginaceae and Cactaceae is called circinotropous.

Bocquet (1959) has discussed the morphology of campylotropous ovules. He is of the opinion that it is important to consider the mode of development and vascular supply to the ovule. According to him, the basic types are orthotropous and anatropous, and there are two basic series—the orthotropous and anatropous. During growth these undergo curvature giving rise to the campylotropous condition. Depending upon the series to which the ovule belongs, the ovules are called orthocampylotropous (Caryophyllaceae) in the ortho series, and anacampylotropous (Leguminosae) in the ana series. If the curvature is much greater, the ovules are designated as orthoamphitropous, as in *Atriplex hortensis,* or ana-amphitropous as in *Pisum sativum.*

In the circinotropous type the young ovule is in line with the axis. Subsequent unilateral growth results in anatropy. The curvature continues so that the ovule takes a complete turn and, eventually, the micropyle faces upward.

Davis (1966) has collected information on ovule morphology from 315 families of which 266 are dicotyledons. The morphology is constant for 248 families, and the anatropous condition is represented in 204 families. Orthotropous ovules are met with in 20 families, and hemianatropous ones in 13 families.

A. Nucellus

Depending upon the extent of development of the nucellus, the ovule may be tenuinucellate or crassinucellate. In the tenuinucellate ovule, which is characteristic of Sympetalae, the archesporial cell develops directly into a megaspore mother cell. The archesporial cell is surrounded by the epidermis only. The delimitation of crassinucellate ovules is not at all satisfactory. Normally, a crassinucellate ovule is one where the archesporial cell cuts off a parietal cell, and its derivatives make the megaspore mother cell deep-seated. The tissue formed by the division of the nucellar epidermis is called nucellar cap (Fig. 2A).

According to Davis (1966), if the parietal cell is not cut off, the ovule should be referred to as tenuinucellate. The difficulty arises in those cases where, although a parietal cell is not formed, nucellar cells surround the megaspore mother cell, e.g., in *Anemone, Aquilegia vulgaris* (Bhandari, 1968; Bhandari and Vijayaraghavan, 1970), and *Clematis* (Vijayaraghavan, 1962). In the terminology suggested by Davis, such ovules would strictly be called tenuinucellate. However, this matter requires a fresh appraisal, and it is time that we redefine the terms tenuinucellate and crassinucellate. In our opinion only such ovules as are seen in Sympetalae, where the megaspore mother cell is surrounded by nucellar epidermis alone, should be referred to as tenuinucellate. If nucellar cells surround the megaspore mother cell, irrespective of whether a parietal cell is or is not cut off, such ovules should be called crassinucellate.

When the two terms tenuinucellate and crassinucellate can encompass all the types of ovules in relation to nucellus, the use of a new term, pseudocrassinucellar is unwarranted. According to Davis (1966), who coined this term, pseudocrassinucellar ovules are those where the cells of the nucellar epidermis divide periclinally giving rise to a nucellar cap. In such cases a parietal cell is not cut off as in *Ficaria ranunculoides* (Souèges, 1913). In the description of pseudocrassinucellar ovules, Davis excludes those crassinucellate ovules that show formation of a nucellar cap.

According to data collected by Davis (1966), the condition of nucellus is known for 314 families of which 260 are dicots. Crassinucellate ovules are characteristic of 179 families, and tenuinucellar ones of 105. Eleven families possess pseudocrassinucellar ovules. In the remaining 19 families, the nucellus forms a generic or specific character.

The nucellus is mostly consumed by the developing embryo sac and endosperm. In several plants, however, it persists and acts as a storage tissue and is called perisperm.

Instances are known in which nuclei of the nucellar cells migrate into the embryo sac and become incorporated in it. This has been observed in *Cocos nucifera* (Quisumbing and Juliano, 1927) and *Tamarix* (B. M. Johri and Kak, 1954). Madge (1934) reported a peculiar situation in *Hedychium gardnerianum*. The nuclei move from cell to cell and ultimately reach the hypostase. From here they enter the embryo sac and are probably nutritive in function.

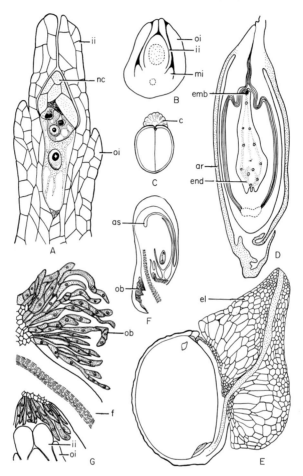

FIG. 2. Ovule and appendages. (A) Longisection of ovule of *Wolffia microscopica* showing integuments and nucellar cap. (B) Longisection of ovule of *Canangium odoratum* showing incipient middle integument. (C) Mature carunculate seed of *Swartzia pinnata;* front
view. (D). Ovule of *Zingiber macrostachyum* showing a third covering. (E) Nearly-mature
seed of *Trillium ovatum* with seed appendage or elaiosome. (F) Anacampylotropous ovule
of *Tetragonia tetragonioides.* (G) Obturator at organized embryo sac stage. (ar) Aril; (as)
air space; (c) caruncle; (el) elaiosome; (emb) embryo; (end) endosperm; (f) funiculus; (ii)
inner integument; (mi) middle integument; (nc) nucellar cap; (ob) obturator; (oi) outer
integument. [(A) after S. C. Maheshwari, 1954; (B) after Corner, 1949; (C) after Corner,
1951; (D) after Panchaksharappa, 1960; (E) after R. Y. Berg, 1958; (F, G) after Prakash,
1967.]

In *Asplundia* (Harling, 1958) the cells of the inner layer of nucellus
enlarge and protrude into the embryo sac. These cells later acquire rich
cytoplasm, and the nuclei become hypertrophied. In one of the species,

Asplundia polymera, the nuclei divide and fuse to form irregularly-shaped synkaryons.

Podostemaceae are unique in having a pseudoembryo sac formed as a result of the breakdown of nucellar cells below the megaspore mother cell. This sac is supposed to play a nutritive role because of the absence of endosperm (Mukkada, 1962).

Ethridge and Herr (1968) reported formation of an indentation or "pore" at the apex of the nucellus in *Rhexia marina*. A nucellar cap is formed consisting of two layers of cells and completely covers the micropylar end of the nucellus. Moreover, it extends downward, on all sides, as irregular lobes. The formation of the pore is probably due to suppression of anticlinal divisions in the nucellar epidermis. As a result, the epidermis is pulled apart due to nucellar enlargement. A nucellar pore has not been reported earlier in angiosperms.

Sometimes, in the lower part of the nucellus, just below the embryo sac, one often finds a group of cells which may or may not contain dense cytoplasm. The walls of these cells are somewhat thickened, and they constitute the hypostase. Johansen (1928), who studied Onagraceae, stated that plants growing in dry areas show a hypostase; plants of damp areas are devoid of this. Such a contention is no longer tenable since hypostase has been recorded in hydrophytes such as *Jussieua* (Khan, 1942) and *Trapa* (Ram, 1956).

The hypostase may persist in the mature seed as in *Aristolochia* (B. M. Johri and Bhatnagar, 1955) or it may be consumed as in *Yucca* (Wunderlich, 1938). According to Dnyansagar (1954), in some Leguminosae the cells of the hypostase (barrier tissue) become thick-walled and probably contain tannin.

Epistase is organized in the upper region of the nucellus. It is a caplike structure formed from the nucellar epidermis or its derivatives. Its cells become cutinized and remain distinguishable even during advanced stages of embryo.

The family Zingiberaceae is characterized by several structures associated with the nucellus. In addition to the hypostase and epistase, the ovules also show a nucellar pad and a nucellar cap. The nucellar pad of *Costus* (Panchaksharappa, 1962b) and *Hedychium* (Sachar and Arora, 1963) consists of unilayered, radially elongated cells in the epidermal region below the micropyle. The nucellar cap, as seen in *Zingiber, Hitchenia* (Panchaksharappa, 1962a, 1966), and *Hedychium* (Harling, 1949), is multilayered and forms as a result of divisions in the nucellar epidermis.

The hypostase and epistase probably serve to "stabilize the water balance of the resting seed over the long period of dormancy during the hot dry season" (Johansen, 1928). Perhaps they also check the growth of the embryo sac. According to Venkata Rao (1953), the hypostase acts as a connecting link between the vascular supply and the embryo sac, and helps in the transport of food materials.

B. Integuments

The ovules may be uni- or bitegmic. In the former there is a single integument, whereas in the latter there are two. The integument is a rim-like envelope. The Sympetalae are characterized by unitegmic ovules; Polypetalae and monocotyledons generally possess bitegmic ovules. Exceptions to this generalization are, of course, available in these groups of plants. Another interesting point is the association of unitegmic ovules with tenuinucellate condition, and bitegmic ovules with crassinucellate condition. In Ranunculaceae, Icacinaceae, Rosaceae, and Piperaceae, both uni- and bitegmic ovules occur. The number of integuments has been useful for taxonomic considerations.

In the bitegmic ovules the inner integument differentiates earlier than the outer, but the latter overgrows the former. The opening in the integument(s) above the nucellus is called the micropyle. The micropyle may be formed by one or both integuments, and it may be straight or zigzag. The micropyle is wide and shallow when the integuments fall short of the nucellus. However, when the integuments overgrow the nucellus, the micropyle is long and narrow.

To a greater or lesser extent, the various parts of the ovule may show fusion. The integuments between themselves, and with the nucellus, may fuse up to various lengths. The line of demarcation between these parts may or may not be discernible.

Although rare, the nucellus is free from the integuments in *Casuarina, Juglans, Myrica, Cleome,* and *Cananga.* In some of the primitive dicotyledons, such as Amentiferae, Ranunculaceae, Leguminosae, Rosaceae, and Capparidaceae, the integuments are free from each other. These, however, lie appressed. The outer integument in *Quercus, Alchemilla,* and *Sibbaldia* fuses above the nucellus closing the micropyle. According to Eames (1961), "This closure has been interpreted as the result of loss of function under chalazogamy or parthenogenesis."

It is a matter of common observation that the outer integument, on the funicular side in anatropous ovules, is not discernible, the fusion of outer integument and funiculus being congenital.

The bitegmic condition has been considered primitive, and the unitegmic one derived. The latter condition may result either from fusion of the two integuments, as in some Lecythidaceae and Myrtaceae (Mauritzon, 1939), or by suppression of one integument (P. Maheshwari, 1950).

In a bitegmic ovule the primordia of both the integuments arise independently, at the base of the young nucellus. In *Lannea* and *Rhus,* and some other members of Anacardiaceae, there is a splitting of the primordium of the single integument resulting in a bitegmic condition (Kelkar, 1958a,b).

In the ovules that possess a single integument and are tenuinucellate, the nucellar epidermis becomes disorganized early in its development. Here the inner epidermis of the integument becomes specialized and takes up the function of nutrition. The cells that surround the archesporium or the embryo sac become enlarged radially and densely cytoplasmic. These cells sometimes contain starch and fat and have been referred to as endothelium or integumentary tapetum, because of similarity to the tapetum of anthers.

The cells of the endothelium often become polyploid. Steffen (1955) reported 32-ploidy by endomitosis in *Pedicularis palustris.* Nair and Jain (1956) observed multinucleate endothelial cells in *Balanites.* In Solanaceae, particularly *Datura,* the cells of the endothelium absorb the outer adjacent layers of the integument which, eventually, itself gets absorbed by the endosperm (Souèges, 1907). The endothelium in compatible and incompatible crosses of *Datura* shows an interesting behavior (Satina *et al.,* 1950). In compatible crosses the size and shape of cells vary during different stages of development. These, however, begin to degenerate when the embryo is heart-shaped. In contrast, in incompatible crosses, in which the endothelium, instead of providing nutrition to the embryo, grows inward and invades the endosperm and embryo, the cells enlarge considerably, multiply, and form tumors.

In sunflower, Savchenko (1960) observed some capitula with abnormal florets resembling short shoots with numerous leaflike scales. The endothelium, in the ovules borne on these florets, divided actively and became eight or nine-layered. The latter could be distinguished into groups of ten to twelve cells separated by transverse cuticular septa. In these ovules the embryo sacs had degenerated. According to Savchenko, the food material could not be utilized by the embryo and, instead, accumulated in the tapetal cells, stimulating them to divide and grow.

Although a character mainly of tenuinucellate ovules, in Acanthaceae and Droseraceae the endothelium is present only in certain members. It is completely absent in Diapensiaceae in spite of the early degeneration of the nucellus. Begoniaceae and Droseraceae are characterized by a false endothelium, because the nucellar cells elongate radially and form a sort of jacket around the embryo sac (Davis, 1966).

Corner (1949) reported a middle integument in some of the Annon-aceae, namely, *Canangium* (Fig. 2B), *Mezzettia,* and *Xylopia.* The addi-tional integument arose as an intercalary structure after fertilization. Since an aril is also present in these plants, the ovules possess four integuments. The middle integument in *Cananga* shows a slightly different behavior in that it develops during prefertilization stages (Periasamy and Swamy, 1961). A rudimentary third integument has also been reported by Vi-jayaraghavan (1964) in *Sarcandra.*

A prominent space between the two integuments, in the region of the chalaza, is an important characteristic of ovules in Centrospermae and more so in Cactaceae (P. Maheshwari and Chopra, 1955; Kapil and Prak-ash, 1969). Cocucci (1961) in *Trianthema argentina* and Prakash (1967) in *Tetragonia tetragonioides* (Fig. 2F) also observed a similar air space in the chalazal end of the ovule.

The presence of stomata on the outer integument has been reported in *Cleome, Isomeris* (Orr, 1921), *Nerine curvifolia* (Schlimbach, 1924), *Argemone mexicana* (Sachar, 1955), *Michelia champaca,* and *Magnolia stellata* (Paliwal and Bhandari, 1962). It is not clear what the probable rolé of stomata could be.

In *Gossypium* (Joshi *et al.,* 1967), stomata differentiate in the outer integument at the chalazal end. They appear 2 days before anthesis, and remain healthy up to 14 to 16 days after pollination. The guard cells are gorged with starch grains. According to Ayyangar (1948), the stomata are presumably associated with respiration and production of hairs on seeds.

In addition to stomata, abundant chlorophyll is present in the integu-ments of *Hymenocallis occidentalis* (Flint and Moreland, 1943). Chloro-phyll has also been reported in *Amaryllis belladonna, Brunsvigia minor* (Hofmeister, 1861), *Sobralia micrantha* (Treub, 1879), *Gladiolus com-munis, Lilium martagon* (O. Berg, 1898), and *Moringa oleifera* (Puri, 1941). In some members of Caesalpinioideae the micropylar end of the ovule becomes red or deep pink because of anthocyanin pigments. Dur-ing postfertilization stages, the outer integument develops chlorophyll and the ovules appear deep green (Corner, 1951).

Some other structures are also associated with the ovules: aril (Fig. 2D), arillode, caruncle (Fig. 2C), and sarcotesta. The aril has been con-sidered as the third integument. P. Maheshwari (1950) stated that the aril "is a new structure arising from the base of the ovule," forming a third integument. Kapil and Vasil (1963) have defined it "as a secondary collarlike outgrowth which develops from the funiculus and surrounds the ovule more or less completely in postfertilization stages." According to Eames (1961), the term aril has been used loosely for fleshy parts of the ovule and includes proliferation from chalaza and integuments almost enveloping the ovule proper. It may extend from the base upward or from the tip of integument downward. This term has also been applied to outgrowth from the chalazal end or to fleshy funiculus as seen in *Magnolia* and some species of *Acacia*. Eames considers the use of aril as a third integument "unfortunate," particularly in those ovules where the aril is an elaboration of either a part or whole of one integument.

In Passifloraceae the aril is thin-walled and contains oil, starch, and yellowish-red chromoplasts (Raju, 1956). Kapil and Vani (1963) reported an annular outgrowth from the funiculus in *Crossosoma californicum*. This outgrowth, or aril, develops numerous fingerlike projections and appears fimbriate. *Myristica* and *Paeonia* also show aril. In *Myristica* the aril is highly vascularized, whereas in *Paeonia* a branch of the vascular bundle partly enters the ovule (Camp and Hubbard, 1963). In *Turnera ulmifolia* (Vijayaraghavan and Kaur, 1966) the aril is white, sickle-shaped, and nonvascularized. It extends to the base of the seed only on the side of the raphe.

When the outgrowth arises from the tip of outer integument and grows downward, partly or completely covering the seed, the structure is called a caruncle or arillode. A caruncle is a common feature of Sapindaceae (Van der Pijl, 1957) and Euphorbiaceae (Kapil, 1958). According to Eames (1961), when the fleshy outgrowths of the integument are small, they are called caruncle, and when large, aril.

In Zingiberaceae (Panchaksharappa, 1960) the covering on the seed is derived both from the funiculus and outer integument.

Sarcotesta is the fleshy or juicy outer integument. In Magnoliaceae it contains fat and is often colored (see also Van der Pijl, 1955).

Since the aril, arillode, and sarcotesta are attractive to animals and birds, their main function is dissemination of seeds.

Elaiosome is an outgrowth from the raphe or hilum (Fig. 2E), and is known in plants belonging to Amaryllidaceae, Liliaceae, and some other families. A conspicuous elaiosome is found in *Trillium ovatum* (R. Y. Berg, 1958) where two meristematic zones, near the raphe, result in a prominent excrescence. Its cells are rich in starch and drops of fatty oils. By some authors (see Netolitzky, 1926), elaiosome has been considered synonymous with caruncle or strophiole. Since its cells are frequently filled with food materials, the seeds are often picked and dispersed by ants.

The obturator is a special structure formed as a swelling of the placenta that grows toward the micropyle fitting like a hood over the nucellus. The obturator may also be formed from various parts of the pistil. It arises as an outgrowth from the funiculus in *Combretum* (Nagaraj, 1955), *Areca* (Venkata Rao, 1959), and *Trianthema* (Cocucci, 1961). In *Poivrea coccinea* (Venkateswarlu, 1952) the long funicle shows large papillate cells, whereas in *Opuntia dillenii* (P. Maheshwari and Chopra, 1955), and some other Cactaceae (Kapil and Prakash, 1969), the inner surface is covered with hairs directed toward the micropyle. In *Tetragonia tetragonioides* (Prakash, 1967) the long funiculus bears an obturator of glandular cells (Fig. 2F and G) which are richly cytoplasmic. Both placental and funicular obturators are known in Aizoaceae (Prakash, 1967).

Leucosyke and *Myriocarpa* (Fagerlind, 1944), *Rhus mysurensis* (Kelkar, 1958b), and *Coleonema* (Desai, 1962) show an integumentary obturator. Cells of the stylar canal and ovary wall are also known to give rise to an obturator. Kapil (1956, 1958) recorded two types of obturators in Euphorbiaceae. In *Chrozophora* it consists of small, thin-walled and compactly arranged cells, whereas in *Euphorbia* and *Acalypha* the cells of the obturator are loose, elongated, and hairlike.

The obturator is devoid of vascular supply and acts as a bridge for the pollen tube. Its main function is to facilitate the entry and growth of the pollen tube. The obturator disorganizes during postfertilization stages.

C. Abnormal and Reduced Ovules

The achene-bearing genera of Ranunculaceae usually possess uniovulate carpels. However, in *Adonis* (Bhandari, 1962, 1966) and *Clematis* (Vijayaraghavan, 1962), in addition to a fertile ovule, two to four accessory sterile ovules are also present. In *Anemone obtusiloba* (Bhandari, 1968), besides the fertile ovule, there are two marginally elongated "cylindrical structures." These represent the sterile ovules. In *Adonis chrysocyathus* and *Anemone rivularis* the sterile ovules are absent. The fertile ovule is anatropous, bitegmic *(Adonis)* or unitegmic *(Anemone, Clematis)*, tenui- or crassinucellate, and is provided with a vascular supply. In contrast, the sterile ovules lack differentiation into nucellus and integument, and do not have a vascular supply. It has been suggested that such ovules probably represent an intermediate stage in the reduction series from a multiovulate to a uniovulate carpel.

In some families of parasitic plants the ovules are not at all well-defined and there is no demarcation between the nucellus and integument (P. Maheshwari *et al.,* 1957; B. M. Johri and Bhatnagar, 1960; Kuijt, 1969). In Loranthaceae, there is a placental–ovular complex, also called mamelon or placenta. It may be lobed or unlobed. In some forms the ovary is completely devoid of placenta. When the placenta is lobed, the archesporium differentiates hypodermally in each lobe. In an unlobed placenta the entire subepidermal tissue functions as archesporium.

When the placenta is absent, the archesporial cells differentiate in the hypodermal layer below the ovarian cavity. Figure 3 gives a diagrammatic representation of the reduction of placenta in Loranthaceae.

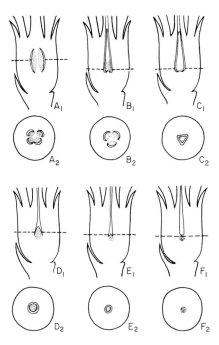

FIG. 3. Longisection and transection of ovary showing reduction of placenta in Loranthaceae. (A$_1$, A$_2$) *Lysiana;* four-chambered ovary showing four-lobed placenta. (B$_1$, B$_2$) *Nuytsia;* three-chambered ovary with three-lobed placenta. (C$_1$, C$_2$) *Macrosolen;* one-chambered ovary with three-lobed placenta. (D$_1$, D$_2$) *Helicanthes;* one-chambered ovary with unlobed conical placenta. (E$_1$, E$_2$) *Amyema;* one-chambered ovary showing unlobed placenta formed as a result of elongation of sporogenous cells. (F$_1$, F$_2$) *Helixanthera;* complete reduction of placenta; archesporial cells differentiate at the base of ovarian cavity. (Modified after P. Maheshwari *et al.*, 1957.)

III. Female Gametophyte

The archesporial cell, in the ovule, with or without cutting off a parietal cell, functions as the megaspore mother cell or embryo sac mother cell (Fig. 4A and B). The megaspore mother cell undergoes reduction division, and its further fate varies in different types of embryo sac development. If both the meiotic divisions are followed by wall formation, a linear tetrad of megaspores is formed (Fig. 4C and D). It is usually the chalazal megaspore that functions, whereas the upper three megaspores

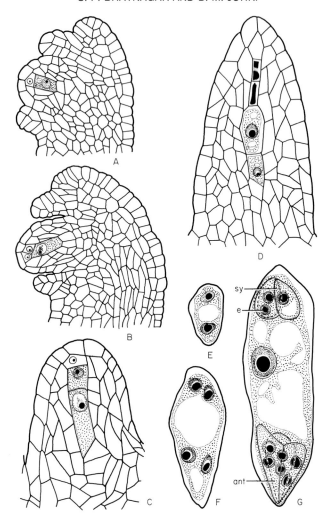

Fig. 4. Megasporangium, megasporogenesis, and female gametophyte in *Aquilegia vulgaris*. (A, B) Longisections of ovules showing megaspore mother cell and parietal cells. (C) Dyad cells. (D) Linear tetrad of megaspores. (E, F) Two-, and four-nucleate embryo sacs. (G) Mature embryo sac; the antipodal cells are binucleate. (ant) Antipodal cells; (e) egg; (sy) synergid. (After Bhandari and Vijayaraghavan, 1970.)

degenerate. The functional megaspore undergoes three mitotic divisions (Fig. 4E and F) resulting in the formation of eight nuclei. Where only one megaspore is involved in the formation of the female gametophyte, the development is referred to as monosporic.

The eight nuclei of the gametophyte organize into two quartets — micropylar and chalazal (Fig. 4G). The micropylar nuclei form the egg apparatus consisting of two synergids and an egg cell, and the upper polar nucleus. The chalazal quartet gives rise to the lower polar nucleus and three antipodal cells. This course of development occurs in the majority of flowering plants and is designated as the *Polygonum* type. In another type of monosporic development, called the *Oenothera* type, it is the micropylar megaspore, sometimes the chalazal, that undergoes only two mitotic divisions forming four nuclei. These nuclei organize into an egg apparatus and upper polar nucleus. *Oenothera* type (Fig. 5) is characteristic of the family Onagraceae. There is, however, an exception. In *Schisandra chinensis,* Swamy (1964) reported both monosporic and bisporic development of gametophyte. The organization of mature embryo sac, which is four-nucleate in both, is similar to the *Oenothera* type. According to Swamy, this is the only authentic report of a bisporic four-nucleate embryo sac.

When, instead of one, two megaspores take part, the development is bisporic. In this category the megaspore mother cell undergoes reduction division giving rise to two dyad cells. The upper dyad cell degenerates, and the lower dyad cell, by two mitotic divisions, produces an eight-nucleate gametophyte. The organized embryo sac is similar to that of the *Polygonum* type. This is the *Allium* type of development (Fig. 5). In *Endymion* type (Fig. 5) the embryo sac is produced from the upper dyad cell.

When both the meiotic divisions in the megaspore mother cell are not followed by wall formation, and all the four nuclei thus formed take part in formation of the embryo sac, the development is referred to as tetrasporic. Several types belonging to this category are represented in Fig. 5.

Only the *Fritillaria* type of the tetrasporic embryo sacs will be discussed here. After the formation of coenomegaspore the nuclei show a 1+3 arrangement (one nucleus at micropylar end and three at chalazal region) — primary four-nucleate stage. During the course of division the lower three nuclei fuse; simultaneously the upper nucleus also divides. This results in the formation of two upper haploid nuclei and two lower triploid nuclei — secondary four-nucleate stage. The two upper and two lower nuclei divide once again so that the eight nuclei organize into a haploid egg apparatus, one haploid polar nucleus, one triploid polar nucleus, and three triploid antipodal cells.

A very good example of the coexistence of mono-, bi-, and tetrasporic embryo sacs is *Delosperma cooperi* of Aizoaceae (Kapil and Prakash, 1966). The authors recorded 14% ovules showing *Polygonum* type, 18% *Endymion* type (Fig. 6A–C), and 68% tetrasporic type. Of the latter, 47% conform to *Drusa* type (Fig. 6D–H), 35% *Penaea* type (Fig. 6I),

and 18% *Adoxa* type (Fig. 6J). Different types of embryo sacs are frequently met with in the same ovary.

Special mention may be made of embryo sacs with unusual character-

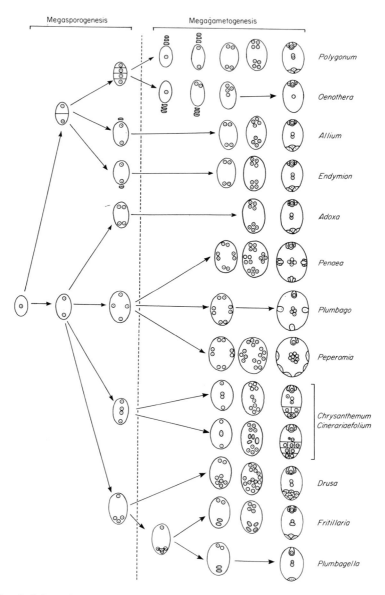

FIG. 5. Schematic representation showing the origin and development of different types
of embryo sacs. (After B. M. Johri, 1963.)

istics in parasitic angiosperms belonging to the group Santalales (see B. M. Johri, 1962) and some other plants.

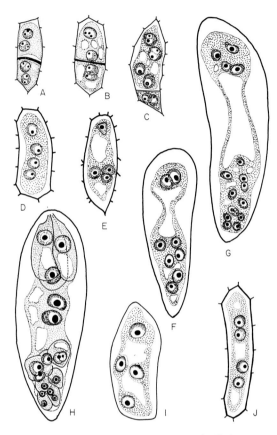

FIG. 6. Different types of embryo sac development in *Delosperma cooperi*. (A, B) Binucleate dyad cells. (C) Four-nucleate embryo sac derived from upper dyad cell (*Endymion* type). (D) Coenomegaspore with nuclei arranged in a linear row. (E) Same as figure D, showing 1+3 arrangement. (F, G) Subsequent stages showing 2+6 and 4+12 arrangements. (H) Organized embryo sac (*Drusa* type). (I) Four megaspore nuclei arranged crosswise (*Penaea* type). (J) Coenomegaspore with 2+2 arrangement of nuclei (*Adoxa* type). (After Kapil and Prakash, 1966.)

In the suborder Santalineae, order Santalales, the embryo sac remains confined to the ovule as in *Strombosia* (Agarwal, 1963b), *Cansjera* (Swamy, 1960), *Opilia* (Swamy and Dayanand Rao, 1963), *Buckleya* (Raj, 1964), *Comandra* (Ram, 1957), and *Iodina* (Bhatnagar and Sabharwal, 1969). However, in *Olax* (Agarwal, 1963a), *Leptomeria* (Ram, 1959b), *Mida* (Bhatnagar, 1960), *Santalum* (Bhatnagar, 1959, 1965), and

Quinchamalium (B. M. Johri and Agarwal, 1965) the tip of the embryo sac extends beyond the ovule and comes to lie in the ovarian cavity.

An interesting feature of the family Loranthaceae is the formation of multiple gametophytes in the same ovary. The embryo sacs develop concurrently and reach to various heights in the ovary, style, and stigma (see B. M. Johri and Bhatnagar, 1960). In *Helixanthera* (B. M. Johri, 1963), the tip of the embryo sac extends as far as the stigmatic epidermis. *Moquiniella* is still more interesting (B. M. Johri and Raj, 1969). The embryo sac grows as in *Helixanthera,* then it curves backward forming an inverted J (Fig. 7A and B). The maximum length is 48 mm which is probably the longest embryo sac known in angiosperms.

In *Pentaphragma* (Kapil and Vijayaraghavan, 1962) and Acanthaceae (Mohan Ram and Sehgal, 1958; P. Maheshwari and Negi, 1955; B. M. Johri and Singh, 1959; Bhatnagar and Puri, 1970) the tip of the embryo

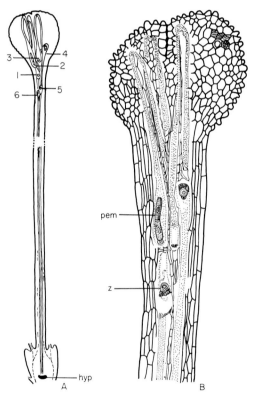

Fig. 7. Embryo sac in *Moquiniella rubra.* (A) Longitudinal section of gynoecium with six embryo sacs of which three (1, 2, and 3) have curved tips (diagrammatic). (B) Magnified view of stigmatic region from figure A showing four embryo sacs, three with curved tips. (hyp) Hypostase; (pem) proembryo; (z) zygote. (After B. M. Johri and Raj, 1969.)

sac grows beyond the ovule and extends into the micropyle or a schizogenous cavity formed in the region of the funiculus. A similar extraovular growth of the female gametophyte is also known in some other angiosperms (Yamazaki, 1954; Khan, 1954).

At its tip, a normally organized gametophyte shows an egg apparatus comprising three cells – an egg cell, and two synergids. One nucleus from the micropylar quartet and one from the chalazal quartet migrate to the center of the gametophyte and function as polar nuclei. There are three antipodal cells situated at the base of the embryo sac.

Rarely, there may be only one synergid or the synergid may even be absent. The number of polar nuclei and antipodal cells may also vary. There is a single polar nucleus in the *Oenothera* type, four in *Penaea* and *Plumbago* types, and eight or more in *Peperomia* type.

There are three antipodal cells in *Polygonum, Allium, Adoxa,* and *Fritillaria* types. The *Peperomia* and *Drusa* types show six and eleven antipodals, respectively. In *Plumbago* and *Plumbagella* types, there is a single antipodal cell which, in the latter, is triploid. There is no antipodal cell in the *Oenothera* type.

Of the various components of the embryo sac, synergids and antipodals are ephemeral.

A. Synergid

The synergid may be hooked or beaked and usually shows a vacuole at the lower end, and nucleus at the upper end. Sometimes, the synergids also show a filiform apparatus.

Quite often one of the synergids may persist in postfertilization stages. It may even become hypertrophied and its nucleus polyploid. Because of endomitosis, in *Allium angulosum* and *Allium pulchellum* the synergid nucleus becomes octoploid (Hasitschka-Jenschke, 1958).

A very peculiar behavior of synergids has been recorded in some members of Crassulaceae, Compositae, and Santalaceae. In the latter two families the synergids elongate considerably. The tips grow beyond the micropyle and sometimes reach the funiculus. This is especially true of *Calendula* and *Ursinea* (Dahlgren, 1924). Also, in Crassulaceae the synergids are haustorial (Subramanyam, 1967). The synergid haustoria are most extensive in *Quinchamalium*. Here, in prepollination stages, the tips of synergids elongate and travel along the surface of the placenta; eventually, the haustoria reach the base of the style.

Jensen (1965a) studied the ultrastructure and histochemistry of the synergids of cotton. The cells are surrounded by a partial wall which gives positive reaction both for carbohydrates and proteins. The reaction is negative for nucleic acids. There is, however, a significant amount of pectin. The synergids of *Capsella* are unique in possessing a chalazal wall (Schulz and Jensen, 1968a).

The filiform apparatus is the extension of the wall at the micropylar end of the cell. Large amounts of endoplasmic reticulum (ER), plastids, and mitochondria are associated with the filiform apparatus. In the synergid, ER is maximum near the filiform apparatus and its concentration decreases in the lower portion of the cell; it is oriented parallel to the long axis of the cell.

The maximum concentration of dictyosomes is in the middle portion of cell. As is characteristic of synergids, the vacuoles are located below the nucleus and are rich in inorganic compounds. Sphaerosome-like bodies are distributed throughout the cell. Plastids and mitochondria show evidence of division. Ribosomes are seen free as well as associated with ER, nucleus, mitochondria, and plastids. Several functions have been attributed to synergids: first, there is the opening of the tip of the pollen tube; second, they help place the male gametes and the discharged material deep into the embryo sac, in between the egg and secondary nucleus (Gerassimova-Navashina and Korobova, 1959; Gerassimova-Navashina, 1961); third, they probably secrete some chemotropic substances; and, last, they may help in absorption, storage, and transport of compounds from the nucellus (Jensen, 1965a) to the developing embryo and endosperm.

B. Egg

In *Gossypium* (Jensen, 1965b) the cytoplasm of the egg contains large amounts of RNA and protein. It shows a negative PAS reaction for carbohydrates, and a definite reaction for DNA is given by the nucleus. The latter also shows the presence of proteins. The nucleolus is rich in RNA and protein.

It is interesting that, just as the synergid, the egg is also partially surrounded by a wall which is seen distinctly at the micropylar region. It becomes thinner toward the chalazal end, and the lower one-third of the cell is surrounded only by the plasma membrane. In *Capsella* (Schulz and Jensen, 1968b) the wall is incomplete at the chalazal end.

The ER is frequently seen close to the plasma membrane and is arranged parallel to the long axis of the egg cell. In the micropylar region the ER forms an expanded network and here, frequently, appear mitochondria, Golgi apparatus, and plastids. The fingerlike projections of the ER into the cytoplasm are another unique feature. This ER is associated with the plasma membrane. Such projections are mostly seen in the micropylar region of the egg. Free ribosomes as well as those associated with ER are found in the cell. Endoplasmic reticulum containing small tubes is occasionally seen in the immediate vicinity of the nucleus. Although scarce, ER was found in all the egg cells examined by Jensen; it increased in the zygote and embryos. The nuclear membrane shows some projections into the cytoplasm which connect with the ER. Clusters of ribosomes are present on the outer membrane of the nucleus.

There are numerous mitochondria with relatively few and short cristae. There are few plastids of uniform shape and size, and they may contain one or two starch grains. Ribosome-like bodies of unknown nature are also present. Like plastids, dictyosomes are also few and consist of three or four cisternae. Vesicles are seen with the dictyosomes.

C. Polar Nuclei

Usually, the polar nuclei are of the same size. In *Dipteracanthus patulus* (P. Maheshwari and Negi, 1955), however, one polar nucleus is smaller than the other. Additional nuclei are known in *Cuscuta reflexa* (B. M. Johri and Tiagi, 1952) and *Utricularia flexuosa* (Khan, 1954). Such a situation may arise either from supernumerary division of one or more nuclei of the unorganized embryo sac or from failure of one or more nuclei to organize into synergids or antipodal cells. The free nuclei move to the center of the embryo sac and fuse with the polar nuclei.

According to Gerassimova-Navashina (1957), the fusion of polar nuclei is possible only when the polars are in a state of division. If the polars fuse before fertilization, the resting phase is assumed after the secondary nucleus has been formed. If already in resting state, the polars just come together and fuse only after having been activated by the male gamete.

The polar nuclei give a positive test for deoxyribonucleic acid (DNA). The cytoplasm of the central cell and nucleoli contain large amounts of ribonucleic acid (RNA) and protein. The wall surrounding the central cell and bordering the nucellus is fairly thick and is believed to be rich in pectic substances.

D. Antipodals

The antipodal cells exhibit great variation in their morphological structure. These may degenerate without being delimited into cells. They may become large multinucleate cells or even polyploid. Sometimes, their number may increase and as many as 300 cells are found in *Sasa* (Yamaura, 1933). In *Aconitum* (Osterwalder, 1898), *Actinophloeus, Areca* and *Howea* (Venkata Rao, 1959), *Argemone* (Sachar, 1955), *Eschscholtzia* (Sachar and Mohan Ram, 1958), and *Caltha* (Kapil and Jalan, 1962) the antipodals enlarge considerably. Supernumerary antipodal cells, sometimes binucleate, are seen in several members of Malvaceae (Venkata Rao, 1954). Multinucleate cells have been recorded in *Amianthium* (Eunus, 1951), *Pennisetum* (Narayanaswamy, 1953), *Stackhousia* (Narang, 1953), and *Tagetes* (Venkateswarlu and Maheswari Devi, 1955). Maheswari Devi (1957) observed only two antipodal cells, each two- to twelve-nucleate, in *Gerbera jamesonii.*

In *Quinchamalium chilense* (B. M. Johri and Agarwal, 1965) the antipodal nuclei do not organize into cells. Instead, they are cut off from the rest of the embryo sac by a wall resulting in an antipodal chamber. This grows through the funiculus and reaches the tip of the placenta and branches profusely. The nuclei become hypertrophied and may even divide, giving rise to five to seven nuclei.

The cytology of antipodal cells has revealed a high degree of polyploidy in several plants. Endoduplication of chromosomes takes place usually during postfertilization stages. It is 8-ploid in *Allium angulosum, Allium pulchellum,* and *Heleborus niger,* 16-ploid in *Othonna crassifolia,* 32-ploid in *Anemone hepatica* and *Clivia miniata,* 64-ploid in *Eranthis hiemalis* and *Kleinia ficoides,* and 64- or 128-ploid in *Aconitum* and *Papaver rhoeas* (Hasitschka, 1956; Hasitschka-Jenschke, 1958, 1959; Tschermak-Woess, 1956, 1957).

Not much information is available on histochemical studies on antipodals. *Taraxacum, Calanthe veitchii, Cypripedium insigne,* and *Dendrobium nobile* have been studied by Poddubnaya-Arnoldi *et al.* (1964) and Zinger and Poddubnaya-Arnoldi (1966), *Stellaria* by Pritchard (1964), and *Vanda* by Alvarez and Sagawa (1965). In *Stellaria,* high content of Feulgen-stainable DNA and histones, high concentration of proteins, low RNA content, and no periodic acid schiff (PAS)-positive material have been reported. In *Dendrobium,* there is a high concentration of peroxidases, cytochrome oxidase, ascorbic acid content, and high concentration of SH compounds as compared to the egg.

Diboll and Larsen (1966) made an electron-microscopic study of the antipodals of *Zea mays.* According to them the inner face of the antipodal walls adjacent to the nucellus is papillate. Due to incomplete formation of cell walls in many antipodals, there is protoplasmic continuity between contiguous cells which results in the appearance of multinucleate protoplasts.

Organelles are in abundance. The mitochondria are oval-to-circular in profile. The internal membranes of the plastids are so arranged that their profiles appear as a system of interconnected rings of varying complexities. Each ring is composed of two appressed unit membranes. Starch grains, whenever present, are small. The dictyosomes are characterized by a moderately loose association of six to eight cisternae and by relatively small vesicles. In KMnO₄-fixed material, elements of ER vary in size from units approximately twice the length of dictyosome cisternae to smaller ones near the size of vesicles. The gradation in size as well as the similarities in the membrane image of the ER, dictyosome cisternae, and nuclear envelope make positive identification of these membrane systems difficult. Numerous minute granules are present throughout the nucleoplasm, often in clusters of several hundred granules.

The granulity of nucleoplasm, large surface area of ER, numerous multicisternal dictyosomes and mitochondria suggest high rates of respiration and synthetic activity. The latter may be related to papillate cell walls, or it may be associated with the movement of large amounts of nutritive materials from adjacent cells.

IV. Pollination and Fertilization

When the anthers are mature, they dehisce and release the pollen grains. The latter are disseminated by several agents such as wind, water, birds, and insects. From the time the pollen grains land on the stigma until the entry of pollen tube inside the ovule, there are a series of important events leading to fertilization. In this connection the growth of the pollen tube through the style is a significant process.

Based upon the distribution of the transmitting tissue, the style is either hollow (Fig. 8A) or solid (Fig. 8C). The hollow style, which is common in monocotyledons, is characterized by the presence of a stylar canal lined with a layer of transmitting tissue. *Allium ursinum* (Guéguen, 1901), *Zephyranthes ajax* (Vasil and M. M. Johri, 1964), and *Nothoscordum fragrans* (M. M. Johri, 1966b) are a few monocots that show a solid style. As seen in most dicotyledons, the solid style lacks the stylar canal and shows one or more strands of transmitting tissue (Bhatnagar and Uma, 1969). A third type of style, "half-closed," has been reported in some members of Cactaceae (Hanf, 1935) and *Artabotrys* (Rao and Gupte, 1951). In these plants the style is hollow, and the transmitting tissue develops only on one side of the stylar canal.

The stigma, which is the expanded portion of the style, receives the pollen grains. The stigmatic surface is smooth in *Smilacina* (M. M. Johri, 1966a). The stigma is generally papillate or hairy. In *Nothoscordum* and *Tulbaghia* (M. M. Johri, 1966b) the apex of the style is papillate and the stigma is not very prominent. In *Ophiopogon* (M. M. Johri, 1966a), a distinct stigma does not differentiate. The stigmatic papillae may be covered with a cuticle as in *Aloe* (M. M. Johri, 1966b; see also Kroh, 1964).

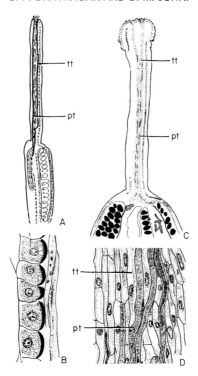

FIG. 8. Style, stigma, and pollen tube. (A) Longisection of gynoecium of *Lloydia alpina* showing course of pollen tubes. (B) Cells of transmitting tissue (longisection) with thickened outer tangential walls; note portions of two pollen tubes. (C) Pollinated stigma (longisection) with upper part of ovary of *Nicandra physaloides;* the ovules have been blackened. (D) Transmitting tissue (longisection) showing pollen tubes. (pt) Pollen tube; (tt) transmitting tissue in style. [(A, B) after M. M. Johri, 1966; (C, D) after Vasil and M. M. Johri, 1964.]

Polyploidy in the stigmatic papillae seems to be of special significance in nourishing the pollen tube as shown in *Spironema fragrans* (Tschermak-Woess, 1959) and *Lilium tigrinum* (Vasil and M. M. Johri, 1964). In *Aloe vera* the cells of the transmitting tissue become tetraploid (M. M. Johri, 1966a).

A detailed study of the morphology and anatomy of the stigma of *Petunia hybrida* has been made by Konar and Linskens (1966). The stigma can be separated into two zones — an upper or secretory zone consisting of epidermis, and the lower or storage zone comprising one to three layers of laterally extended cells.

The stigmatic secretion is highly unspecific. Lipids have been observed in the stigmatic exudate of *Lotus* (Bubar, 1958), *Aegle* and *Pavonia* (Vasil and M. M. Johri, 1964), *Petunia* (Konar and Linskens, 1966), and *Sola-*

num melongena, Solanum nigrum, and *Heliotropium eichwaldi* (Bhatnagar and Uma, 1969). A day prior to pollination, the stigmatic papillae secrete mucilage in *Aloe* (M. M. Johri, 1966a). Resinous secretion is exuded in *Koelreuteria paniculata* (Baum, 1950). The stigmatic secretion helps to protect the stigma from desiccation. It also plays some role in receiving the pollen and providing optimal conditions for pollen germination.

The pollen grains, after germination, put forth tubes which creep over the stigmatic papillae and enter the transmitting tissue. Only rarely do the tubes pierce the papillae and grow intracellularly, e.g., in *Agrostemma* (Hanf, 1935), *Dianthus* (Buell, 1952), and some members of Malvaceae. As a consequence of pollination, various changes occur in the stigmatic cells of Gramineae. They show better permeability and stainability, change in shape and size of nucleus, and a gradual withering and collapse. These changes were designated *stigma reaction* (Kato and Watanabe, 1957).

As the pollen tube grows between the stylar cells, the enzymes secreted by it soften the middle lamellae and the walls appear as if converted to mucilage. The passage of pollen tube into the gynoecium is believed to be mainly determined by the disposition of transmitting tissue (Schnarf, 1929; Renner and Preuss-Herzog, 1943; Bhatnagar and Uma, 1969). The superficial growth of pollen tubes over the transmitting tissue (Fig. 8B) in hollow styles has been termed ectotrophic, and the intrusive growth through the intercellular spaces (Fig. 8D) in the solid style endotrophic (Vasil and M. M. Johri, 1964; M. M. Johri, 1966a,b; Bhatnagar and Uma, 1969). Iwanami (1959) and Mascarenhas and Machlis (1962a,b) reported that growth of the pollen tube and its entry into the ovule is probably a chemotropic phenomenon. Mascarenhas and Machlis (1964) showed that in *Antirrhinum majus* the pollen tubes grow chemotropically toward the calcium source and that the ovules had a very high concentration of calcium. According to Welk *et al.* (1965), the ovules and transmitting tissue are sources of the chemotropic factor responsible for directional growth of the pollen tube.

Schoch-Bodmer and Huber (1947) believed that pollen tubes traversing the styles secrete pectinase which dissolves the intercellular substances in the transmitting tissue. From pollen germination and electron-microscopic studies of pollen tubes in *Lilium,* Iwanami (1959) concluded that the tube elongates at its tip.

Intracarpellary pollen grains were first observed by B. M. Johri (1936a) in *Butomopsis lanceolata.* Since then this feature has been reported in several other plants, such as *Butomus umbellatus, Boottia cordata, Limnocharis emarginata, Hydrocleis nymphoides, Trillium* (B. M. Johri and Bhatnagar, 1957), *Elettaria cardamomum* (Panchaksharappa, 1962a), *Fritillaria roylei,* and *Lilium tigrinum* (Vasil and M. M. Johri, 1964).

Germinating pollen grains have been observed in the ovaries of *Butomopsis lanceolata* (B. M. Johri, 1936a), *Limnocharis emarginata* and *Trillium sessile* var. *giganteum* (B. M. Johri and Bhatnagar, 1957), and *Fritillaria roylei* (Vasil and M. M. Johri, 1964). The diameter of intracarpellary pollen is slightly larger than of pollen on the stigma or in anthers. The slight increase appears to be due to swelling caused by mucilage in the style (B. M. Johri and Bhatnagar, 1957).

After traversing the style, the pollen tube reaches the embryo sac through the micropyle. This is referred to as porogamous fertilization. However, if the pollen tube enters through the chalaza, as in *Casuarina,* the fertilization is chalazogamous. In *Pistacia* (Grundwag and Fahn, 1969) and *Eucommia* (Eckardt, 1963) the pollen tube rarely enters the embryo sac through the antipodal end, and this is termed intravesicular chalazogamy. Irrespective of the place of entry of the pollen tube in the ovule, it invariably travels up to the tip of the ovules and enters the embryo sac through the micropyle.

There are three possible mechanisms of pollen tube entry in the embryo sac: (*a*) between the egg and one synergid, (*b*) between the embryo sac wall and one synergid, and (*c*) directly into one of the synergids. That synergids play a role in the entry of pollen tube inside the embryo sac, has been shown by electron-microscopic studies.

The pollen tube enters the embryo sac at the apex of the filiform apparatus which facilitates the entry. It is believed that some chemotropic substance in the synergid attracts the pollen tube. Jensen (1965a) reported a high concentration of calcium in synergids of cotton. A concentration gradient is probably established as a result of the collapse of synergid vacuoles resulting in the diffusion of these materials. The preferential entry of the pollen tube into one of the synergids has been attributed to the different concentration of calcium in the two synergids.

Mascarenhas (1966), however, is of the opinion that in *Antirrhinum majus* there is no increasing gradient or high concentration of calcium in the micropylar part of the embryo sac. Pollen tubes exhibit chemotropic responses toward ovules when grown *in vitro.* According to Mascarenhas (1966), factors other than calcium may also be responsible for directional growth of the pollen tube. Recently (Van Went, 1970), the synergids of *Petunia* have been reported to produce and secrete substances that direct the growth of pollen tubes.

Earlier, it was believed that one or both the synergids are destroyed as a result of the entry of the pollen tube inside the embryo sac. Recent studies of Jensen and D. B. Fisher (1968) on *Gossypium* and of Cass

and Jensen (1970) on *Hordeum* have revealed that one of the synergids begins to degenerate after pollination but before the pollen tube
reaches the embryo sac. The pollen tube enters this degenerated synergid
and discharges its contents (see Diboll, 1968; Schulz and Jensen, 1968a;
Jensen and D. B. Fisher, 1968; Kaplan, 1969; Cocucci and Jensen, 1969;
Cass and Jensen, 1970). In cotton (Jensen and D. B. Fisher, 1968), a
lateral pore develops in the pollen tube. In *Epidendrum,* however, the
pore is terminal (Cocucci and Jensen, 1969). It is through this pore that
the sperms are discharged.

The discharge of the pollen tube, as seen in *Gossypium hirsutum* (D. B.
Fisher and Jensen, 1969), *Capsella bursa-pastoris* (Schulz and Jensen,
1968a), and *Epidendrum scutella* (Cocucci and Jensen, 1969), is of a
similar pattern and is fountainlike. Depending upon the force with which
the pollen tube or synergid bursts, the discharged material comes to lie
deep in the embryo sac, usually between the egg and the central cell.

According to Jensen and D. B. Fisher (1968), in cotton, the degenerated
synergid does not burst. The plasma membrane of this synergid disappears — one sperm comes in contact with the egg plasma membrane,
and the other sperm with the plasma membrane of the central cell. A
small opening in the fused plasma membranes of egg and sperm can allow
the entry of sperm nucleus into the egg. In *Capsella* also, Schulz and
Jensen (1968a) reported a rupture in the common wall of the egg and
synergid. It is through this rupture that the sperm nucleus enters the egg.
In *Petunia,* Van Went (1970) reported the bursting of the synergid and
failed to observe any ruptures or pores in the walls of the zygote or primary endosperm cell.

Branching of the tip of the pollen tube has been reported in plants such
as *Peperomia* and *Portulaca.* One of the branches is directed toward the
egg and the other toward the central cell. It is through these branches
that the sperm nuclei reach their respective partners.

The sperm nuclei are carried passively through the cytoplasm of the
egg or central cell. The sperms do not show any evidence of amoeboid
movement.

One of the sperm nuclei first comes in contact with the egg nucleus and
begins to fuse before the other sperm and the polar nuclei. This probably
reflects (*a*) a difference in the rate of movement of the two sperm nuclei,
(*b*) the difference in distance the male gametes have to travel to reach the
egg and polar nuclei, and (*c*) the difference in time the male gametes leave
the synergid to fuse with the respective nuclei (Jensen and D. B. Fisher,
1967).

Although both the male gametes are identical, their fusion behavior
with the egg and polars is different, particularly in respect of time which is

much longer between syngamy than between triple fusion. This probably could be due to differences between the egg and central cell (Jensen, 1965b; Jensen and D. B. Fisher, 1967) including their cytoplasm and nuclei. The latter seem less active in the egg than in the central cell. This is also borne out by ultrastructural and histochemical studies. According to Jensen and D. B. Fisher (1967), the metabolic status of the central cell is high as compared to the egg, and "would thus appear to be a contributing factor in determining the rate of nuclear fusion." This probably explains why fusion of polar nuclei with one of the sperms is accomplished earlier than fusion of the egg and other sperm, even if the latter starts before the former.

It is believed that it is only the sperm nuclei that fuse with the egg and polar nuclei. There is no evidence of the pollen tube or sperm cytoplasm either in the egg or in the central cell (see also Van Went, 1970).

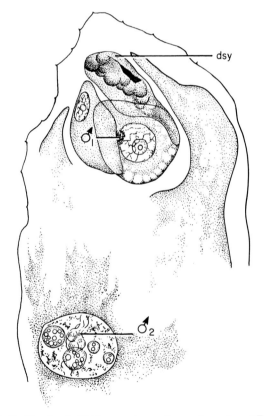

Fig. 9. Double fertilization in *Triticum vulgare*. One male nucleus ($\male_1$) has fused with egg nucleus, and the second ($\male_2$) with secondary nucleus. The remaining contents of pollen tube can be seen in the degenerated synergid (dsy). (Modified after Batygina, 1962.)

Once the sperm nuclei have been discharged into the synergid, one of them moves to the egg (Fig. 9) and the other migrates to the central cell to fuse with the polar nuclei (Figs. 9 and 10A–D).

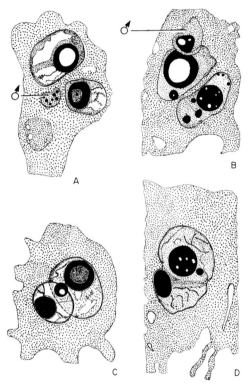

FIG. 10. Triple fusion in *Adonis aestivalis*. (A) Portion of embryo sac enlarged to show two polar nuclei and a male gamete. (B–D) Stages during triple fusion. (After Bhandari, 1966.)

There are three types of karyogamy involved in the fusion of male and female nuclei (Gerassimova-Navashina, 1960, 1969). The classification is based on differences of the cyclic state of the nuclei. Whereas female nuclei are in a state of rest, the male nuclei are still in mitotic cycle of telophase.

In the first type the male gamete fuses immediately on coming in contact with the female nucleus and reaches the ultimate phase of mitotic cycle. This type of fusion is called *premitotic,* because the union of the two nuclei takes place before the zygote mitosis.

In the *postmitotic* type the male gamete does not sink into the female nucleus at once, but undergoes a resting period while still in contact with the female nucleus. Here the fusion of sexual nuclei occurs after the zygote has entered mitosis.

In the third or *intermediate* type the male nucleus, after completing its mitotic cycle and in a state of rest, enters the female nucleus. There is incomplete mixing of nuclei, and chromosomes are seen apart and can be observed in prophase in the zygote nucleus.

The nature of X-bodies has been elusive for a long time. These were once considered as the sperm cytoplasm (Jensen and D. B. Fisher, 1968). D. B. Fisher and Jensen (1969) now consider these bodies as the nuclei of the degenerated synergid and pollen tube (see also Cass and Jensen, 1970; Van Went, 1970).

V. Endosperm

The endosperm is a tissue unique to angiosperms. The primary endosperm nucleus results from fusion of the polar nuclei and one of the male nuclei. Normally, all the three nuclei are haploid and, therefore, the fusion product is triploid. The endosperm tissue is formed as a result of repeated divisions of the primary endosperm nucleus. There are three types of endosperm development: Nuclear, Cellular, and Helobial.

A. *Nuclear Endosperm*

In the Nuclear-type development (Fig. 11A–E) the primary endosperm nucleus undergoes a series of free-nuclear divisions. The latter may be synchronous or nonsynchronous. When many nuclei have been formed, they arrange themselves peripherally in a layer of cytoplasm. Later, centripetal wall formation takes place. The cellularization first begins in the micropylar region and extends downward (Fig. 11F–H). The embryo sac eventually becomes completely cellular. Initially, the cells are polygonal and vacuolate (Fig. 11I), and may subsequently become thick-walled and filled with reserve food material (Fig. 11J).

Wall formation does not take place in plants such as *Oxyspora* (Subramanyam, 1951) and *Limnanthes* (Mathur, 1956). In several families, wall formation is limited only to the upper and middle region of the embryo sac so that the chalazal region remains free-nuclear. This region often elongates and functions as a haustorium. It was through the technique of dissected wholemounts that Kausik (1941) discovered, for the first time, a "vermiform appendage" in the endosperm of *Grevillea robusta*. Since then similar observations have been made in several members of Proteaceae, Leguminosae, and Cucurbitaceae. There is much variation in the extent and behavior of the chalazal haustorium

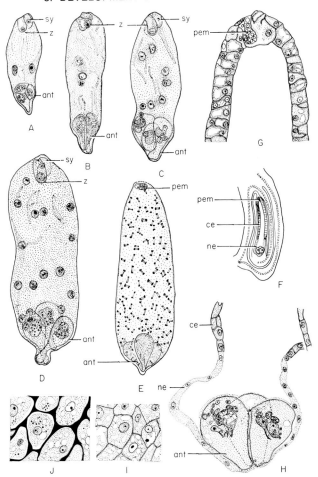

Fɪɢ. 11. Nuclear endosperm in *Anemone rivularis*. (A) Two-nucleate endosperm (whole-mount). (B–D) Four-, eight-, and sixteen-nucleate endosperm (wholemount). (E) Embryo sac showing synchronous divisions of endosperm nuclei. (F) Index figure for figures G and H; note wall formation in upper portion of endosperm and free-nuclear chalazal region. (G) Enlarged view of micropylar portion of embryo sac from figure F showing initiation of wall formation in endosperm. (H) Lower portion in figure F to show free-nuclear endosperm at chalazal end. (I) Endosperm cells from a young seed. (J) Endosperm cells from mature seed showing thick walls and reserve food material. (ant) Antipodal cells; (ce) cellular endosperm; (ne) nuclear endosperm; (pem) proembryo; (sy) synergid; (z) zygote. (After Bhandari, 1968.)

which is usually tubular and coenocytic with dense cytoplasm. The nuclear size is variable, and hypertrophy and polyploidy are common. The lower portion of the haustorium in *Mimosa pudica* (B. M. Johri and

Garg, 1959) forms several processes, which enter the nucellus; the haustorium is coiled in *Calliandra* (Dnyansagar, 1958). In *Cyamopsis psoralioides* and *Desmodium pulchellum* (Rau, 1953), the haustorium eventually becomes cellular.

In *Desmodium,* there is considerable variation in the structure of the haustorium (B. M. Johri and Garg, 1959). It is free-nuclear in *Desmodium floribundum* (Fig. 12A). In *Desmodium gangeticum,* the haustorium is similar to *D. floribundum* except that there is cell formation in the endosperm up to the tubular portion while the vesicular tip remains free-nuclear (Fig. 12B and C). The haustorium, to begin with, is free-

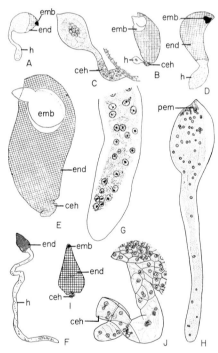

Fig. 12. Endosperm in Leguminosae (A–E) and Cucurbitaceae (F–J). (A) Endosperm of *Desmodium floribundum* at early heart-shaped stage of embryo; the haustorium is free-nuclear. (B) Endosperm of *Desmodium gangeticum* at early dicotyledonous stage of embryo. (C) Enlarged view of haustorium from figure B showing cellular condition in the upper part and free-nuclear in the lower portion. (D, E) Endosperm of *Desmodium laburnaefolium;* note cellular condition of haustorium in figure E. (F) Endosperm of *Cucurbita pepo* showing coenocytic haustorium. (G) Tip of haustorium enlarged from figure F. (H) *Benincasa cerifera;* free-nuclear endosperm and bicelled proembryo; note the elongation of lower part of embryo sac into a tubular process. (I) Outline diagram for figure J. (J) Haustorium marked in figure I showing cell formation. (ceh) Cellular haustorium; (emb) embryo; (end) endosperm; (h) haustorium; (pem) proembryo. [(A–E) after B. M. Johri and Garg, 1959; (F, G) after Chopra and Basu, 1965; (H–J) after Chopra and Agarwal, 1958.]

nuclear in *Desmodium laburnaefolium,* but later becomes completely cellular (Fig. 12D and E).

A chalazal endosperm haustorium, either coenocytic or cellular (Fig. 12F–J), is of common occurrence in Cucurbitaceae (see Chopra and Agarwal, 1958; Chopra and Basu, 1965). In *Citrullus fistulosus* (Chopra, 1955), the haustorium becomes cellular with multinucleate cells. Subsequently, due to subdivisions, the cells become uninucleate. *Echinocystis lobata* (Seth, 1962) has the longest haustorium in the family. The haustorium, when studied in fresh material, shows protoplasmic streaming. Similar observations have been made in the living materials of *Grevillea robusta* by Kausik (1941) and in *Desmodium triflorum* by Rau (1953).

The endosperm of coconut *(Cocos nucifera)* is very interesting. When the fruits are about 50 mm long, the embryo sac is filled with a watery fluid, also called milk. The latter has numerous free nuclei and cytoplasmic particles, and is referred to as liquid syncytium. The ploidy of the nuclei in the syncytium varies from $2n$ to $10n$ (Dutt, 1953). The size of the nuclei is also variable and ranges from 10 to 90 μ. With increase in the size of the embryo sac, the nuclei also multiply. When the fruit is approximately 100 mm in length, numerous large spherical cells or vesicles appear in the syncytium (Cutter and Freeman, 1955). These vesicles probably arise as a result of coalescence of cytoplasm around the free nuclei, and the diameter of vesicles varies from 10 to 300 μ, each enclosing 1–40 nuclei.

B. Cellular Endosperm

The Cellular endosperm is characterized by absence of a free-nuclear phase, and division of the primary endosperm nucleus and subsequent divisions are invariably followed by wall formation (Fig. 13A–F). A noteworthy feature of Cellular endosperm, in several plants, is the formation of haustoria which are much more varied than in the Nuclear type.

The haustoria may develop at the micropylar or chalazal end, or both. A micropylar haustorium is produced in *Impatiens roylei* (Dahlgren, 1934) and *Hydrocera triflora* (Venkateswarlu and Lakshminarayana, 1957). Kapil and Bhandari (1964) reported a two-celled chalazal endosperm haustorium in *Magnolia obovata* (Fig. 14A–D). This is the first report of the occurrence of an endosperm haustorium in the family Magnoliaceae which, incidentally, is a primitive taxon.

The parasitic angiosperms show rather interesting features. B. M. Johri and Bhatnagar (1969) have reviewed the endosperm in Santalales (Fig. 15A–E). A chalazal haustorium is common, particularly in the suborder Santalineae. Division of the primary endosperm nucleus is followed by

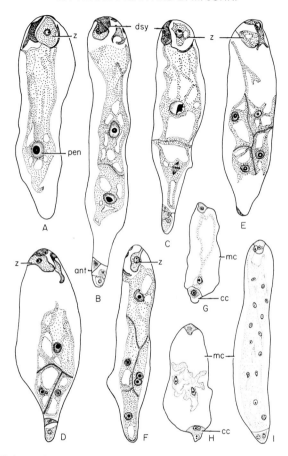

FIG. 13. Cellular and Helobial endosperm in *Parrotiopsis jacquemontiana* (A–F) and *Philydrum lanuginosum* (G–I). (A) Primary endosperm nucleus lying near chalazal region. (B) Division of primary endosperm nucleus. (C, D) Division of nucleus of chalazal chamber followed by oblique-vertical wall. (E, F) Four- and five-celled endosperm. (G) Delimitation of a small chalazal chamber. (H) Two-nucleate micropylar chamber. (I) Cell formation in chalazal chamber; the micropylar chamber is free-nuclear. (ant) Antipodal cells; (cc) chalazal chamber; (dsy) degenerated synergid; (mc) micropylar chamber; (pen) primary endosperm nucleus; (z) zygote. [(A–F) after Kaul, 1969; (G–I) after Kapil and Walia, 1965.]

formation of a micropylar and a chalazal chamber. The micropylar chamber gives rise to the endosperm proper, whereas the chalazal chamber functions as the haustorium. The latter shows considerable variation in its extension, behavior and activity. The prevalent condition is a unicellular and uninucleate haustorium; it is four-nucleate in *Olax*. In some plants, it becomes multicellular. Secondary haustoria are reported in

Cansjera (Swamy, 1960), *Mida* (Bhatnagar, 1960), and *Comandra* (Ram, 1957). A very aggressive haustorium is known in *Cansjera* and *Iodina* (Bhatnagar and Sabharwal, 1969).

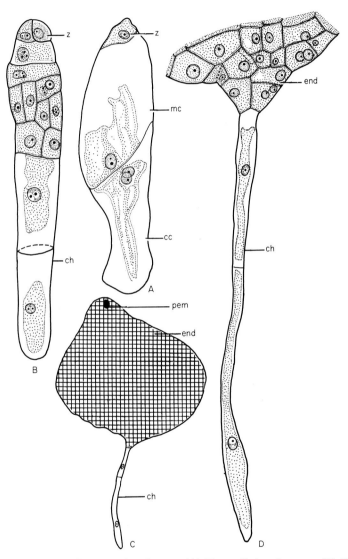

FIG. 14. Endosperm in *Magnolia obovata*. (A) Two-celled endosperm. (B) Thirteen-celled endosperm. (C) Diagram of endosperm at globular stage of embryo to show chalazal haustorium. (D) Enlarged view of chalazal haustorium from figure C. (cc) Chalazal chamber; (ch) chalazal haustorium; (end) endosperm; (mc) micropylar chamber; (pem) proembryo; (z) zygote. (After Kapil and Bhandari, 1964.)

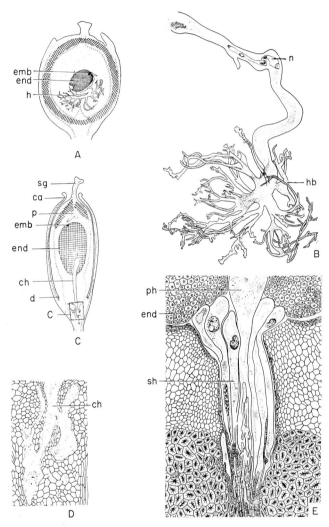

FIG. 15. Endosperm in Santalales. (A) Longisection of fruit of *Iodina rhombifolia* at globular stage of embryo; note the aggressive nature of chalazal endosperm haustorium (diagrammatic). (B) Haustorium with branched lower end and hypertrophied nucleus. (C) Longisection of fruit of *Olax stricta* at globular stage of embryo; the chalazal haustorium extends into ovarian tissue. (D) Magnified view of portion marked C in figure C; note the ramifications of chalazal haustorium. (E) Longisection of portion of fruit of *Cansjera rheedii* showing primary and secondary endosperm haustoria. (ca) Calyx; (ch) chalazal haustorium; (d) disc; (emb) embryo; (end) endosperm; (h) haustorium; (hb) haustorial branches; (n) nucleus; (p) placenta; (ph) primary haustorium; (sg) stigma; (sh) secondary haustoria. [(A, B) after Bhatnagar and Sabharwal, 1969; (C, D) after Agarwal, 1963a; (E) after Swamy, 1960.]

Scrophulariaceae and Orobanchaceae are other interesting parasitic families. In the former, with certain exceptions, both micropylar and chalazal haustoria are formed. Moreover, the haustoria show considerable variation in their structure and organization.

The micropylar haustorium is two-celled with one or two nuclei in each cell. The chalazal haustorium is one-celled and binucleate in *Euphrasia, Orthocarpus* (Arekal, 1963), *Pedicularis* (R. Y. Berg, 1954), and *Striga* (Tiagi, 1956). In *Euphrasia,* the haustorium often forms two lobes one of which sometimes becomes extra ovular. *Ellisiophyllum* shows only micropylar endosperm haustorium, and *Chaenorrhinum* only chalazal haustorium. The micropylar haustorium in *Melampyrum, Pedicularis,* and *Rhinanthus* is one-celled and tetranucleate. The haustoria are aggressive and invade the tissues of the integument and even the pericarp. In *Melampyrum lineare,* one of the micropylar tubular processes (Fig. 16A) of the haustorium enlarges and enters the funiculus (Fig. 16B) where it branches extensively (Arekal, 1963). The chalazal haustorium is binucleate and broader above and narrow below. It comes in contact with the vascular supply and persists longer than the micropylar haustorium.

An excellent account of the endosperm in Acanthaceae has been furnished by Mohan Ram and Wadhi (1965). Development of the endosperm is asymmetric. Initially, three cells are formed and, whereas the upper and lower cells function as haustoria, the central chamber gives rise to the endosperm proper. In *Thunbergia alata* (Mohan Ram and Wadhi, 1964), however, the micropylar chamber forms a branched, coenocytic haustorium, whereas the chalazal chamber gives rise to the endosperm proper and several elongated, tubular secondary haustoria (Fig. 17A).

In Loasaceae (Garcia, 1963), the micropylar haustorium is free-nuclear with hyphalike protuberances which may enter the funiculus or even reach the placenta. There is variation with regard to the chalazal haustorium. In *Loasa,* it is balloonlike, in *Cajophora* it is elongated with a multicellular narrow base and much-branched apical portion, and in *Mentzelia* it is elongated but unbranched. In *Blumenbachia* (Fig. 17B), the micropylar haustorium consists of two superimposed cells, each binucleate. The chalazal haustorium is uninucleate and profusely branched. The branches invade the cells of the integument.

The development of endosperm in *Klugia notoniana* (Gesneriaceae) is very unusual (Arekal, 1961). Both micropylar and chalazal haustoria are organized. The chalazal haustorium is binucleate, but the two nuclei fuse producing a large nucleus. The haustorium extends laterally and grows upward consuming the cells of the integument lying between the outer epidermis and endothelium (Fig. 17C). With the decline in the

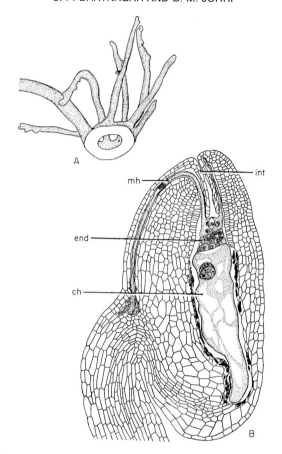

FıG. 16. Endosperm haustorium in *Melampyrum lineare*. (A) Micropylar haustorium (dissection). (B) Young seed showing extension of one of the micropylar haustorial tubes into the funiculus. (ch) Chalazal haustorium; (end) endosperm; (int) integument; (mh) micropylar haustorium. (After Arekal, 1963.)

activity of the chalazal haustorium, the micropylar haustorium becomes active.

In Loranthaceae, several endosperms develop in different embryo sacs in an ovary and, finally, fuse with each other forming a composite structure. This feature is not known in any other family.

C. *Helobial Endosperm*

The Helobial endosperm in angiosperms has been reviewed by Swamy and Parameswaran (1963). The primary endosperm nucleus divides and two unequal chambers are formed, the micropylar one being larger than

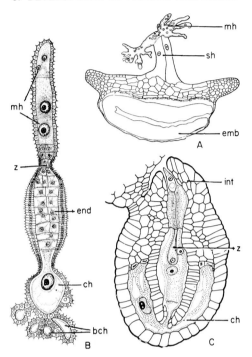

FIG. 17. Endosperm haustoria. (A) Inverted hat-shaped endosperm of *Thunbergia alata;* note the tubelike secondary haustoria and micropylar haustorium with its ramifications. (B) Endosperm proper with micropylar and chalazal haustoria in *Blumenbachia insignis;* note two binucleate cells of micropylar haustorium. (C) Longisection of ovule of *Klugia notoniana* with chalazal haustorium extending laterally upward. (bch) Branches of chalazal haustorium; (ch) chalazal haustorium; (emb) embryo; (end) endosperm; (int) integument; (mh) micropylar haustorium; (sh) secondary haustoria; (z) zygote. [(A) after Mohan Ram and Wadhi, 1964; (B) after Garcia, 1963; (C) after Arekal, 1961.]

the chalazal one (Fig. 13G). In the micropylar chamber, several free-nuclear divisions (Fig. 13H and I) take place but, ultimately, it becomes cellular. The fate of the chalazal chamber is different; its nucleus may or may not divide. It usually remains coenocytic. However, it may sometimes become cellular.

In *Phylidrum lanuginosum* (Kapil and Walia, 1965), the chalazal chamber frequently divides earlier than the micropylar chamber; the division may be followed by wall formation (Fig. 13H and I). The chalazal chamber becomes two- to six-celled. The activity of this chamber ceases and the cells degenerate. Such a precocious activity of the chalazal chamber in a Helobial endosperm is not reported in other monocotyledons where wall formation in the chalazal chamber is much delayed (see Swamy and Parameswaran, 1963).

D. Ruminate Endosperm

Sometimes the endosperm may be uneven or irregular and is referred
to as ruminate. It cannot be classed as a separate type of endosperm as
was done by Linskens (1969) because rumination does not start in the
initial stages of development. It is the ultimate irregular shape that may
be due to the uneven inner surface of the testa. Rumination is caused by
activity of either the seed coat or the endosperm. Periasamy (1962) has
discussed the various types of rumination. If rumination is caused by
the seed coat, it may be due to ingrowth or infolding of the seed coat as
seen in Annonaceae, Aristolochiaceae, Palmae, and Vitaceae. The unequal radial elongation of the cells of the seed coat may also result in
rumination as in *Passiflora.*

In those plants where the endosperm itself causes rumination, the
endosperm is aggressive and digests the nucellus. The irregularities thus
arising on the inner face of the testa make the endosperm ruminate, e.g.,
in *Coccoloba, Diospyros,* and *Myristica.* In *Degeneria* and *Tiliacora* the
endosperm remains quiescent, and the irregularities of seed coat affect
the nucellus first. Later, when the endosperm has become enlarged, it
becomes sandwiched between the preformed ingrowths of seed coat giving the endosperm a ruminate appearance.

Periasamy (1962) has distinguished, on a morphological basis, seven
types of rumination. These are present in the following genera: *Passiflora,
Verbascum, Annona, Myristica, Coccoloba, Spigelia,* and *Elytraria.*
Ruminate endosperm is known in thirty-two families of angiosperms.

E. Cytology of Endosperm

Depending upon the number of nuclei that fuse with the male gamete,
the ploidy of endosperm cells varies. The primary endosperm nucleus
may be diploid, as in Onagraceae, or as much as $15n$ as in *Peperomia
hispidula* (P. Maheshwari, 1950).

Giant nuclei have been reported in endosperms and, because of their
large size, several workers have investigated their nuclear structure and
behavior. Duncan and Ross (1950) have studied this in maize and noted
that, due to endomitosis, the nuclei in the central part of the endosperm
became 1000 times their original volume 24 days after pollination. This
involved increase in the number of strands per chromosome (polyteny),
and the length and breadth of individual chromosomes also increased.
There was considerable increase in the amount of nucleolar material.
However, the number of knobs and nucleoli remained the same, indicating the constancy of chromosomes per nucleus. Duncan and Ross con-

sider endomitosis in endosperms a derived and secondary phase of growth in a period of special physiological activity. According to Punnett (1953), however, hexaploid cells are formed as a result of two duplications during the resting stage subsequent to normal mitosis.

At first, in *Gagea,* the endosperm nuclei were very large, and there was an actual increase in the number of chromosomes. Later, the size of the nuclei appeared to decrease, but looked denser. This is probably due to either depolyploidization or dehydration of maturing endosperms (Geitler, 1948). In another member of the Liliaceae, *Allium ursinum* (Geitler, 1955), polyploid nuclei are formed in the chalazal region of endosperm. They reach up to the 6-ploid and 12-ploid state. The volume of the nuclei varies from 4 to 7 times the original triploid nuclei. Polyploidization does not correspond with endomitosis. Steffen (1956) reported that, in *Pedicularis palustris,* the nuclei in the upper region of the endosperm are hexaploid, whereas in the chalazal region the nuclei are dodekaploid. The ploidy in the chalazal haustorium is $96n$; in the micropylar haustorium the ploidy varies from $192n$ to $384n$. With increase in the number of nuclei, the number of nucleoli also increases.

Enzenberg (1961) has made extensive investigations on the cytology of endosperm. He studied members of several families and reported that, due to endomitosis, the ploidy reached up to $384n$. The haustorial organs showed a still higher degree of ploidy. Erbrich (1965) made similar observations on the cytology of endosperm and endosperm haustoria. The ploidy of the nucleus of the chalazal haustorium in *Thesium alpinum* and *Thesium lineare* is $384n$, whereas in the micropylar haustorium of *Codonopsis clematidea* the ploidy varied from $96n$ to $192n$. The highest ploidy is seen in *Arum maculatum* where the nucleus becomes $24576n$. At the time of degeneration of the haustorium, the endosperm proper becomes dodekaploid due to endomitosis. In *Melampyrum lineare,* each nucleus of the micropylar haustorium attains a ploidy of $1536n$. Such high degrees of polyploidy seem to be closely linked with the aggressive nature of endosperm haustoria.

The division of nuclei by amitosis is not very frequent. However, division of endosperm nuclei, in addition to normal mitosis, by amitosis or fragmentation is usual and has been recorded by several workers (Datta, 1955; Scott, 1953; Tandon and Kapoor, 1961, 1962, 1963). The nuclei first enlarge and become constricted in the middle; with deepening of the constriction, the nucleus eventually fragments into two or more unequal or equal parts.

Chlorophyll has been reported in the endosperm of certain plants. In *Mathiola* and *Raphanus,* Yoffe (1952) observed chloroplasts both in the nuclear and cellular phases of endosperms. Similar reports have been made by Kuijt (1960) in certain mistletoes, and by Rigby (1959) in *Lysiana* of Loranthaceae.

VI. Embryo

Depending upon the number of cotyledons in the embryo, the angiosperms have been divided into two major groups — monocotyledons and dicotyledons. Instances are on record where the embryo is without cotyledons (acotyledonous) or there are more than two cotyledons (polycotyledonous).

The two main stages in development of the embryo are (*a*) proembryo, and (*b*) embryo proper. Incidentally, in both monocotyledons and dicotyledons, the earlier stages in embryogeny up to the proembryo stage are similar.

The zygote undergoes a resting period which is of a longer duration when the endosperm is of the Nuclear type, as compared to the Cellular type. The zygote divides later than the primary endosperm nucleus. An axial symmetry resulting in establishing polarity is brought about by internal differentiation of the zygote. Potentialities for further development of the embryo are located in the apical pole which is directed toward the center of the embryo sac.

According to Souèges (1935), the proembryo possesses an axial symmetry, a feature it inherits from the egg. In dicotyledons, this symmetry continues until cotyledonary primordia appear. In monocotyledons the axial symmetry is maintained until the lateral differentiation of shoot apex; later, bilateral symmetry is attained.

The division of the zygote is transverse; rarely vertical or oblique. In Loranthaceae, the zygote invariably divides longitudinally. As a result of transverse division, two cells are produced. The smaller cell near the cavity of the embryo sac is the apical cell (ca); the larger cell is referred to as the basal cell (cb) and faces the micropyle.

A. Types of Embryogeny

The following main types of embryogeny have been distinguished (P. Maheshwari, 1950):

1. The apical cell (ca) divides by a longitudinal wall.
    a. The basal cell (cb) plays only a minor role or none in the development of the embryo — Crucifer type.
    b. Both the cells (ca and cb) take part in development of the embryo — Asterad type.

2. The apical cell undergoes transverse division.
   a. The basal cell plays a minor role or none in development of the embryo.
      i. The basal cell usually forms a suspensor of two or more cells — Solanad type.
      ii. The basal cell remains undivided, and the suspensor, if present, is derived from terminal cell — Caryophyllad type.
   b. Both the cells take part in development of the embryo — Chenopodiad type.

B. *Embryo Development in Dicots*

The development in *Ceratocephalus falcatus* (Bhandari and Asnani, 1968) is a good example of a dicot (Fig. 18A–Q). The zygote divides transversely giving rise to a large basal (cb) and a small apical cell (ca). Cell cb undergoes a transverse division forming two superposed cells, ci and m; cell ca divides vertically forming two juxtaposed cells. This results in a perpendicularly-shaped four-celled proembryo. Of the two daughter cells of cb, cell ci divides transversely giving rise to n and n'. These two cells divide further forming a row of three or four cells. The latter constitute the forerunners of suspensor. Cell m and its derivatives mostly undergo vertical divisions forming a group of four to six cells. The latter divide by oblique-periclinal walls resulting in the inner cells (iec) and outer cells (pco). Of these, cells iec form the initials of root apex, and cells pco are the progenitors of root cap.

The two juxtaposed cells, formed as a result of division of ca, divide vertically giving rise to a quadrant q. Each cell of the quadrant divides transversely resulting in an octant. The cells of the octant are arranged in two tiers (l and l'). Each tier is composed of four cells. The octant stage is followed by vertical divisions in both tiers l and l' resulting in a globular proembryo. Now periclinal divisions in the peripheral cells of the globular proembryo delimit an outer layer, de, the dermatogen. Cotyledons and the shoot apex are derived from the tier l, and the hypocotyl-radicle axis from tier l'.

C. *Embryo Development in Monocots*

Najas lacerata (Swamy and Lakshmanan, 1962) illustrates the development of embryo in monocotyledons (Fig. 19A–L). The zygote divides to form a large cell cb and a small cell ca. Cell cb remains undivided and develops into a haustorial cell of the suspensor. Cell ca undergoes a transverse division giving rise to two cells, c and d. The latter (d) divides by a transverse wall resulting in ci and m. In cells c and

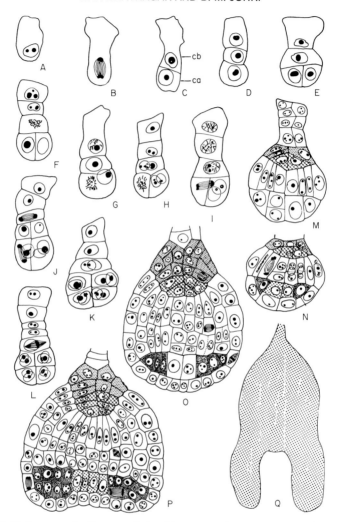

Fig. 18. Embryogeny in *Ceratocephalus falcatus*. (A–C). Transverse division of zygote to form cells ca and cb. (D, E) Three- and four-celled proembryos. (F–J) Stages in the development of quadrant. (K, L) Proembryo showing octant stage. (M–O) Globular stages of embryo; note the delimitation of dermatogen, cotyledonary initials, and shoot apex initials. (P) Longisection of early heart-shaped embryo. (Q) Longisection of mature dicotyledonous embryo (diagrammatic). (After Bhandari and Asnani, 1968.)

m, two vertical divisions at right angles to each other, produce two superposed tiers (q and m) of four cells each. Cell ci undergoes a transverse division forming n and n'. Whereas n divides vertically, cell n' on transverse division gives rise to cells o and p. The latter (p) undergoes another transverse division producing u and s. Cell o divides vertically.

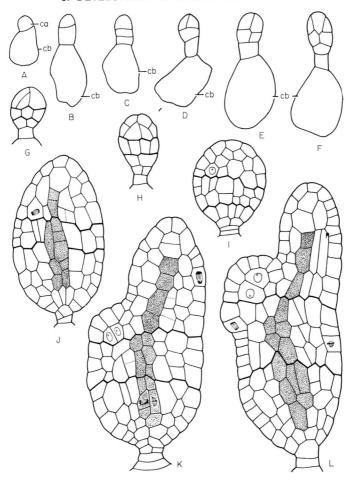

FIG. 19. Successive stages (A–L) of embryogenesis in *Najas lacerata*. (After Swamy and Lakshmanan, 1962.)

The cells of the tier q undergo a periclinal division demarcating an epidermis and an inner (axial) group of four cells. The proembryo now appears spherical. Further divisions in tiers m and n are mostly transverse. This facilitates the lengthwise elongation of the embryo. When the embryo has become somewhat oval, initiation of "plerome" takes place. The central cells of tiers q to n become densely cytoplasmic (stippled cells). Of the eight cells comprising the tier q, three of the inner cells divide whereas the remaining cell is quiescent. This results in the tip of the embryo becoming notched. The rapidly growing region develops into the cotyledon. The quiescent segment of q, after division, forms initials of the epicotyl. Thus, both the cotyledon and epicotyl are derived from the terminal tier q of the proembryo.

Swamy and his collaborators (Swamy, 1962; Swamy and Padmanabhan, 1962) are of the opinion that in monocotyledons the single cotyledon as well as the shoot apex are terminal and arise as a result of the first vertical division in the terminal cell of the proembryo. This contention is opposed to the earlier views that the cotyledon is terminal and the shoot apex lateral in monocotyledons. Moreover, according to Swamy, it is not true that the earlier stages of the embryo development in dicotyledons and monocotyledons are comparable.

Swamy (1962) concluded:

> . . . the essential differences between the dicotyledonous and monocotyledonous embryos will have to be looked for neither in the number of cotyledons nor in their topographical relationships, but in the morphogenetic potentialities of the primary terminal meristems, especially the one at the shoot pole. In the dicotyledons these tissues develop into the functional root and shoot systems. In the monocotyledons, on the other hand, the primary shoot meristem should be assumed to have become sterile and reduced to a stublike termination in which are consolidated the derivatives of the two cotyledons as well; the activity of the terminal root meristem is also similarly suppressed or is only transitory.

Batygina (1969) has suggested the possibility of a new type of embryogenesis in Gramineae — the Graminad type. According to her, in *Triticum,* a dorsiventral symmetry is established as a result of a peculiar oblique position of cell walls early in embryogeny. In addition, there is an early formation of scutellum (see also Poddubnaya-Arnoldi, 1964).

Associated with the different type of embryogeny, the embryo of grasses is strikingly different from the embryos of other monocotyledons and needs a special mention. In maize, for example, the embryo has a single cotyledon attached laterally to the embryonal axis. The secretory epithelial epidermis of the cotyledon, also called scutellum, is adjacent to the endosperm. The radicle is covered by a special tissue, the coleorhiza. Around the epicotyl is a leaflike membranous structure, the coleoptile. There is a pore at the apex of the coleoptile, and through this pore emerges the first foliage leaf.

D. *Suspensor*

In several plants the suspensor of the embryos exhibits great variation with regard to its size, shape, and number of cells. In Orchidaceae itself there is great diversity (Swamy, 1949). In *Cypripedium,* the suspensor is single-celled saclike, conical, or tubular. The suspensor in *Ophrys* consists of five to ten cells which grow beyond the micropyle. On reaching the placenta, the suspensor sends out haustorial branches. In *Epi-*

dendrum, the cells of the suspensor look like a bunch of grapes. In *Vanda,* the cells elongate downward and envelope more than half of the embryo. In *Dendrobium,* the suspensor forms a much-branched haustorium (Poddubnaya-Arnoldi, 1967).

A variety of modifications of the suspensor are known in Leguminosae (see P. Maheshwari, 1950). It may consist of a row of several cells which may be binucleate; sometimes the cells are large and coenocytic. In *Cytisus,* the cells appear as a bunch of grapes.

In Podostemaceae also there may be a single large coenocytic haustorial cell as in *Indotristicha* (Mukkada, 1963); or several thin-walled haustorial branches may grow in between the integuments, e.g., in *Dicraea* (Mukkada, 1962).

Suspensor haustoria have also been reported in Rubiaceae, Halorrhagaceae, Fumariaceae, and Crassulaceae. Loranthaceae, perhaps, show the longest suspensors in angiosperms.

E. Undifferentiated Embryos

Normally, the mature embryo consists of a radicle, a plumule, and one or two cotyledons. However, in some plants the embryos are reduced and lack differentiation into organs. The coiled embryo of *Cuscuta* shows a shoot apex but is devoid of the cotyledons and radicle. In Orobanchaceae (Tiagi, 1951, 1963, 1965; Rangan and Rangaswamy, 1968), a family of total root parasites, the embryo lacks a radicle, hypocotyl, cotyledon, epicotyl, and plumule. The embryos of Orchidaceae are minute and undifferentiated (Withner, 1959; Arditti, 1967; Poddubnaya-Arnoldi, 1967).

In several other families, such as Balanophoraceae, Hydnoraceae, Lentibulariaceae, Pyrolaceae, Rafflesiaceae, Pandanaceae, and Xyridaceae the embryos lack differentiation of tissues, organs, or both (see Rangaswamy, 1967).

F. Embryogeny in Paeonia

In 1951, Yakovlev published his observations on the development of the embryo in *Paeonia,* which is unique and unknown in angiosperms. The division of the zygote nucleus and subsequent divisions of daughter nuclei are not accompanied by wall formation resulting in a coenocytic structure. Yakovlev and Yoffe (1957, 1961, 1965) described two phases in the embryogenesis of *Paeonia.* In the first phase, a coenocytic structure is formed, and the nuclei become delimited by walls. In the second phase, a few embryonal primordia differentiate peripherally and give rise to embryos. However, only one embryo attains maturity and is typically dicotyledonous (see also Yakovlev, 1969). This observation has been confirmed by Cave *et al.* (1961), Carniel (1967), Matthiessen (1962), and Moskov (1964) in various species of *Paeonia.*

Murgai (1959, 1962), who also studied several species of *Paeonia,* does not agree with the finding of Yakovlev and Yoffe. She observed a transverse wall, after division of the zygote, separating a smaller apical and a much larger basal cell. The basal cell becomes coenocytic whereas the terminal cell divides vertically. Further development of the apical tier is delayed until cell formation occurs in the basal cell. During later stages, according to Murgai, it is difficult to ascertain the derivatives of the apical and basal cells, but the development of the embryo conforms to the observations of Yakovlev and Yoffe. This intriguing problem is still under investigation at the University of Delhi.

VII.   Polyembryony

Since 1719 when Leeuwenhoek observed the presence of more than one embryo in the seeds of orange, many other plants have been reported to show the phenomenon of polyembryony. P. Maheshwari and Sachar (1963) have classified polyembryony into (*a*) true polyembryony, and (*b*) false polyembryony.

a. True polyembryony. The embryos arise within the embryo sac, either by the budding or cleavage of the zygotic proembryo, or from the synergids and antipodal cells. The embryos may also originate from the cells of the nucellus or integument and, eventually, grow into the embryo sac. When the embryos arise from the somatic cells of the ovule, the phenomenon is referred to as adventive embryony.

b. False polyembryony. Supernumerary embryos may arise from different embryo sacs growing in the same nucellus or by fusion of two or more nucelli with independent embryo sacs.

Polyembryony, as a result of splitting of the zygotic proembryo, is of common occurrence in gymnosperms. In angiosperms, however, it is less frequent. Proliferations arise from the lower end of the proembryonal mass of cells which later develop into independent embryos, e.g., in *Primula auricula* (Veillet-Bartoszewska, 1957) and *Cocos nucifera* (Whitehead and Chapman, 1962).

In orchids, such as *Cymbidium, Eulophia,* and *Habenaria* (Swamy, 1942, 1943, 1946b, 1949), the zygotic embryo gives rise to additional embryos by budding, proliferation, or cleavage. Suspensor polyembryony is a common feature in *Exocarpus* (Fig. 20A), a member of Santalaceae (Ram, 1959a; Bhatnagar and Joshi, 1965). As many as five or six pro-embryos may develop simultaneously but, eventually, only one of them takes the lead and reaches maturity. In *Zygophyllum fabago*, sometimes suspensor embryos are formed and may develop up to the heart-shaped stage (Masand, 1963).

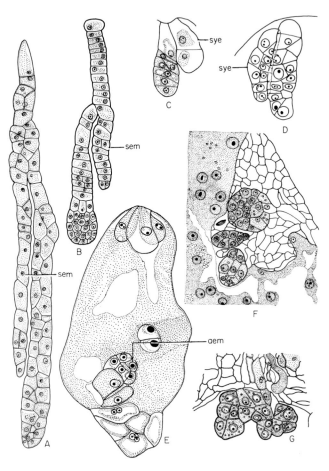

FIG. 20. Polyembryony. (A, B) Suspensor polyembryony in *Exocarpus spartea* and *Garrya veatchii*. (C, D) Synergid polyembryony in *Peganum harmala* and *Argemone mexicana*. (E) Abnormal embryo sac of *Paspalum scrobiculatum* showing antipodal embryo. (F, G) Nucellar polyembryony in *Mangifera odorata* and *Opuntia dillenii*. (aem) Antipodal embryo; (sem) suspensor embryo; (sye) synergid embryo. [(A) after Ram, 1959a; (B) after Kapil and Mohana Rao, 1966; (C) after Kapil and Ahluwalia, 1963; (D) after Sachar, 1955; (E) after Narayanaswami, 1954; (F) after Sachar and Chopra, 1957; (G) after P. Maheshwari and Chopra, 1955.]

In several other plants, cleavage polyembryony occurs only as an abnormality. Crété (1938, 1944) reported additional embryos which arose from cells of the suspensor in *Actinidia chinensis* and *Lobelia syphilitica*.

Similar observations have been made by Subramanyam (1944) in *Sonerila wallichii,* by Kausik and Subramanyam (1946) in *Isotoma longifolia,* by P. Maheshwari and Negi (1955) in *Dipteracanthus patulus,* by Vijayaraghavan (1965) in *Actinidia polygama,* and by Kapil and Mohana Rao (1966) in *Garrya veatchii* (Fig. 20B).

Besides the zygote, synergids commonly give rise to additional embryos. In *Sagittaria graminea* (B. M. Johri, 1936b), *Poa alpina* (Håkansson, 1943), and *Aristolochia bracteata* (B. M. Johri and Bhatnagar, 1955), the synergids are fertilized and then develop into embryos. This can be brought about either by the entry of more than one pollen tube into the embryo sac or by additional sperms in the same pollen tube.

Instances are known when unfertilized synergids also develop into embryos; these are haploid and the examples are *Phaseolus vulgaris* (Belikova, 1952), *Argemone mexicana* (Fig. 20D; Sachar, 1955), and *Lannea coromandelica* (Kelkar, 1961). Additional embryos may also arise as a result of hybridization, e.g., *Nicotiana glutinosa* ($n = 12$), when pollinated with *Nicotania tabacum* ($n = 24$), produces twin embryos — one from the zygote which is triploid and the other from the synergid which is haploid (Cooper, 1943).

Synergids develop into embryos (Fig. 20C) in *Sicyos* (Crété, 1958), *Peganum* (Kapil and Ahluwalia, 1963), *Melia* and *Azadirachta* (Nair, 1970), and several orchids (Poddubnaya-Arnoldi, 1967).

It may be recalled that antipodal cells are generally ephemeral and degenerate shortly before or after fertilization. Therefore, the formation of embryos from antipodal cells is rather rare. However, in several plants, such as *Paspalum scrobiculatum* (Narayanaswami, 1954) and *Rudbeckia sullivantii* (Battaglia, 1955), the antipodal cells not only persist but also divide producing "embryos" (Fig. 20E). Yakovlev and Solntzeva (1965) observed in *Stipa* that all the three antipodal cells start dividing forming a tetrad of linearly arranged cells. This simulates a young proembryo produced by the zygote. The three antipodal embryos do not develop into normal embryos, but degenerate. In *Ulmus campestris* (Guignard and Mestre, 1966), additional embryos may develop from the antipodal cells, but they do not seem to survive.

Reports about embryos arising from the endosperm have proved to be erroneous and so far not a single authentic case is on record.

In adventive embryony the embryos develop from the cells of the nucellus and integument lying outside the embryo sac. In bitegmic ovules, additional embryos arise from the cells of the inner integument. Nucellar polyembryony has been reported in *Trillium undulatum* (Swamy, 1948b), *Cucumis melo* var. *pubescens* (D. Singh, 1955), *Momordica charantia* (Agrawal and Singh, 1957), and *Aphanamixis polystachya* (Ghosh, 1962).

Citrus and *Mangifera* (Sachar and Chopra, 1957) are naturally poly-embryonic, and the accessory embryos develop from the nucellar cells (Fig. 20F). In *Citrus* the nucellar cells in the micropylar region, destined to develop into supernumerary embryos, develop at the expense of neighboring cells. They grow into the embryo sac cavity and undergo further development reaching various stages up to maturity. Although somewhat difficult, the adventive embryos may be distinguished from the zygotic embryo by their lateral position, irregular shape, and lack of suspensor.

There is a high percentage of polyembryonic seeds in *Zeuxine* and *Spiranthes* (Swamy, 1946a, 1948a, 1949). In *Spiranthes,* the embryos arise from the inner epidermis of the inner integument, whereas in *Zeuxine* they develop from cells of the nucellar epidermis.

In *Opuntia* (P. Maheshwari and Chopra, 1955) and *Aegle marmelos* (B. M. Johri and Ahuja, 1957), the egg fails to divide and degenerates, and polyembryony is due to the activity of nucellar cells (Fig. 20G).

Endosperm formation is normal in plants showing adventive embryony. The only exception is *Opuntia aurantiaca* (Archibald, 1939) where all the components of the embryo sac degenerate. This, however, appears doubtful since other species of *Opuntia* which show adventive embryony also develop abundant free-nuclear endosperm (P. Maheshwari and Chopra, 1955).

Production of adventive embryos may be autonomous or independent of the stimulus of pollination and fertilization. In *Nigritella nigra* (Afzelius, 1928), the presence of pollen tubes in the ovary accelerates the formation of adventive embryos. In most plants, pollination, or both pollination and fertilization, are necessary for initiation of such embryos.

False polyembryony is quite common in Loranthaceae (Narayana, 1954). An ovule, in the conventional sense, is absent. Many embryo sacs start developing concurrently in the same ovary and reach various lengths in the ovary, style, and stigma. Following fertilization, several biseriate proembryos grow downward and enter the composite endosperm (Fig. 21). Later, only one embryo reaches maturity and the others degenerate. Sometimes, two embryos were observed in the same fruit.

False polyembryony may also be brought about by fusion of two or more ovules as in *Rhizophora mucronata* (Kumar and Joshi, 1942) and *Chaenostoma foetidum* (Crété, 1948). Another method is by formation of multiple embryo sacs which may arise either by (*a*) derivatives of one, two, or more megaspore mother cells, or (*b*) from nucellar cells by apo-spory. Twin embryos formed in different embryo sacs have been reported in *Citrus* (Bacchi, 1943), *Poa pratensis* (Nielson, 1946), and *Casuarina equisetifolia* (Swamy, 1948c). According to Solntzeva (1957), in the

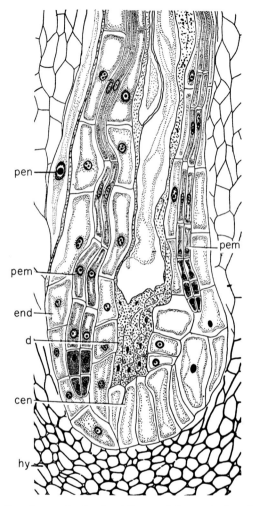

FIG. 21. False polyembryony in *Dendrophthoe neelgherrensis*. Longisection of central part of ovary showing two embryos embedded in composite endosperm. (cen) Composite endosperm; (d) degenerated tissue between embryo sacs; (end) endosperm; (hy) hypostase; (pem) proembryo; (pen) primary endosperm nucleus. (After Narayana, 1954.)

nucellus of *Fragaria grandiflora,* two or more gametophytes develop which fuse forming a common embryo sac containing several embryos.

In *Pennisetum ciliare* and *Cenchrus setigerus* (W. D. Fisher *et al.,* 1954), the normal embryo sac develops up to the four-nucleate stage but is crowded out by the aposporous embryo sacs. This results in the formation of twin embryos observed in nearly 22% seeds (see also Snyder *et al.,* 1955).

In some plants, polyembryony is known to occur by more than one method. In *Allium odorum,* supernumerary embryos originate both from synergids and antipodal cells (Hegelmaier, 1897). According to Haberlandt (1923, 1925), cells of the inner integument also produce embryos. Håkansson (1951) reported that in *Allium odorum* and *Allium nutans* the embryo sacs are generally diploid and embryos develop parthenogenetically. An additional embryo developing from the synergid has been recorded by Guttenberg *et al.* (1951) in *Allium giganteum.* Only rarely the zygotic and synergid embryos fuse forming a common structure.

Synergids, antipodals, and nuceller cells produce embryos in some species of *Elatostema* (Fagerlind, 1944). Several embryo sacs in the same ovule may also result in polyembryony.

Bouman and Boesewinkel (1969) have proposed a new classification for polyembryony dividing it into three categories:

1. Additional embryos arising from sporophytic cells of the parental generation. This includes cells of the nucellus (nucellar polyembryony) and integument (tegumentary polyembryony).

2. Supernumerary embryos arising from gametophytic cells either (*a*) by formation of two or more embryo sacs in one ovule (plurisaccal polyembryony), or (*b*) from cells of a single embryo sac (unisaccal polyembryony).

3. Additional embryos arising from the new (filial) sporophyte (zygote) or "unreduced" unfertilized egg cell in three possible ways: (*a*) derivatives of the zygote developing into embryos, (*b*) cleavage of the proembryo, and (*c*) division of the embryo.

Further, according to these authors, false polyembryony must be restricted to such examples where polyembryony results by the secondary fusion of two or more ovules or seeds.

The causes leading to polyembryony are practically unknown. Haberlandt's (1921, 1922) "necrohormone theory" is no longer tenable, and subsequent workers failed to confirm his findings. Polyembryony may be due to genetic causes, and hybridization probably plays some role in this phenomenon.

Polyembryony is of great significance in plant breeding and horticulture. Homozygous diploids can be raised from haploids. One can also obtain genetically pure plants by making use of adventive embryos.

VIII. Seed Coat

As the ovule matures into a seed, the integuments undergo conspicuous changes. Mostly there is reduction in thickness and disorganization but, sometimes, additional layers may be formed increasing the thickness. The funiculus usually abscises leaving a scar, called the hilum; that portion of the funiculus which is adnate in anatropous ovules appears as a longitudinal ridge on the seed and is termed raphe.

In order to have a clear picture of the structure of the seed coat, the best course is to study its development from the time when the ovule is mature till it ripens into a seed. There are several reports about the structure of mature seeds, but ontogenetic development has not been investigated in all cases.

Detailed studies have been carried out in Malvaceae, Euphorbiaceae, Cucurbitaceae, Acanthaceae, Leguminosae, and some other families (B. Singh, 1964). As examples, we will refer to the seed coat structure in cotton and Cucurbitaceae.

In cotton, which has a bitegmic ovule (Fig. 22A), both integuments contribute toward formation of the seed coat (Ramchandani et al., 1966; Joshi et al., 1967). The outer integument consists of four to eight layers (Fig. 22B) and is distinguishable into three zones: (a) outer epidermis, (b) outer pigmented zone of two to five layers of cells filled with tannin and starch, and (c) inner epidermis (Fig. 22C). With maturation of the seed, there is considerable enlargement of cells. The inner epidermis may remain single-layered as in *Gossypium arboreum,* or it may divide producing two or three layers as in *Gossypium barbadense, Gossypium herbaceum,* and *Gossypium hirsutum.* It accumulates abundant starch and is usually referred to as the colorless layer. The cells of the outer epidermis become tanniniferous.

The inner integument comprises eight to fifteen layers (Fig. 22B). In earlier stages there is enlargement of cells, and starch is deposited in three or four layers below the outer epidermis. These cells, in addition to accumulation of tannin, become lignified. The cells of the outer epidermis elongate radially, many times their original size, become thick-walled, and nuclei and cytoplasm become restricted to the outer wall (Fig. 22D–F). These cells, referred to as palisade cells, are broader on the inner side. Inner epidermal cells also elongate radially and develop platelike thickenings on the walls (Fig. 22G).

In a mature seed the outer integument is differentiated into an outer epidermis, a pigmented zone of four or five layers, and a zone of two or three colorless layers. The inner integument is distinguishable into a palisade layer, inner pigmented zone of fifteen to twenty layers, and the fringe layer (Fig. 22F).

In the innermost layer of the seed coat, which develops into the fringe layer, special type of thickening appears which becomes conspicuous in older seeds (Fig. 22F).

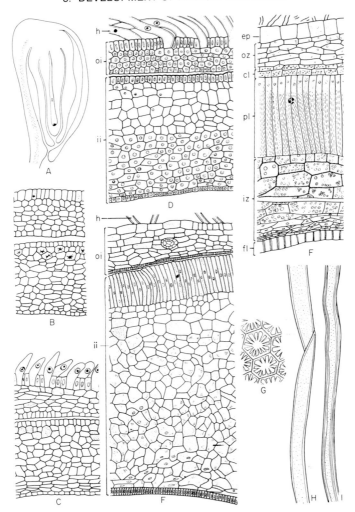

FIG. 22. Development of seed coat in *Gossypium herbaceum*. (A) Longisection of ovule at mature embryo sac stage. (B) Portion of ovule at anthesis. (C) Two or three days after anthesis. (D) Five or six days after anthesis. (E) Fifteen days after anthesis. (F) Portion of mature seed. (G) Palisade cells in transection; note the peculiar thickenings on walls. (H, I) Lint and fuzz hairs. (cl) Colorless zone; (ep) epidermis; (fl) fringe layer; (h) hair; (ii) inner integument; (iz) inner pigmented zone; (oi) outer integument; (oz) outer pigmented zone; (pl) palisade layer. (After Ramchandani *et al.*, 1966.)

The nucellus is absorbed. The endosperm is also consumed by the embryo.

In cotton the fibers are distributed all over the surface of the seed. They arise from the epidermal cells of the outer integument. At maturity, the fibers are distinguishable into lint and fuzz hairs (Fig. 22H and I). The lint hairs are long with a thin wall, broader lumen, and conspicuous spiral twist, whereas the fuzz hairs are small with a thick wall and narrow lumen.

Development of the seed coat in Cucurbitaceae (Fig. 23A–J) also presents some interesting features (B. Singh, 1952, 1953; D. Singh, 1961, 1964, 1965, 1967). The ovule is bitegmic, but only the outer integument takes part in development of the seed coat, and the inner integu-

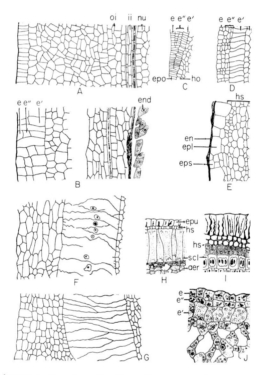

FIG. 23. Development of seed coat in Cucurbitaceae. *Sechium edule* (A, B, J); *Echinocystis wrightii* (C–F); *Sicyos angulata* (G, H); *Luffa cylindrica* (I). (A) Longisection of portion of integuments. (B) Longisection of integument from seed coat. (C) Part of outer integument showing division of ovular epidermis. (D) Same as figure C; note radial elongation of e′ and periclinal divisions in e and e″. (E) Longisection of outer part of outer integument. (F) Longisection of inner part of outer integument. (G) Part of outer integument (lateral view). (H, I) Outer integument from seed coat. (J) Longisection of part of seed coat (lateral view). (aer) Aerenchyma; (e, e′, e″) layers of cells formed by tangential divisions of ovular epidermis; (en) carpel endothelium; (end) endosperm; (epl, eps) large- and small-celled outer epidermis of seed coat; (epo) ovular epidermis; (epu) epidermis of seed coat formed of uniform cells; (ho) ovular hypodermis; (hs) seed hypodermis; (ii) inner integument; (nu) nucellus; (oi) outer integument; (scl) sclerenchymatous layer. [(A, B, J) after D. Singh, 1965; (C–F) after Singh, 1952; (G–I) after B. Singh, 1953.]

ment degenerates. The cells of the outer integument divide periclinally and, at maturity, the seed coat is differentiated into five zones: (*a*) seed epidermis consisting of cells radially or tangentially elongated, homogeneous or heterogeneous, thin-walled or with rodlike, tubular, spiral, or reticulate thickenings on the radial walls; (*b*) hypodermis is one- to many-layered with uniform or sinuate and thin or thick-walled cells; (*c*) main sclerenchymatous layer comprising brachy-, osteo-, or branched types of sclereids; (*d*) aerenchyma derived from hypodermis, comprising stellate cells that have prominent air spaces; and (*e*) chlorenchyma of thin-walled cells formed from the remaining layers of integument.

In Acanthaceae, which possess unitegmic ovules, the massive integument is consumed by the developing endosperm and only the epidermis persists in the mature seed (see Wadhi, 1970). The seed epidermis develops various types of hairs or thickening (Bhatnagar and Puri, 1970). *Elytraria* (B. M. Johri and Singh, 1959), *Andrographis* (Mohan Ram, 1960; Mohan Ram and Masand, 1962), and *Haplanthus* (Phatak and Ambegaokar, 1961) are devoid of seed coat at maturity, and the outer layers of endosperm take up the functions of the seed coat.

Mistletoes (Loranthaceae), members of Santalaceae, and related families either lack integuments or they are poorly developed. If present, the integuments are consumed by the developing endosperm. *Crinum* (Amaryllidaceae) has well-developed integuments, but they are also completely digested by the endosperm by the time the seed reaches maturity. Seeds devoid of testa are usually referred to as "naked."

The chief functions of the seed coat are protection against attack by microorganisms or insects, mechanical injury, and desiccation, and to help in dispersal.

IX. Mature Seed

For describing the morphology of seed, it is important to consider its position in the fruit and its form and surface.

Where the seed is attached to the funiculus or the placenta forms the base of the seed, the apex (top) is the extremity of the line drawn through the axis of seed. When the seed is derived from an orthotropous, basal ovule, it is called erect. If the base of the seed is toward the tip of fruit, it is reversed. In a horizontal seed, its base is at right angles to that of fruit.

The surface of the seed may be smooth, wrinkled, ribbed, punctate, reticulate, glabrous, pulpy, hairy, or have markings resembling fingerprints.

In *Plantago*, the external morphology of seeds is quite variable and has been used to identify the seeds of various species (Misra, 1964, 1970). The seeds may be boat-shaped or flat with further modifications in these two types. There is diversity in the internal morphology too. In *Plantago lagopus, Plantago ovata*, and *Plantago pumila*, the narrow lateral wings of the seed show ventrally curved ends. The wings are more-or-less straight in *Plantago amplexicaulis*. There is also a conical ridgelike structure on the ventral side. The characteristic lateral wings are absent in *Plantago coronopus* and *Plantago major*. The cotyledons are disposed laterally, except in *Plantago major* where they are placed anteroposteriorly.

Usually, the external features of seed are used for its identification. Martin (1946) has shown that the internal morphology is equally important. With the help of hand sections, he studied the gross internal morphology of seeds of 1287 genera of angiosperms.

For purposes of classification, Martin (1946) has taken into account (*a*) size of embryo in relation to endosperm, and (*b*) differences in size, shape, and position of embryo in seed. With regard to the first point (embryo–endosperm ratio), there are five size-designations represented volumetrically in quarter units. The units are small, quarter, half, dominant, and total (Fig. 24).

Depending upon the size, shape, and position of the embryo in seed, the seed is described as "peripheral" and "axile." These are further subdivided into twelve types. According to Martin, the "rudimentary" type is an exception. He also refers to the phylogeny and evolutionary tendencies among seeds.

X. Conclusions

The processes involved in the formation of seeds have interested man for ages. From the time the flower opens until the seed reaches maturity, several interesting events intervene. The first is pollination which ensures a proper supply of pollen on the stigma. Several agents help in the transfer of pollen to the stigma.

Fertilization, which follows pollination, is the most significant event in the life history of a plant. In angiosperms, fertilization comprises the union of one male gamete with the egg (syngamy), and union of the other male gamete with two polar nuclei (triple fusion). The latter phenomenon is unique to flowering plants.

Recent electron-microscopic studies have revealed several interesting aspects of fertilization. The pollen tube invariably enters the degenerate synergid and discharges its contents. It would be rather unwise to generalize such a statement, because instances are on record where both

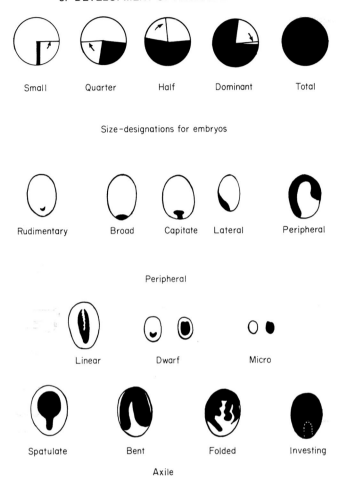

FIG. 24. Types of seed according to Martin (1946.)

synergids degenerate before entry of the pollen tube. Sometimes, gameto-phytes may lack synergids, e.g., in *Plumbago* and *Plumbagella*.

Since it is the degenerate synergid that is involved in the entry of the pollen tube, the factors leading to synergid degeneration are unknown. It has, however, been shown that the pollen tube itself is not responsible for this behavior. Whether it is pollination, or changes in the ovule, or both, that trigger synergid degeneration, requires further study.

The last word has not yet been said on the role of male cytoplasm in fertilization. The lack of data seems essentially due to the difficulty in obtaining proper stages of syngamy. Moreover, a suitable technique must be worked out to trace the male cytoplasm before, during, and after syngamy.

Study of the female gametophyte presents many interesting features, and several unsolved problems. Factors causing variable behavior of nuclei during megasporogenesis and megagametogenesis are not fully understood. We are quite ignorant as to causes that lead to high degrees of ploidy of synergids and antipodal cells. Nutrition of the embryo sac, and DNA and RNA correlations at various stages of development of female gametophyte, also require detailed investigations.

The development of endosperm in angiosperms may be Nuclear, Cellular, or Helobial. Whereas the first two types occur in both dicotyledons and monocotyledons, the Helobial type is predominant among monocotyledons. Although Swamy and Parameswaran (1963) consider the Helobial type exclusive to monocots, this type has conclusively been shown in some dicotyledons also. The phylogenetic relationships of the three types of endosperm are obscure.

In several taxa of angiosperms, endosperm haustoria are formed. These haustoria may be micropylar or chalazal, or both. The haustoria contain dense cytoplasm and prominent nuclei which may become polyploid. The function of haustoria seems to be to supply nourishment to the developing embryo. To solve some of the problems regarding the role of haustoria associated with embryo sac, endosperm, and embryo, it would be worth while to apply histochemical, biochemical, and ultrastructural techniques. As to the significance of haustoria, Kaplan (1969) remarks: "Because of their idioblastic character, haustorial cells are of fundamental interest not only in their systematic distribution, but as prime examples of cellular specialization in relation to seed development."

We do not have adequate information on development of the seed coat, and an intensive and exhaustive study of the histological changes undergone by the integument(s) should prove very rewarding.

To quote Martin and Barkley (1961): "A quick check on a seed's internal morphology can often indicate its family or genus and thus help the investigator avoid serious mistakes about seeds which look surprisingly alike from the outside, but internally they are very distinct." A study of anatomy, and an analysis of viability of seeds by chemicals and electrical conductivity would be of immense help to the geneticist, economic botanist, and agronomist.

ACKNOWLEDGMENTS

We are grateful to Dr. S. S. Bhojwani and Miss Vatsala for checking the manuscript; Mr. Krishan Lal and Mr. R. A. Saini for preparing the illustrations; and Mr. Sat Pal for typing the manuscript.

REFERENCES

Afzelius, K. (1928). Die Embryobildung bei *Nigritella nigra*. *Sv. Bot. Tidskr.* **22**, 82–91.

Agarwal, S. (1963a). Morphological and embryological studies in the family Olacaceae. I. *Olax* L. *Phytomorphology* **13**, 185–196.

Agarwal, S. (1963b). Morphological and embryological studies in the family Olacaceae. II. *Strombosia* Blume. *Phytomorphology* **13**, 348–356.

Agrawal, J. S., and Singh, S. P. (1957). Nucellar polyembryony in *Momordica charantia* Linn. *Sci. Cult.* **22**, 630–631.

Alvarez, M. R., and Sagawa, Y. (1965). A histochemical study of embryo sac development in *Vanda* (Orchidaceae). *Caryologia* **18**, 241–249.

Amici, G. B. (1824). Observations microscopiques sur diverses espèces des plantes. *Ann. Sci. Nat. Bot.* **2**, 41–70 and 211–248.

Amici, G. B. (1830). Note sur la mode d'action du pollen sur le stigmate. Extrait d'une lettre d'Amici à Mirbel. *Ann. Sci. Nat. Bot.* **21**, 329–332.

Amici, G. B. (1847). Sur la fécondation des Orchidées. *Ann. Sci. Nat. Bot.* **7/8**, 193–205.

Archibald, E. E. A. (1939). The development of the ovule and seed of jointed cactus (*Opuntia aurantiaca* Lindley). *S. Afr. J. Sci.* **36**, 195–211.

Arditti, J. (1967). Factors affecting the germination of orchid seeds. *Bot. Rev.* **33**, 1–83.

Arekal, G. D. (1961). Embryology of *Klugia notoniana*. *Bot. Gaz.* **123**, 144–150.

Arekal, G. D. (1963). Embryological studies in Canadian representatives of the tribe Rhinantheae, Scrophulariaceae. *Can. J. Bot.* **41**, 267–302.

Ayyangar, G. S. (1948). Some observations on stomata found on cotton ovules. *Indian Cotton Grow. Rev.* **2**, 187–192.

Bacchi, O. (1943). Cytological observations in *Citrus*. III. Megasporogenesis, fertilization and polyembryony. *Bot. Gaz.* **105**, 221–225.

Battaglia, E. (1955). Unusual cytological features in the apomictic *Rudbeckia sullivantii* Boynton et Beadle. *Caryologia* **8**, 1–32.

Batygina, T. B. (1962). Fertilization process in wheat. *Tr. Botan. Inst. Acad. Nauk. SSSR* **9**, 260–293 (in Russian).

Batygina, T. B. (1969). On the possibility of separation of a new type of embryogenesis in Angiospermae. *Rev. Cytol. Biol. Veg.* **32**, 335–341.

Baum, H. (1950). Das Narbensekret von *Koelreuteria paniculata*. *Oesterr. Bot. Z.* **97**, 517–519.

Belikova, N. L. (1952). Polyembryony in beans (*Phaseolus*). *Byull. Mosk. Obshchest. Ispyt. Prir., Otd. Biol.* **57**, 65–66.

Berg, O. (1898). Beitrag zur Kenntnis der Entwicklung des Embryosackes der Angiospermen. Dissertation, Erlangen.

Berg, R. Y. (1954). Development and dispersal of the seed of *Pedicularis sylvatica Nytt. Mag. Bot.* **2**, 1–60.

Berg, R. Y. (1958). Seed dispersal, morphology, and phylogeny of *Trillium*. *Skr. Norske Videnskapsakad. Oslo, I.: Mat.-Naturv. Kl. No. 1*.

Bhandari, N. N. (1962). Studies in the family Ranunculaceae. III. Development of the female gametophyte in *Adonis annua* L. *Phytomorphology* **12**, 70–74.

Bhandari, N. N. (1966). Studies in the family Ranunculaceae. IX. Embryology of *Adonis* Dill. ex Linn. *Phytomorphology* **16**, 578–587.

Bhandari, N. N. (1968). Studies in the family Ranunculaceae. X. Embryology of *Anemone* L. *Phytomorphology* **18**, 487–497.

Bhandari, N. N., and Asnani, S. (1968). Studies in the family Ranunculaceae. XI. Morphology and embryology of *Ceratocephalus falcatus* Pers. *Beitr. Biol. Pflanz.* **45**, 271–290.

Bhandari, N. N., and Vijayaraghavan, M. R. (1970). Studies in the family Ranunculaceae. XII. Embryology of *Aquilegia vulgaris. Beitr. Biol. Pflanz.* **46**, 337–354.

Bhatnagar, S. P. (1959). Some observations on the post-fertilization development of the embryo sac of *Santalum. Phytomorphology* **9**, 87–91.

Bhatnagar, S. P. (1960). Morphological and embryological studies in the family Santalaceae. IV. *Mida salicifolia* A. Cunn. *Phytomorphology* **10**, 198–207.

Bhatnagar, S. P. (1965). Studies on angiospermic parasites. II. *Santalum album.* The sandalwood tree. *Bull. Nat. Bot. Gard., Lucknow* **112**, 1–90.

Bhatnagar, S. P., and Joshi, P. C. (1965). Morphological and embryological studies in the family Santalaceae. VII. *Exocarpus bidwillii* Hook. f. *Proc. Nat. Inst. Sci. India, Part B* **31**, 34–44.

Bhatnagar, S. P., and Puri, S. (1970). Morphology and embryology of *Justicia betonica* Linn. *Oesterr. Bot. Z.* **118**, 55–71.

Bhatnagar, S. P., and Sabharwal, G. (1969). Morphology and embryology of *Iodina rhombifolia* Hook. & Arn. *Beitr. Biol. Pflanz.* **45**, 464–479.

Bhatnagar, S. P., and Uma, M. C. (1969). The structure of style and stigma in some Tubiflorae. *Phytomorphology* **19**, 99–109.

Bocquet, G. (1959). The campylotropous ovule. *Phytomorphology* **9**, 222–227.

Bouman, F., and Boesewinkel, F. D. (1969). On a case of polyembryony in *Pterocarya fraxinifolia* (Juglandaceae) and on polyembryony in general. *Acta Bot. Neer.* **18**, 50–57.

Bubar, J. S. (1958). An association between variability in ovule development within ovaries and self-incompatibility in *Lotus* (Leguminosae). *Can. J. Bot.* **36**, 65–72.

Buell, K. M. (1952). Developmental morphology in *Dianthus.* I. Structure of the pistil and seed development. *Amer. J. Bot.* **39**, 194–210.

Camp, W. H., and Hubbard, M. M. (1963). Vascular supply and structure of the ovule and aril in peony and of the aril in nutmeg. *Amer. J. Bot.* **50**, 174–178.

Carniel, K. (1967). Ueber die Embryobildung in der Gattung *Paeonia. Oesterr. Bot. Z.* **114**, 4–19.

Cass, D. D., and Jensen, W. A. (1970). Fertilization in barley. *Amer. J. Bot.* **57**, 62–70.

Cave, M. S., Arnott, H. J., and Cook, S. A. (1961). Embryogeny in the California peonies with reference to their taxonomic position. *Amer. J. Bot.* **48**, 397–404.

Chopra, R. N. (1955). Some observations on endosperm development in the Cucurbitaceae. *Phytomorphology* **5**, 219–230.

Chopra, R. N., and Agarwal, S. (1958). Some further observations on the endosperm haustoria in the Cucurbitaceae. *Phytomorphology* **8**, 194–201.

Chopra, R. N., and Basu, B. (1965). Female gametophyte and endosperm of some members of the Cucurbitaceae. *Phytomorphology* **15**, 217–223.

Cocucci, A. E. (1961). Embriologia de *Trianthema argentina. Kurtziana* **1**, 105–122.

Cocucci, A. E., and Jensen, W. A. (1969). Orchid embryology: Megagametophyte of *Epidendrum scutella* following fertilization. *Amer. J. Bot.* **56**, 629–640.

Cooper, D. C. (1943). Haploid-diploid twin embryos in *Lilium* and *Nicotiana. Amer. J. Bot.* **30**, 408–413.

Corner, E. J. H. (1949). The annonaceous seed and its four integuments. *New Phytol.* **48**, 332–364.

Corner, E. J. H. (1951). The leguminous seed. *Phytomorphology* **1**, 117–152.

Crété, P. (1938). La polyembryonie chez le *Lobelia syphilitica* L. *Bull. Soc. Bot. Fr.* **85**, 580–583.

Crété, P. (1944). Polyembryonie chez l'*Actinidia chinensis* Planch. *Bull. Soc. Bot. Fr.* **91**, 89–92.

Crété, P. (1948). Recherches embryologiques chez les Scrofulariacées. Développement de l'embryon chez le *Chaenostoma foetidum* (Jacq.) Benth. *Bull. Soc. Bot. Fr.* **95**, 142–146.

Crété, P. (1958). La Parthenogenèse chez le *Sicyos angulata. Bull. Soc. Bot. Fr.* **105**, 18–19.

Cutter, V. M., Jr., and Freeman, B. (1955). Nuclear aberrations in the syncytial endosperm of *Cocos nucifera. J. Elisha Mitchell Sci..Soc.* **71**, 49–58.

Dahlgren, K. V. O. (1924). Studien über die Endospermbilung der Kompositen. *Sv. Bot. Tidskr.* **18**, 177–203.

Dahlgren, K. V. O. (1934). Die Embryologie von *Impatiens*'roylei. *Sv. Bot. Tidskr.* **28**, 103–125.

Datta, M. (1955). The occurrence and division of free nuclei in the endospermal milk in some Palmae. *Trans. Bose Res. Inst., Calcutta* **19**, 117–125.

Davis, G. L. (1966). "Systematic Embryology of the Angiosperms." New York.

Desai, S. (1962). Cytology and embryology of the Rutaceae. *Phytomorphology* **12**, 178–184.

Diboll, A. G. (1968). Fine structural development of the megagametophyte of *Zea mays* following fertilization. *Amer. J. Bot.* **55**, 797–806.

Diboll, A. G., and Larson, D. A. (1966). An electron microscopic study of the mature megagametophyte in *Zea mays. Amer. J. Bot.* **53**, 391–402.

Dnynansagar, V. R. (1954). Embryological studies in the Leguminosae. X. Supplementary observations on the development of endosperm and embryo in *Leucaena glauca* Benth. and *Mimosa hamata* Willd. *J. Indian Bot. Soc.* **33**, 433–442.

Dnyansagar, V. R. (1958). Embryological studies in the Leguminosae. VIII. *Acacia auriculaeformis* A. Cunn., *Adenanthera pavonina* Linn., *Calliandra hematocephala* Benth. *Lloydia* **21**, 1–25.

Duncan, R. E., and Ross, J. G. (1950). The nucleus in differentiation and development. III. Nuclei of maize endosperm. *J. Hered.* **41**, 259–268.

Dutt, M. (1953). Dividing nuclei in coconut milk. *Nature (London)* **171**, 799.

Eames, A. J. (1961). "Morphology of the Angiosperms." New York.

Eckardt, T. (1963). Some observations on the morphology and embryology of *Eucommia ulmoides* Oliv. Maheshwari Comm. Vol. *J. Indian Bot. Soc.* **42A**, 27–34.

Enzenberg, U. (1961). Beiträge zur Karyologie des Endosperm. *Oesterr. Bot. Z.* **108**, 245–285.

Erbrich, P. (1965). Über endopolyploidie und Kernstrukturen in Endospermhaustorien. *Oesterr. Bot. Z.* **112**, 197–262.

Ethridge, A. L., and Herr, J. M., Jr. (1968). The development of the ovule and megagametophyte in *Rhexia marina. Can. J. Bot.* **46**, 133–139.

Eunus, A. M. (1951). Contributions to the embryology of the Liliaceae. IV. Gametophytes of *Smilacina stellata. New Phytol.* **49**, 269–273.

Fagerlind, F. (1944). Die Samenbildung und die Zytologie bei agamospermischen und sexuellen Arten von *Elatostema* und einigen nahestehenden Gattungen nebst Beluchtung einiger damit Zusammenhängender Probleme. *Kgl. Sv. Vetenskapsakad., Handl.* **21**, 1–130.

Fisher, D. B., and Jensen, W. A. (1969). Cotton embryogenesis: The identification, as nuclei, of the X-bodies in the degenerating synergid. *Planta* **84**, 122–133.

Fisher, W. D., Bashaw, E. C., and Holt, E. C. (1954). Evidence for apomixis in *Pennisetum ciliare* and *Cenchrus setigerus. Agron. J.* **46**, 401–404.

Flint, L. H., and Moreland, C. F. (1943). Notes on photosynthetic activity in seeds of the spider lily. *Amer. J. Bot.* **30**, 315–317.

Freeman, Orville, L. (1961). *In* "Seeds. The Yearbook of Agriculture" (A. Stefferud, ed.), pp. v–vi. USDA, Washington, D.C.

Garcia, V. (1963). Embryological studies on the Loasaceae. Doctor Biol. Sci. Thesis, University of Buenos Aires.

Geitler, L. (1948). Notizen zur endomitotischen Polyploidisierung in Trichozyten und Elaisosomen sowie über Kernstrukturen bei *Gagea lutea*. *Chromosoma* **3**, 271–281.

Geitler, L. (1955). Riesenkerne in Endosperm von *Allium ursinum*. *Oesterr. Bot. Z.* **102**, 460–475.

Gerassimova-Navashina, H. (1957). On some cytological principles underlying double fertilization. *Phytomorphology* **7**, 150–167.

Gerassimova-Navashina, H. (1960). A contribution to the cytology of fertilization in flowering plants. *Nucleus* **3**, 111–120.

Gerassimova-Navashina, H. (1961). Fertilization and events leading up to fertilization, and their bearing on the origin of angiosperms. *Phytomorphology* **11**, 139–146.

Gerassimova-Navashina, H. (1969). Some cytological aspects of double fertilization. *Rev. Cytol. Biol. Veg.* **32**, 301–308.

Gerassimova-Navashina, H., and Korobova, S. N. (1959). On the role of synergids in fertilization. *Dokl. Akad. Nauk SSSR* **124**, 223–226.

Ghosh, R. B. (1962). A note on the nucellar polyembryony in *Aphanamixis polystachya*, (Wall.) Parker. *Curr. Sci.* **31**, 165.

Grundwag, H., and Fahn, A. (1969). The relation of embryology to the low seed set in *Pistacia vera* (Anacardiaceae). *Phytomorphology* **19**, 225–235.

Guéguen, F. (1901). Anatomie comparée du tissu conducteur du style et du stigmate des Phanérogames (I. Monocotylédones, Apétales et Gamopétales). Thesis, Faculty of Science, Paris.

Guignard, J. L., and Mestre, J. C. (1966). On the development of embryos from antipodal cells of *Ulmus campestris* L. *Bull. Soc. Bot. Fr.* **113**, 227–228.

Guttenberg, H. V., Heydel, H., and Pankow, H. (1951). Embryologische Studien an Monokotyledonen. II. Die Entwicklung des Embryon von *Allium giganteum* Regel. *Flora (Jena)* **141**, 476–500.

Haberlandt, G. (1921). Über experimentelle Erzeugung von Adventivembryonen bei *Oenothera lamarckiana*. *Sitzungsber. Preuss. Akad. Wiss., Phys.-Math. Kl.* **40**, 695–725.

Haberlandt, G. (1922). Über Zellteilungshormone und ihre Beziehungen zur Wundeheilung, Befruchtung, Parthenogenesis und Adventivembryonie. *Biol. Zentralbl.* **42**, 145–172.

Haberlandt, G. (1923). Zur Embryologie von *Allium odorum*. *Ber. Deut. Bot. Ges.* **41**, 174–179.

Haberlandt, G. (1925). Zur Embryologie und Cytologie von *Allium odorum*. *Ber. Deut. Bot. Ges.* **43**, 559–564.

Håkansson, A. (1943). Die Entwicklung des Embryosackes und die Befruchtung bei *Poa alpina*. *Hereditas* **29**, 25–61.

Håkansson, A. (1951). Parthenogenesis in *Allium*. *Bot. Notis.* pp. 143–179.

Hanf, M. (1935). Vergleichende und entwicklungsgeschichtliche Untersuchungen über Morphologie und Anatomie der Griffel und Griffeläste. *Beih. Bot. Zentralbl.* **54A**, 99–141.

Harling, G. (1949). Zur Embryologie der Gattung *Hedychium*. *Sv. Bot. Tidskr.* **43**, 357–364.

Harling, G. (1958). Monograph of the Cyclanthaceae. *Acta Horti Bergiani* **18**, 1–428.

Hasitschka, G. (1956). Bildung von Chromosomenbünden nach Art der Speicheldrüsenchromosomen, spiralisierte Ruhekernchromosomen und andere Struktureigentümlichkeiten in den endopolyploiden Riesenkernen der Antipoden von *Papavar rhoeas*. *Chromosoma* **9**, 87–113.

Hasitschka-Jenschke, G. (1958). Zur Karyologie der Samenanlage dreier *Allium*-Arten. *Oesterr. Bot. Z.* **105**, 71–82.

Hasitschka-Jenschke, G. (1959). Vergleichende karyologische Untersuchungen an Antipoden. *Chromosoma* **10**, 229–267.

Hegelmaier, F. (1897). Zur Kenntnis der Polyembryonie von *Allium odorum. Bot. Z.* **55**, 133–170.

Hill, J. B., Popp, H. W., and Grove, A. R., Jr. (1967). "Botany," 4th ed. McGraw-Hill, New York.

Hofmeister, W. (1849). "Die Entstehung des Embryo der Phanerogamen." Leipzig.

Hofmeister, W. (1861). Neue Beiträge zur Kenntnis der Embryobildung der Phanerogamen. II. Monocotyledonen. *Abh. Saechs. Akad. Wiss. Leipzig, Math.-Naturw. Kl.* **7**, 629–760.

Iwanami, Y. (1959). Physiological studies of pollen. *J. Yokohama Munic., Univ., Ser. C* **116**, 1–137.

Jensen, W. A. (1965a). The ultrastructure and histochemistry of the synergids of cotton. *Amer. J. Bot.* **52**, 238–256.

Jensen, W. A. (1965b). The ultrastructure and composition of the egg and central cell of cotton. *Amer. J. Bot.* **52**, 781–798.

Jensen, W. A., and Fisher, D. B. (1967). Cotton embryogenesis: Double fertilization. Panchanan Maheshwari Mem. Vol. *Phytomorphology* **17**, 261–269.

Jensen, W. A., and Fisher, D. B. (1968). Cotton embryogenesis: The entrance and discharge of the pollen tube into the embryo sac. *Planta* **58**, 158–183.

Johansen, D. A. (1928). The hypostase: Its presence and function in the ovule of Onagraceae. *Proc. Nat. Acad. Sci. U.S.* **14**, 710–713.

Johri, B. M. (1936a). The life history of *Butomopsis lanceolata* Kunth. *Proc. Indian Acad. Sci., Sect. B* **4**, 139–162.

Johri, B. M. (1936b). Studies in the family Alismaceae. IV. *Alisma plantago* L., *A. plantago-aquatica* L., and *Sagittaria graminea* Mich. *Proc. Indian Acad. Sci., Sect. B* **4**, 128–138.

Johri, B. M. (1962). Female gametophyte of the Santalales. *In* "Plant Embryology — A Symposium," pp. 192–198. Council Sci. Ind. Res., New Delhi.

Johri, B. M. (1963). Female gametophyte. *In* "Recent Advances in the Embryology of Angiosperms" (P. Maheshwari, ed.), pp. 69–104. Int. Soc. Plant Morphol., Delhi.

Johri, B. M., and Agarwal, S. (1965). Morphological and embryological studies in the family Santalaceae. VIII. *Quinchamalium chilense* Lam. *Phytomorphology* **15**, 360–372.

Johri, B. M., and Ahuja, M. R. (1957). A contribution to the floral morphology and embryology of *Aegle marmelos* Correa. *Phytomorphology* **7**, 10–24.

Johri, B. M., and Bhatnagar, S. P. (1955). A contribution to the morphology and life history of *Aristolochia. Phytomorphology* **5**, 123–137.

Johri, B. M., and Bhatnagar, S. P. (1957). Intra-carpellary pollen grains in angiosperms. *Phytomorphology* **7**, 292–296.

Johri, B. M., and Bhatnagar, S. P. (1960). Embryology and taxonomy of the Santalales. I. *Proc. Nat. Inst. Sci. India, Part B* **26**, 199–220.

Johri, B. M., and Bhatnagar, S. P. (1969). Endosperm in the Santalales. *Rev. Cytol. Biol. Veg.* **32**, 353–369.

Johri, B. M., and Garg, S. (1959). Development of endosperm haustoria in some Leguminosae. *Phytomorphology* **9**, 34–46.

Johri, B. M., and Kak, D. (1954). The embryology of *Tamarix. Phytomorphology* **4**, 230–247.

Johri, B. M., and Raj, B. (1969). Morphological and embryological studies in the family Loranthaceae. XII. *Moquiniella rubra* (Spreng. f.) Balle. *Oesterr. Bot. Z.* **116**, 475–485.

Johri, B. M., and Singh, H. (1959). The morphology, embryology and systematic position of *Elytraria acaulis* (Linn.) Lindau. *Bot. Notis.* **112**, 227–251.

Johri, B. M., and Tiagi, B. (1952). Floral morphology and seed formation in *Cuscuta reflexa* Roxb. *Phytomorphology* **2**, 162–180.

Johri, M. M. (1966a). The style, stigma and pollen tube. II. Some taxa of the Liliaceae and Trilliaceae. *Phytomorphology* **16**, 92–109.

Johri, M. M. (1966b). The style, stigma and pollen tube. III. Some taxa of the Amaryllidaceae. *Phytomorphology* **16**, 142–157.

Joshi, P. C., Wadhwani, A. M. (née Ramchandani, S.), and Johri, B. M. (1967). Morphological and embryological studies of *Gossypium* L. *Proc. Nat. Inst. Sci. India, Part B* **33**, 37–93.

Kapil, R. N. (1956). A further contribution to the morphology and life history of *Chrozophora* Neck. *Phytomorphology* **6**, 278–288.

Kapil, R. N. (1958). Morphological and embryological studies in some Euphorbiaceae together with a discussion on the systematic position of the family. Ph.D. Thesis, University of Delhi.

Kapil, R. N., and Ahluwalia, K. (1963). Embryology of *Peganum harmala* Linn. *Phytomorphology* **13**, 127–140.

Kapil, R. N., and Bhandari, N. N. (1964). Morphology and embryology of *Magnolia* Dill. ex Linn. *Proc. Nat. Inst. Sci. India, Part B* **30**, 245–262.

Kapil, R. N., and Jalan, S. (1962). Studies in the family Ranunculaceae. I. The embryology of *Caltha palustris* L. *In* "Plant Embryology – A Symposium," pp. 205–214. Council Sci. Ind. Res., New Delhi.

Kapil, R. N., and Mohana Rao, P. R. (1966). Studies of the Garryaceae. II. Embryology and systematic position of *Garrya* Douglas ex Lindley. *Phytomorphology* **16**, 564–578.

Kapil, R. N., and Prakash, N. (1966). Co-existence of mono-, bi-, and tetrasporic embryo sacs in *Delosperma cooperi* Hook. f. (Aizoaceae). *Beitr. Biol. Pflanz.* **42**, 381–392.

Kapil, R. N., and Prakash, N. (1969). Embryology of *Cereus jamacaru* and *Ferocactus wislizeni* and comments on the systematic position of the Cactaceae. *Bot. Notis.* **122**, 409–426.

Kapil, R. N., and Vani, R. S. (1963). Embryology and systematic position of *Crossosoma californicum* Nutt. *Curr. Sci.* **32**, 493–495.

Kapil, R. N., and Vasil, I. K. (1963). Ovule. *In* "Recent Advances in the Embryology of Angiosperms" (P. Maheshwari, ed.), pp. 41–67. Int. Soc. Plant Morphol., Delhi.

Kapil, R. N., and Vijayaraghavan, M. R. (1962). Embryology and systematic position of *Pentaphragma horsefieldii* (Miq.) Airy Shaw. *Curr. Sci.* **31**, 270–272.

Kapil, R. N., and Walia, K. (1965). The embryology of *Philydrum lanuginosum* Banks, ex Gaertn. and the systematic position of the Phylidraceae. *Beitr. Biol. Pflanz.* **41**, 381–404.

Kaplan, D. R. (1969). Seed development in *Downingia*. *Phytomorphology* **19**, 253–278.

Kato, K., and Watanabe, K. (1957). The stigma reaction. II. The presence of the stigma reaction in the intra-specific and the inter-generic pollination in Gramineae. *Bot. Mag. (Tokyo)* **70**, 76–101.

Kaul, U. (1969). Endosperm in *Parrotiopsis jacquemontiana*. *Phytomorphology* **19**, 197–199.

Kausik, S. B. (1941). Development of the vermiform appendage in *Grevillea robusta* Cunn. *Proc. Indian Acad. Sci., Sect. B* **14**, 137–140.

Kausik, S. B., and Subramanyam, K. (1946). A case of polyembryony in *Isotoma longifolia* Presl. *Curr. Sci.* **15**, 257–258.

Kelkar, S. S. (1958a). A contribution to the embryology of *Lannea coromandelica*. *J. Univ. Bombay* **26**, 152–159.

Kelkar, S. S. (1958b). Embryology of *Rhus mysurensis*. *J. Indian Bot. Soc.* **37**, 114–122.

Kelkar, S. S. (1961). The development of endosperm and embryo in *Lannea coromandelica* (Houtt.) Merr. *J. Univ. Bombay* **29**, 1–5.

Khan, R. (1942). A contribution to the embryology of *Jussieua repens*. *J. Indian Bot. Soc.* **21**, 267–282.

Khan, R. (1954). A contribution to the embryology of *Utricularia flexuosa* Vahl. *Phytomorphology* **4**, 80–117.

Konar, R. N., and Linskens, H. F. (1966). The morphology and anatomy of the stigma of *Petunia hybrida*. *Planta* **71**, 356–371.

Kroh, M. (1964). An electron microscopic study of the behavior of Cruciferae pollen after pollination. *In* "Pollen Physiology and Fertilization" (H. F. Linskens, ed.), pp. 221–224. Amsterdam.

Kuijt, J. (1960). Morphological aspects of parasitism in the dwarf mistletoes (*Arceuthobium*). *Univ. Calif., Berkeley, Publ. Bot.* **30**, 337–436.

Kuijt, J. (1969). "The Biology of Parasitic Flowering Plants." Univ. of California Press, Berkeley.

Kumar, L. S. S., and Joshi, M. V. (1942). False polyembryony in viviparous *Rhizophora mucronata*. *Curr. Sci.* **11**, 242.

Linskens, H. F. (1969). Fertilization mechanisms in higher plants. *In* "Fertilization: Comparative Morphology, Biochemistry and Immunology" (C. B. Metz and A. Monroy, eds.), Vol. 2. pp. 189–253. Academic Press, New York.

Madge, M. (1934). Nuclear migrations in *Hedychium*. *Proc. Linn. Soc. London* **146**, 108–109.

Maheshwari, P. (1950). "An Introduction to the Embryology of Angiosperms." McGraw-Hill, New York.

Maheshwari, P., and Chopra, R. N. (1955). The structure and development of the ovule and seed of *Opuntia dillenii*. *Phytomorphology* **5**, 112–122.

Maheshwari, P., and Negi, V. (1955). The embryology of *Dipteracanthus patulus* (Jacq.) Nees. *Phytomorphology* **5**, 456–472.

Maheshwari, P., and Sachar, R. C. (1963). Polyembryony. *In* "Recent Advances in the Embryology of Angiosperms" (P. Maheshwari, ed.), pp. 265–296. Int. Soc. Plant Morphol., Delhi.

Maheshwari, P., Johri, B. M., and Dixit, S. N. (1957). The floral morphology and embryology of the Loranthoideae (Loranthaceae). *J. Madras Univ., Sect. B* **27**, 121–136.

Maheshwari, S. C. (1954). The embryology of *Wolffia*. *Phytomorphology* **4**, 355–365.

Maheswari Devi, H. (1957). Embryological studies in Compositae. III. *Gerbera jamesonii* Bohs. *Proc. Indian Acad. Sci., Sect. B* **46**, 68–74.

Martin, A. C. (1946). The comparative internal morphology of seeds. *Amer. Midl. Natur.* **36**, 513–660.

Martin, A. C., and Barkley, W. D. (1961). "Seed Identification Manual." California.

Masand, P. (1963). Embryology of *Zygophyllum fabago* Linn. *Phytomorphology* **13**, 293–302.

Mascarenhas, J. P. (1966). The distribution of ionic calcium in the tissues of the gynoecium of *Antirrhinum majus*. *Protoplasma* **62**, 53–58.

Mascarenhas, J. P., and Machlis, L. (1962a). The hormonal control of the directional growth of pollen tubes. *Vitam. Horm. (New York)* **20**, 347–372.

Mascarenhas, J. P., and Machlis, L. (1962b). The pollen tube chemotropic factor from bioassay, extraction and partial purification. *Amer. J. Bot.* **49**, 482–489.

Mascarenhas, J. P., and Machlis, L. (1964). Chemotropic response of the pollen of *Antirrhinum majus* to calcium. *Plant Physiol.* **39**, 70–77.

Mathur, N. (1956). The embryology of *Limnanthes*. *Phytomorphology* **6**, 41–51.

Matthiessen, A. (1962). A contribution to the embryology of *Paeonia*. *Acta Horti Bergiani* **20**, 57–61.

Mauritzon, J. (1939). Contributions to the embryology of the orders Rosales and Myrtales. *Acta Univ. Lund.* **35**, 1–120.

Misra, R. C. (1964). Development and structure of angiosperm seed. III. *Plantago* L. *Bull. Nat. Bot. Gard., Lucknow* **105**, 1–14.

Misra, R. C. (1970). Plantaginaceae. *In* "Symposium on Comparative Embryology of Angiosperms," pp. 298–304. Indian Nat. Sci. Acad., New Delhi.

Mohan Ram, H. Y. (1960). The development of the seed in *Andrographis serpyllifolia*. *Amer. J. Bot.* **47**, 215–219.

Mohan Ram, H. Y., and Masand, P. (1962). Endosperm and seed development in *Andrographis echioides* Nees. *Curr. Sci.* **31**, 7–8.

Mohan Ram, H. Y., and Sehgal, P. P. (1958). The life history of *Justicia simplex* Don. *Phytomorphology* **8**, 124–136.

Mohan Ram, H. Y., and Wadhi, M. (1964). Endosperm in Acanthaceae. *Phytomorphology* **14**, 388–413.

Mohan Ram, H. Y., and Wadhi, M. (1965). Embryology and the delimitation of the Acanthaceae. *Phytomorphology* **15**, 201–205.

Moskov, I. V. (1964). The development of the embryo in some *Paeonia* varieties. *Bot. Zh.* **49**, 887–894.

Mukkada, A. J. (1962). Some observations on the embryology of *Dicraea stylosa*. *In* "Plant Embryology—A Symposium," pp. 139–145. Council Sci. Ind. Res., New Delhi.

Mukkada, A. J. (1963). A contribution to the embryology of *Indotristicha ramosissima* (Wight) van Royen. *Proc. Indian Sci. Congr., 50th.* Part III, Abstr. 387.

Murgai, P. (1959). The development of the embryo in *Paeonia*—A reinvestigation. *Phytomorphology* **9**, 275–277.

Murgai, P. (1962). Embryology of *Paeonia* together with a discussion on its systematic position. *In* "Plant Embryology—A Symposium," pp. 215–223. Council Sci. Ind. Res., New Delhi.

Nagaraj, M. (1955). Some aspects of floral morphology in *Combretum ovalifolium* Roxb. *J. Mysore Univ., Ser. B* **14**, 77–85.

Nair, N. C. (1970). Meliaceae. *In* "Symposium on Comparative Embryology of Angiosperms," pp. 151–155. Indian Nat. Sci. Acad., New Delhi.

Nair, N. C., and Jain, P. K. (1956). Floral morphology and embryology of *Balanites roxburghii*. *Lloydia* **19**, 269–279.

Narang, N. (1953). The life history of *Stackhousia linariaefolia* A. Cunn. with a discussion on its systematic position. *Phytomorphology* **3**, 485–493.

Narayana, R. (1954). Contribution to the embryology of *Dendrophthoe* Mart. *Phytomorphology* **4**, 173–179.

Narayanaswami, S. (1953). The structure and development of the caryopsis in some Indian millets. I. *Pennisetum typhoideum* Rich. *Phytomorphology* **3**, 98–112.

Narayanaswami, S. (1954). The structure and development of the caryopsis in some Indian millets. II. *Paspalum scrobiculatum*. *Bull. Torrey Bot. Club* **81**, 288–299.

Nawaschin, S. G. (1898). Resultate einer Revision der Befruchtungsvorgänge bei *Lilium martagon* and *Fritillaria tenella*. *Izv. Imp. Akad. Nauk* **9**, 377–382.

Netolitzky, F. (1926). "Anatomie der Angiospermen-Samen." Bornträger, Berlin.

Nielsen, E. L. (1946). The origin of multiple macrogametophytes in *Poa pratensis*. *Bot. Gaz.* **108**, 41–50.

Orr, M. Y. (1921). The structure of the ovular integuments and development of the testa in *Cleome* and *Isomeris*. *Notes Roy. Bot. Gard. Edinburgh* **60**, 243.

Osterwalder, A. (1898). Beiträge zur Embryologie von *Aconitum napellus*. *Flora (Jena)* **85**, 254–292.

Paliwal, G. S., and Bhandari, N. N. (1962). Stomatal studies in some Magnoliaceae. *Phytomorphology* **12**, 409–412.

Panchaksharappa, M. G. (1960). Embryological studies in the Zingiberaceae. Ph.D. Thesis, University of Delhi.

Panchaksharappa, M. G. (1962a). Embryological studies in some members of the Zingiberaceae. *In* "Plant Embryology – A Symposium," pp. 224–228. Council Sci. Ind. Res., New Delhi.

Panchaksharappa, M. G. (1962b). Embryological studies in the family Zingiberaceae. I. *Costus speciosus*. *Phytomorphology* **12**, 418–430.

Panchaksharappa, M. G. (1966). Embryological studies in some members of Zingiberaceae. II. *Elettaria cardamomum, Hitchenia caulina* and *Zingiber macrostachyum*. *Phytomorphology* **16**, 412–417.

Periasamy, K. (1962). The ruminate endosperm. Development and types of rumination. *In* "Plant Embryology – A Symposium," pp. 62–64. Council Sci. Ind. Res., New Delhi.

Periasamy, K., and Swamy, B. G. L. (1961). Studies in the Annonaceae. II. The development of ovule and seed in *Cananga odorata* and *Miliusa wightiana*. *J. Indian Bot. Soc.* **40**, 206–216.

Phatak, V. G., and Ambegaokar, K. B. (1961). Embryological studies in the Acanthaceae. IV. Development of embryo sac and seed formation in *Haplanthus tentaculatus* Nees. *J. Indian Bot. Soc.* **40**, 525–534.

Poddubnaya-Arnoldi, V. A. (1964). Research on fertilization and embryogenesis in some Angiospermae. *In* "Problems of Recent Embryology," pp. 25–32. Moscow Univ.

Poddubnaya-Arnoldi, V. A. (1967). Comparative embryology of the Orchidaceae. Panchanan Maheshwari Mem. Vol. *Phytomorphology* **17**, 312–320.

Poddubnaya-Arnoldi, V. A., Zinger, N. V., and Petrovskaya-Baranova, T. P. (1964). A histochemical investigation of the ovules, embryo sacs and seeds in some angiosperms. *In* "Pollen Physiology and Fertilization" (H. F. Linskens, ed.), pp. 3–7. North Holland Publ. Co., Amsterdam.

Prakash, N. (1967). Life history of *Tetragonia tetragonioides* (Pall.) O. Kuntze. *Aust. J. Bot.* **15**, 413–424.

Pritchard, H. N. (1964). A cytochemical study of embryo sac development in *Stellaria media*. *Amer. J. Bot.* **51**, 371–378.

Punnett, H. H. (1953). Cytological evidence of hexaploid cells in maize endosperm. *J. Hered.* **44**, 257–259.

Puri, V. (1941). Life history of *Moringa oleifera* Lamk. *J. Indian Bot. Soc.* **20**, 263–284.

Quisumbing, E., and Juliano, J. B. (1927). Development of ovule and embryo sac of *Cocos nucifera*. *Bot. Gaz.* **84**, 279–293.

Raj, B. (1964). Female gametophyte of *Buckleya lanceolata* Sieb. et Zucc. *Curr. Sci.* **33**, 348–349.

Raju, M. V. S. (1956). Embryology of the Passifloraceae. I. Gametogenesis and seed development of *Passiflora calcarata* Mast. *J. Indian Bot. Soc.* **35**, 126–138.

Ram, M. (1956). Floral morphology and embryology of *Trapa bispinosa* Roxb., with a discussion on the systematic position of the genus. *Phytomorphology* **6**, 312–323.

Ram, M. (1957). Morphological and embryological studies in the family Santalaceae. I. *Comandra umbellata* (L.) Nutt. *Phytomorphology* **7**, 24–35.

Ram, M. (1959a). Morphological and embryological studies in the family Santalaceae. II. *Exocarpus,* with a discussion on its systematic position. *Phytomorphology* **9,** 4–19.

Ram, M. (1959b). Morphological and embryological studies in the family Santalaceae. III. *Leptomeria* R. Br. *Phytomorphology* **9,** 20–33.

Ramchandani, S., Joshi, P. C., and Pundir, N. S. (1966). Seed development in *Gossypium* Linn. *Indian Cotton J.* **20,** 97–106.

Rangan, T. S., and Rangaswamy, N. S. (1968). Morphogenic investigations on parasitic angiosperms. I. *Cistanche tubulosa* Wight (Family Orobanchaceae). *Can. J. Bot.* **46,** 263–266.

Rangaswamy, N. S. (1967). Morphogenesis of seed germination in angiosperms. Panchanan Maheshwari Mem. Vol. *Phytomorphology* **17,** 477–487.

Rao, V. S., and Gupte, K. (1951). A few observations on the carpels of *Artabotrys. J. Univ. Bombay* **20,** 62–65.

Rau, M. A. (1953). Some observations on the endosperm in Papilionaceae. *Phytomorphology* **3,** 209–222.

Renner, O., and Preuss-Herzog, G. (1943). Der Weg der Pollenschlauche in Fruchtknoten der Oenotheren. *Flora (Jena)* **36,** 215–222.

Rigby, J. F. (1959). Light as a control in the germination and development of several mistletoe species. *Proc. Linn. Soc. N.S.W.* **84,** 335–337.

Sachar, R. C. (1955). The embryology of *Argemone mexicana* L. – A reinvestigation. *Phytomorphology* **5,** 200–218.

Sachar, R. C., and Arora, U. (1963). Some embryological aspects of *Amomum dealbatum* and *Hedychium acuminatum. Bot. Gaz.* **124,** 353–360.

Sachar, R. C., and Chopra, R. N. (1957). A study of the endosperm and embryo in *Mangifera* L. *Indian J. agr. Sci.* **27,** 219–228.

Sachar, R. C., and Mohan Ram, H. Y. (1958). The embryology of *Eschscholtzia californica* Cham. *Phytomorphology* **8,** 114–124.

Satina, S., Rappaport, J., and Blakeslee, A. F. (1950). Ovular tumors connected with incompatible crosses in *Datura. Amer. J. Bot.* **37,** 576–586.

Savchenko, M. I. (1960). Anomalies in the structure of angiosperm ovules. *Dokl. Akad. Nauk SSSR* **130,** 15–17.

Schlimbach, H. (1924). Beiträge zur Kenntnis der Samenanlagen und Samen der Amaryllidaceen mit Berucksichtigung des Wassergehaltes der Samen. *Flora (Jena)* **117,** 41–54.

Schnarf, K. (1929). "Embryologie der Angiospermen." Berlin.

Schoch-Bodmer, H., and Huber, P. (1947). Die Ernährung der Pollenschlauche durch das Leitgewebe. *Vierteljahresschr. Naturforsch. Ges. Zürich* **92,** 43–48.

Schulz, R., and Jensen, W. A. (1968a). *Capsella* embryogenesis: The synergids before and after fertilization. *Amer. J. Bot.* **55,** 541–552.

Schulz, R., and Jensen, W. A. (1968b). *Capsella* embryogenesis: The egg, zygote, and young embryo. *Amer. J. Bot.* **55,** 807–819.

Scott, F. M. (1953). The physical consistency of the endosperm of *Echinocystis macrocarpa. Phytomorphology* **3,** 66–76.

Seth, P. N. (1962). A contribution to the embryology of the Cucurbitaceae. M.Sc. Thesis, University of Delhi.

Singh, B. (1952). Studies on the structure and development of seeds of Cucurbitaceae. I. Seeds of *Echinocystis wrightii* Cogn. *Phytomorphology* **2,** 201–209.

Singh, B. (1953). Studies on the structure and development of seeds of Cucurbitaceae. *Phytomorphology* **3,** 224–239.

Singh, B. (1964). Development and structure of angiosperm seed. I. *Bull. Nat. Bot. Gard., Lucknow* **89,** 1–115.

Singh, D. (1955). Embryological studies in *Cucumis melo* L. var. *pubescens* Willd. *J. Indian Bot. Soc.* **36,** 72–78.

Singh, D. (1961). Studies on endosperm and development of seeds in Cucurbitaceae and some of its related families. *Agra Univ. J. Res., Sci.* **10,** 117–124.

Singh, D. (1964). Seed structure in the classification of the Cucurbitaceae. *Proc. Int. Bot. Congr., 10th, 1964 Abstract,* pp. 516–517.

Singh, D. (1965). Ovule and seed of *Sechium edule* Sw. – A reinvestigation. *Curr. Sci.* **34,** 696–697.

Singh, D. (1967). Structure and development of seed of the Cucurbitaceae. I. Seeds of *Biswarea* Cogn., *Edgaria* Clarke, and *Herpetospermum* Hook. f. *Proc. Indian Acad. Sci., Sect. B* **65,** 267–274.

Snyder, L. A., Hernandez, A. R., and Warmke, H. E. (1955). The mechanism of apomixis in *Pennisetum ciliare. Bot. Gaz.* **116,** 209–219.

Solntzeva, M. P. (1957). Characteristics of the structure of embryo sacs of *Fragaria grandiflora* and occurrence of polyembryony. *Dokl. Acad. Nauk SSSR* **116,** 181–184.

Souèges, R. (1907). Développement et structure du tegument seminal chez les Solanacées. *Ann. Sci. Nat. Bot.* **6,** 1–24.

Souèges, R. (1913). Recherches sur l'embryogénie des Renonculacées. *Bull. Soc. Bot. Fr.* **60,** 150–157.

Souèges, R. (1935). "La cellule embryonnaire." Paris.

Steffen, K. (1955). Kern- und Nucleolenwachstum bei endomitotischer Polyploidisierung. (Ein Beitrag zur karyologischen Anatomie von *Pedicularis palustris* L.) *Planta* **45,** 379–394.

Steffen, K. (1956). Endomitosen im Endosperm von *Pedicularis palustris* L. *Planta* **47,** 625–652.

Strasburger, E. (1884). "Neue Untersuchungen über den Befruchtungsvorgang bei den Phanerogamen." Jena.

Subramanyam, K. (1944). A contribution to the life history of *Sonerilla wallichii* Benn. *Proc. Indian Acad. Sci., Sect. B* **19,** 115–120.

Subramanyam, K. (1951). Flower structure and seed development in *Isotoma fluviatilis* F.v.M. *Proc. Nat. Inst. Sci. India, Part B* **17,** 275–285.

Subramanyam, K. (1967). Some aspects of the embryology of *Sedum chrysanthum* (Boissier) Raymond-Hamlet with a discussion on its systematic position. Panchanan Maheshwari Mem. Vol. *Phytomorphology* **17,** 240–247.

Swamy, B. G. L. (1942). Female gametophyte and embryogeny in *Cymbidium bicolor* Lindl. *Proc. Indian Acad. Sci., Sect. B* **15,** 194–201.

Swamy, B. G. L. (1943). Gametogenesis and embryogeny of *Eulophia epidendra* Fischer. *Proc. Nat. Inst. Sci. India, Part B* **9,** 59–65.

Swamy, B. G. L. (1946a). The embryology of *Zeuxine sulcata* Lindl. *New Phytol.* **45,** 132–136.

Swamy, B. G. L. (1946b). Embryology of *Habenaria. Proc. Nat. Inst. Sci. India, Part B* **12,** 413–426.

Swamy, B. G. L. (1948a). Agamospermy in *Spiranthes cernua. Lloydia* **11,** 149–162.

Swamy, B. G. L. (1948b). On the post-fertilization development of *Trillium undulatum. Cellule* **52,** 7–14.

Swamy, B. G. L. (1948c). A contribution to the life history of *Casuarina. Proc. Amer. Acad. Arts Sci.* **77,** 1–32.

Swamy, B. G. L. (1949). Embryological studies in the Orchidaceae. III. Embryogeny. *Amer. Midl. Natur.* **41,** 202–232.

Swamy, B. G. L. (1960). Contributions to the embryology of *Cansjera rheedii. Phytomorphology* **10,** 397–409.

Swamy, B. G. L. (1962). The embryo of monocotyledons: A working hypothesis from a new approach. *In* "Plant Embryology—A Symposium," pp. 113–123. Council Sci. Ind. Res., New Delhi.

Swamy, B. G. L. (1964). Macrogametophytic ontogeny in *Schisandra chinensis. J. Indian Bot. Soc.* **43,** 391–396.

Swamy, B. G. L., and Dayanand Rao, J. (1963). The endosperm in *Opilia amentacea* Roxb. *Phytomorphology* **13,** 423–428.

Swamy, B. G. L., and Lakshmanan, K. K. (1962). Contributions to the embryology of the Najadaceae. *J. Indian Bot. Soc.* **41,** 247–267.

Swamy, B. G. L., and Padmanabhan, D. (1962). A reconnaissance of angiosperm embryogenesis. *J. Indian Bot. Soc.* **41,** 422–435.

Swamy, B. G. L., and Parameswaran, N. (1963). The Helobial endosperm. *Biol. Rev.* **38,** 1–50.

Tandon, S. L., and Kapoor, B. M. (1961). Amitosis in the endosperm of *Zephyranthes ajax* Sprenger. *Curr. Sci.* **30,** 72–73.

Tandon, S. L., and Kapoor, B. M. (1962). Contribution to the cytology of endosperm in some angiosperms. I. *Zephyranthes ajax* Sprenger. *Caryologia* **15,** 21–41.

Tandon, S. L., and Kapoor, B. M. (1963). Cytological studies in the endosperm of *Nothoscordum fragrans* Kunth. *Curr. Sci.* **32,** 325–326.

Tiagi, B. (1951). Studies in the family Orobanchaceae. III. A contribution to the embryology of *Orobanche cernua* and *O. aegyptiaca. Phytomorphology* **1,** 158–169.

Tiagi, B. (1956). A contribution to the embryology of *Striga orobanchoides* Benth. and *Striga euphrasioides* Benth. *Bull. Torrey Bot. Club* **83,** 154–171.

Tiagi, B. (1963). Studies in the family Orobanchaceae. IV. Embryology of *Boschniackia himalaica* and *B. tuberosa*, with remarks on the evolution of the family. *Bot. Notis.* **116,** 81–93.

Tiagi, B. (1965). Studies in the family Orobanchaceae. VI. Development of the seed in *Conopholis americana* (L. fil.) Walter. *Acta Bot. Szeged* **11,** 253–261.

Treub, M. (1879). Notes sur l'embryogénie de quelques Orchidées. *Natuurk. Verh. Kon. Akad. Amsterdam* **19,** 1–50.

Tschermak-Woess, E. (1956). Notizen über die Riesenkerne und "Riesenchromosomen" in den Antipoden von *Aconitum. Chromosoma* **8,** 114–134.

Tschermak-Woess, E. (1957). Über Kernstrukturen in den endopolyploiden Antipoden von *Clivia miniata. Chromosoma* **8,** 637–649.

Tschermak-Woess, E. (1959). Endopolyploidie der Narbenpapillen bei *Spironema fragrans. Oesterr. Bot. Z.* **106,** 74–80.

van der Pijl, L. (1955). Sarcotesta, aril, pulpa, and the evolution of the angiosperm fruit. I. and II. *Proc., Kon. Ned. Akad. Wetensch., Ser. C* **58,** 154–161 and 307–312.

van der Pijl, L. (1957). On the arilloids of *Nephelium, Euphorbia, Litchi* and *Aesculus,* and the seeds of Sapindaceae in general. *Acta Bot. Neer.* **6,** 618–641.

Van Went, J. L. (1970). The ultrastructure of the fertilized embryo sac of *Petunia. Acta Bot. Neer.* **19,** 468–480.

Vasil, I. K., and Johri, M. M. (1964). The style, stigma and pollen tube. I. *Phytomorphology* **14,** 352–369.

Veillet-Bartoszewka, M. (1957). La polyembryonie chez le *Primula auricula* L. *Bull. Soc. Bot. Fr.* **104,** 473–475.

Venkata Rao, C. (1953). Contribution to the embryology of Sterculiaceae. V. *J. Indian Bot. Soc.* **32,** 208–238.

Venkata Rao, C. (1954). Embryological studies in Malvaceae. I. Development of gametophytes. *Proc. Nat. Inst. Sci. India, Part B* **20,** 127–150.

Venkata Rao, C. (1959). Contributions to the embryology of Palmae. II. Ceroxylineae. *J. Indian Bot. Soc.* **38**, 46–75.

Venkateswarlu, J. (1952). Contributions to the embryology of Combretaceae. I. *Poivrea coccinea* DC. *Phytomorphology* **2**, 231–240.

Venkateswarlu, J., and Lakshminarayana, L. (1957). A contribution to the embryology of *Hydrocera triflora* W. & A. *Phytomorphology* **7**, 194–203.

Venkateswarlu, J., and Maheswari Devi, H. (1955). Embryological studies in Compositae. II. Helenieae. *Proc. Nat. Inst. Sci. India, Part B* **21**, 149–161.

Vijayaraghavan, M. R. (1962). Studies in the family Ranunculaceae. II. The female gametophyte of *Clematis gauriana* Roxb. *Phytomorphology* **12**, 45–49.

Vijayaraghavan, M. R. (1964). Morphology and embryology of a vesselless dicotyledon – *Sarcandra irvingbaileyi,* and systematic position of the Chloranthaceae. *Phytomorphology* **14**, 429–441.

Vijayaraghavan, M. R. (1965). Morphology and embryology of *Actinidia polygama* Franch. & Sav. and systematic position of the family Actinidiaceae. *Phytomorphology* **15**, 224–235.

Vijayaraghavan, M. R., and Kaur, D. (1966). Morphology and embryology of *Turnera ulmifolia* L. and affinities of the family Turneraceae. *Phytomorphology* **16**, 539–553.

Wadhi, M. (1970). Acanthaceae. *In* "Symposium on Comparative Embryology of Angiosperms," pp. 266–271. Indian Nat. Sci. Acad., New Delhi.

Welk, M., Sr., Millington, W. F., and Rosen, W. G. (1965). Chemotropic activity and the pathway of pollen tube in lily. *Amer. J. Bot.* **52**, 774–781.

Whitehead, R. A., and Chapman, G. P. (1962). Twinning and haploidy in *Cocos nucifera* Linn. *Nature (London)* **195**, 1228–1229.

Withner, C. L. (1959). "The Orchids – A Scientific Survey." New York.

Wunderlich, R. (1938). Ein künstlich bestäubter Fruchtknoten von *Yucca filamentosa.* *Oesterr. Bot. Z.* **87**, 109–113.

Yakovlev, M. S. (1951). On the unity of embryogenesis of the angiosperms and gymnosperms. *Proc. Bot. Inst. Acad. Sci. USSR* **7**, 356–365 (in Russian).

Yakovlev, M. S. (1969). Embryogenesis and some problems of phylogenesis. *Rev. Cytol. Biol. Veg.* **32**, 325–330.

Yakovlev, M. S., and Solntzeva, M. P. (1965). Some problems of flower morphology and embryology of feather-grasses. *In* "Flower Morphology and Reproductive Processes in Angiosperms" (M. S. Yakovlev, ed.), pp. 61–73. Leningrad (in Russian).

Yakovlev, M. S., and Yoffe, M. D. (1957). On some peculiar features in the embryogeny of *Paeonia. Phytomorphology* **7**, 74–82.

Yakovlev, M. S., and Yoffe, M. D. (1961). Further studies of the new type of embryogenesis in the Angiosperms. *Bot. J. Sci. USSR* **46**, 1402–1422.

Yakovlev, M. S., and Yoffe, M. D. (1965). The embryology of the genus *Paeonia* L. *In* "Flower Morphology and Reproductive Processes in Angiosperms" (M. S. Yakovlev, ed.), pp. 140–176. Leningrad (in Russian).

Yamaura, A. (1933). Karyologische und embryologische studien über einige *Bambusa-*Arten. *Bot. Mag. (Tokyo)* **47**, 551–555.

Yamazaki, T. (1954). Notes on *Lindernia, Vandellia, Torenia* and their allied genera in Eastern Asia. II. *Bot. Mag. (Tokyo)* **68**, 14–23.

Yoffe, M. D. (1952). The presence of chlorophyll in the endosperm of cruciferous plants. *Dokl. Akad. Nauk SSSR* **82**, 473–476.

Zinger, N. V., and Poddubnaya-Arnoldi, V. A. (1966). Application of histochemical techniques to the study of embryonic processes in certain orchids. *Phytomorphology* **16**, 111–124.

4

ANATOMICAL MECHANISMS OF SEED DISPERSAL

Abraham Fahn and Ella Werker

I. Introduction

For many plants the dispersal of seeds over wide areas is of great advantage for survival of a species. Seed dispersal helps to prevent competition among seedlings, facilitates utilization of suitable sites and occupation of new localities, and also enables new genotypes to find appropriate environmental conditions. For other plants, however, dispersal of seeds to great distances is not advantageous, e.g., when very specific ecological conditions are required, when the mother plant grows in a niche surrounded by an unfavorable locality such as arid zones or on islands, and when specific agents are required for other stages of the plant's life such as pollination.

However, many plants have no particular mechanisms either for distant dispersal or for prevention of dispersal. Seed dispersal in such plants is purely a matter of chance.

Seed dispersal may be classified according to different criteria: the agent of dispersal, e.g., animals, wind, and water; the structure and constitution of the dispersal apparatus, whether fleshy or dry, dehiscent or indehiscent, or the special structural characteristics adapted for dispersal, such as wings, plume, and mucilage. Any one classification method cannot be wholly satisfactory since all classifications overlap. Moreover, more than one mechanism, each in a different location and not necessarily acting by the same force, may be involved in the dispersal of a seed (polychorous plants). There are heterocarpous plants which produce two kinds of diaspores differing in manner and agent of dispersal. In addition, a certain structure or mechanism can be adapted for dispersal in different and sometimes even antagonistic ways; for example, mucilage may aid distant dispersal in some plants and prevent it in others.

In spite of all these reservations, the classification according to the agent of dispersal (cf. van der Pijl, 1969) will be used in this chapter.

Plant organs involved in dispersal may vary considerably. They may consist of the seed only, of part or the whole fruit, and may include different floral parts or vegetative organs. In extreme cases the whole plant may even constitute the diaspore.

The terminology used here for the various types of fruits is that used by Fahn (1967).

Books dealing with dispersal of plants, e.g., Ulbrich (1928), Ridley (1930), and van der Pijl (1969) refer mainly to the gross morphological structure of diaspores. Morphological characteristics result from anatomical changes. Some adaptations for dispersal, however, are not only expressed in changes in outer morphological appearance, but almost entirely in specific structure and arrangement of the various diaspore tissues. Moreover, some mechanisms of diaspore dispersal are based only on the ultrastructure of cell walls. In related taxa the location of a specialized tissue may have shifted during evolution from one place to another in an organ and from one organ of the diaspore to another. Thus, there is a transfer of function. According to Corner (1958), this transfer seems always to be from the central or apical region toward the basal part. Transfer of function requires a close correlation between the new organ which takes part in dispersal and abscission tissue. At present data on anatomical adaptations of diaspores to various dispersal methods are sparse and scattered in the literature.

In this review we shall describe the major types of anatomical aspects of mechanisms involved in seed dispersal and illustrate them by selected examples.

II. Abscission

Seeds may either remain on the mother plant until the latter wilts or until the seeds themselves, or fruits, or other seed dispersal units are picked by some living creature. More often, however, the seed dispersal unit, i.e., the diaspore, develops a special tissue at its base, termed abscission tissue, which makes possible its separation from the mother plant. This separation, in the case of seeds, may follow dehiscence of the fruit, which also occurs with the aid of abscission tissue (Figs. 1 and 2).

Seeds of dehiscent fruits or the more complicated diaspores may develop further specialized tissues which are adapted to the various modes of dispersal. These will be described later.

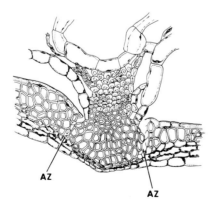

Fig. 1. Cross section of a ripe siliqua at the region of dehiscence of *Aethionema carneum* showing abscission tissue (AZ) with thin-walled cells. (Adapted from Zohary and Fahn, 1950.)

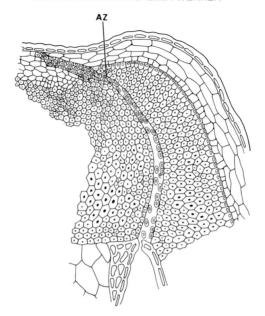

FIG. 2. Cross section of a ripe siliqua at the region of dehiscence of *Leptaleum filifolium* showing abscission tissue (AZ) consisting of cells the walls of which are swollen at this stage. (Adapted from Zohary, 1948.)

The abscission zone is composed of a weak tissue a few layers thick, which may be easily torn. In different species the cells of this weak tissue may be rounded or elongated and thin- or thick-walled. The cell walls may sometimes be lignified or suberized. Often the abscission zone is delineated by an indentation around the organ.

Two main types of abscission are referred to in the literature following Mühldorf (1926) and Pfeiffer (1928): (*1*) disintegration of part or all of the cell wall material as a result of chemical changes (*schizolysis* and *histolysis* according to Pfeiffer) and (*2*) mechanical tearing of the abscission tissue (*rhexolysis* according to Pfeiffer). In the first type either the middle lamellae alone or the middle lamellae together with part or the whole cell wall disintegrate. When the whole cell wall breaks down, naked protoplasts may remain in the mature abscission zone. In the second type of abscission, mechanical tearing of tissue may occur between cells, i.e., along the middle lamellae or across the cells—the cell walls themselves are torn.

A. *Abscission of Diaspores*

McCown (1943) described abscission of the fruit of *Pyrus malus*. In this species, abscission occurs at the base of the pedicel. A constricted zone, already evident in the stage of flowering, persists during the stage of fruit ripening. The cells of all the tissues in this zone, except those of the pith, are smaller than those of the adjacent zones. In the mature pedicel the pith is composed of stone cells, and in the pericycle there are stone cells and fibers. The cortical cells are very thick-walled. At the constriction zone the cells are thick-walled but contain chiefly cellulose. Secondary thickening of the pedicel also occurs. Abscission of mature pedicels is initiated independently in the pith and cortex and is not preceded by cell divisions. The walls of a few tiers of the stone and parenchyma cells of the pith swell and extend. During this process, these walls lose their anisotropic properties, thus indicating a physical change in the cellulose. In the course of cell separation in the pith, dissolution of the pectic compounds of the middle lamellae and disintegration of the primary walls and much of the secondary walls occur. Only the lignified lamellae and thin remnants of the secondary walls persist. Just prior to separation of the swollen cells, there is a change from insoluble to soluble forms of pectic materials. In the separation of cells in tissues other than the pith, less secondary wall disintegration accompanies the dissolution of the middle lamellae and primary wall. Vessels and fibers in the path of separation are ruptured mainly by the weight of the fruit.

Jensen and Valdovinos (1967), using the pedicels of the tobacco and tomato flower at the time of anthesis, demonstrated with the aid of the electron microscope that the groove which delineates the abscission zone extends a small distance into the pedicel with delicate branchings (Figs. 3 and 4). They also found many microbodies with crystalloid cores in the cells of the abscission zone and suggested that these microbodies may have some function in the abscission process.

In *Prunus cerasus,* an abscission zone is formed between the pedicel and the fruit which is composed of small parenchymatous cells. According to Stösser *et al.* (1969a,b), prior to abscission there is a partial breakdown of cell wall constituents, namely pectic materials, noncellulotic polysaccharides, and cellulose. Prior to and/or during the development of the abscission layer, calcium and magnesium were found to move out of the walls. This breakdown is thought to be caused by enzymic action. The degradation of cell wall components then causes cell separation. No abscission layer is formed across the vascular bundles. No apparent abscission layer was observed under normal conditions in the transition zone between fruit and pedicel of sweet cherry (*Prunus avium*). However, in aborting sweet cherry fruits and after application of 2-chloroethylphosphonic acid, similar cell separation and partial dissolution of cell walls occur at the base of the pedicel (Stösser, 1967, 1970; Stösser *et al.,* 1969a).

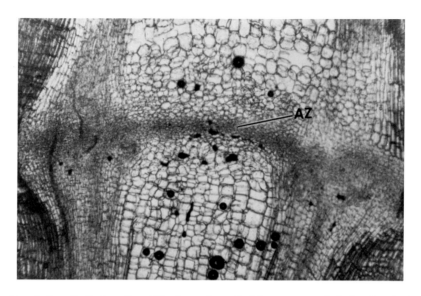

FIG. 3. Longitudinal section of the base of a pedicel from a tomato fruit showing abscission tissue (AZ). Magnification: × 42.

The abscission of the fruit from the cupula in *Corylus* takes place by disintegration of the middle lamellae of parenchymatous cells with many intercellular spaces situated between two sclerenchymatous tissues (Mühldorf, 1926).

Three types of diaspores are known in *Aegilops:* (*1*) the whole spike; (*2*) the spikelet together with part of the rachis; and (*3*) the spikelet itself. The type of diaspore depends on the location of the abscission zone. Markgraf (1925) and Frank (1963) described abscission in different species. According to their descriptions it appears that in most cases abscission is achieved by rhexolysis, i.e., mechanical tearing of cells. In many species of *Aegilops* the only specific anatomical feature observable in the abscission zone is the shortening of the fibers in the periphery of the rachis and of the parenchyma cells in the pith. In the abscission zone of a few species of *Aegilops,* thin-walled and lignified parenchyma cells have been reported to appear below the peripheral fibers. In both types, cell tearing occurs. In still other species, (e.g., *Aegilops squarrosa* L.), a special two-layered zone of small unlignified parenchyma cells crosses both the peripheral fibrous tissue and the pith parenchyma. The cells separate along the middle lamellae of two cell layers in this zone.

The abscission layer that separates the cremocarps (mericarps) in Umbelliferae is not sharply delimited from neighboring tissue. In *Chaero-*

FIG. 4. Electron micrograph of a portion of a longitudinal section of the pedicel of a tomato fruit showing the indentation (i) of the abscission zone. Magnification: × 32,130. (Courtesy of Thomas E. Jensen and Jack G. Valdovinos, Herbert H. Lehman College, Bronx, N.Y.)

phyllum aureum (Mühldorf, 1926), it is composed of round cells loosely arranged with large intercellular spaces which form lacunae. The cells are more densely grouped near the epidermis. The epidermis itself is grooved in this region due to the smaller dimensions of its cells.

Separation at the abscission tissue can occur without the aid of any additional force acting upon the tissue or it may occur only with the aid of such a force. This force can be produced by the neighboring tissue, by weight of the fruit, or it may be external such as by wind and rain. An additional external force or the weight of the diaspore itself aids in the final detachment, especially in those cases in which diaspores remain attached to the plant by strands of vascular tissue. The forces exerted by neighboring tissue may be due either to differential drying of tissues or to a decrease in volume of the fruit coat resulting from overall drying to such an extent that it is less than the volume occupied by the seeds (Holden, 1956) or to turgor pressure. All of these forces occur in dehiscent fruits (see Section VII), and turgor pressure also in diaspore detachment. An example of detachment resulting from turgor pressure is the aggregate fruit of *Rubus* (Reeve, 1954a) the ripening drupelets of which are weakly attached to the receptacle by xylem strands. The surrounding phloem and parenchyma dry up. Detachment of drupelets apparently takes place through suberization of the tissues. In some varieties, enlargement of the conical receptacle, due to both growth and cell turgor pushes the berry upward so that vascular connections of the drupelet are severed.

B.   Abscission of Seeds

Abscission of the seed of *Pisum sativum,* for example, which was described by Mühldorf (1926), takes place in the region of contact between the funicle and the seed where the spongy parenchyma of the funicle is replaced by an abscission tissue built of round cells. As the seed ripens the spongy parenchyma of the funicle dries up and the funicle shrinks, both transversely and longitudinally. The funicle is thus torn from the seed at the zone of the round parenchymatous cells. The seed still remains connected to the funicle in the region of the sclerenchymatous and vascular strand which is torn only when the valves of the fruit are suddenly opened and twisted.

C.   Fruit Dehiscence

There are three varieties of *Linum usitatissimum* (Holden, 1956): (*1*) with distinctly dehiscent capsules; (*2*) with semidehiscent capsules; and (*3*) with indehiscent capsules. In the distinctly dehiscent variety, the capsule dehisces both *septicidally,* i.e., between the carpels, and *loculicidally,* i.e., along the dorsal bundle of each carpel. Septicidal dehiscence occurs through the abscission zones in the fruit wall which consists of parenchymatous cells. These are flanked on either side by fibers. Dehiscence also extends through the median plane of the partitions. In regions where loculicidal dehiscence occurs, no specially differentiated zone of abscission is found. In this case dehiscence involves splitting of a mass of fibers which accompanies the midrib of each carpel through their middle lamellae. Progressing inward, separation and breaking of the weakened parenchyma cells of the false partition occur. The three fruit types differ anatomically in several ways. The greatest dehiscence is associated with a radial elongation, and the least with tangential elongation of the cells of the third layer of the pericarp. In the semidehiscent variety the cells are intermediate in direction of elongation.

Cells in the area of dehiscence contain large amounts of hemicellulose and pectic substances and some cellulose. Prior to dehiscence, there is a softening of the cell walls and of the middle lamellae which results from chemical changes in the hemicellulose and pectic substances. Cells of the second layer of the pericarp (counting from the outside) in the dehiscent variety develop one or two oxalate crystals per cell. Holden suggested that a possible role of oxalic acid in the living cells might be to remove and tie up calcium during periods of growth, thus preventing the calcium from combining with the carboxyl groups of the pectic substances.

Holden describes the final phase of dehiscence:

> the drying-out of the ripened fruit which leads to the opening, or failure of opening, of the valves. The actual tension produced results from an accordion-like action of the cells of the third layer of the pericarp; these expand upon the absorption of water or shrink on loss of water. Changes in cell shape cause, or relieve, a strain on the longitudinally oriented fibers which cap the loculicidal vascular bundle and flank the septicidal dehiscence zone. Because cells of the third layer swell or contract in diameter more than in length, those varieties which have tangentially oriented cells in this layer have less capacity to expand and contract the circumference of the fruit than those which have radially oriented cells. The direction of shrinkage or expansion reflects the architecture of the cell wall of these cells . . . . The tension which results from drying of the capsule in the dehiscent variety is sufficient to cause complete dehiscence and opening of the capsule. It is adequate to cause breaking of the weakened cell walls, but in most cases separation occurs at the middle lamella. In the semidehiscent variety the tension is adequate for complete septicidal, but for only a small amount of loculicidal dehiscence. Tension is so weak in the indehiscent variety that no dehiscence occurs. It is in the last-mentioned capsules that the cutinized epidermis remains intact and thus prevents water absorption, in contrast to those which have had their epidermis ruptured. The result is that seeds of the semidehiscent varieties are more likely to be damaged by water.

Circumscissile dehiscence, i.e., dehiscence by means of a transverse split resulting in the formation of a lid is found in *Portulaca grandiflora,*

Hyoscyamus niger, some species of *Plantago* (Rethke, 1946), and other plants. Separation in *P. grandiflora* is due primarily to the separation of cells at the abscission zone rather than by disruption or disintegration of their walls. In *Hyoscyamus niger* and *Plantago* species, however, separation of cells as well as their rupture occur. The splitting force in all these genera is due to the development of seeds that completely fill the cavity of the ovary and maintain their size while the wall of the fruit shrinks upon drying.

When dehiscence of capsules occurs by means of outwardly flared teeth, it is said to be *valvate.* The epidermal cells of these capsules have very thick outer walls whereas the mesocarp tissue is parenchymatous. Elongated, thick-walled cells may sometimes be present below the epidermis as in *Vaccaria pyramidata* (see Fig. 49). The cells of the inner epidermis may also be thick-walled. In the capsules of certain species of Primulaceae such as *Lysimachia muritiana,* the cells of the inner epidermis have particularly thick walls on the side closest to the mesocarp (von Guttenberg, 1926). The dehiscence of the capsule is brought about by the anisotropic imbibition of the cell walls (see Section VII). The walls that bring about dehiscence in this case are mainly the very thick walls of the outer or inner epidermis. These thick walls determine the direction of the bending of the teeth. The thinner walls of the epidermis, of the sclerenchymatous tissue below the epidermis, or of the sclerenchyma accompanying the vascular bundles usually constitute the resistance tissue. The swelling and, therefore, the shrinkage of these tissues along the axis of bending is relatively restricted. These differences in swelling and shrinkage cause the characteristic opening movements. In this type of capsule, abscission tissue is developed between the teeth.

The opening of the valves of legumes, as described, for example, by Baum (1948) for species of *Wisteria,* takes place along an obvious zone composed of small cells the walls of which thicken before they separate from one another (see also Section VII).

In conclusion, the distinction between *rhexolysis, histolysis,* and *schizolysis,* as suggested by Pfeifer (1928) does not seem to us wholly satisfactory. According to Pfeifer, in the two last abscission types, disintegration of whole cell walls or of the middle lamellae only takes place. However, he also mentions that, in some cases of *rhexolysis,* loosening of cells may be involved. This appears to us to be a more common phenomenon. Therefore it seems that no rigid distinction should be made between these three types of abscission and that there are actually all kinds of transitory states from complete disintegration of walls in the abscission zone to very little disintegration of wall material. When disintegration does not occur, only a strong outer force need be involved in separation.

The problem of abscission is of economic value in agriculture. Prevention of premature abscission of fruit is one of the important problems of fruit growers. The problem is well known in apples, pears, peaches, oranges (Cooper *et al.,* 1968) and tomato (Reinders-Gouwentak and Bing, 1948). Prevention of abscission in cereals and of opening of fruits in sesame (Ashri and Ladijinski, 1964) and flax (Holden, 1956) so that loss of seeds does not occur is another important problem. However, if an abscission layer is not sufficiently well developed, the fruit is too firmly attached to the plant and removal by mechanical harvesting becomes difficult.

Some of these problems can be overcome by spraying with hormones and other chemicals which postpone or facilitate abscission (Cooper *et al.,* 1968) as well as by genetic selection.

III. Dispersion

The diaspores are moved from the mother plant to the place of seed germination either by external agents or by methods originating from the plant itself. The various methods of diaspore transport are classified here as: *zoochory* (dispersal by animals; *chory* from *chorein*—to wander), including endozoochory, epizoochory, and synzoochory; *anemochory* (dispersal by wind), including meteoranemochory (flyers) and chamaechory (rollers); *hydrochory* (dispersal by water); *autochory* (dispersal by the plant itself). All dispersal types based on external agents are classified by Ulbrich (1928) as *allochores* in order to distinguish them from the autochores.

The external agents always promote seed dispersal to great distances *(telechory),* thus enabling a species to spread and occupy new territories. Autochory serves in some species as a method for dispersal to relatively large distances and in others, especially in species adapted to arid zones, ensures the germination of seeds near the mother plant by dispersing the diaspores close to it, generally when water is readily available *(antitelechory).*

IV. Zoochory

The zoochorous plants, i.e., those dispersed by animals, were divided by Sernander (1901, see Nordhagen, 1959; van der Pijl, 1969) into three groups: *(1)* the *endozoochorous* plants, the diaspores of which are eaten by animals and mankind but where the hard seeds or fruit stones pass through the intestinal canal without damage; *(2)* the *epizoochorous* plants, the diaspores of which adhere to the fleece, coat, or feathers of animals or the garments and boots of man but gradually loosen and fall to the ground; (3) the *synzoochorous* plants, the diaspores of which are collected by animals either for making stores before the winter or the dry season comes or for feeding of the young.

In zoochores as in all other groups of telechores it must be remembered that a sharp classification is sometimes very difficult since the same diaspore may be dispersed by different means. The seeds of *Ficus,* for example, are dispersed by animals swallowing the fruits (endozoochory) or biting at the fig and subsequently dropping it or by the figs falling into streams and floating with the current (hydrochory) (Ridley, 1930). Another example of a fruit which is dispersed by several means is the berry of *Viscum album.* It may be carried in the beak, where the viscid seed is immediately deposited when the bird whets its bill on branches (epizoochory or synzoochory). However, most of the seeds are just regurgitated from the gizzard or pass completely through the animal (endozoochory) (van der Pijl, 1969).

A. *Endozoochory*

Diaspores may be eaten by various animals such as snails, fish, reptiles, bats, birds, and rodents. In order to be dispersed in this way the diaspores must have the means to attract the appropriate animals. This can be achieved by different properties such as color, odor, abundance of storage material, and size.

1. COLOR

Color of diaspores is usually the result of pigments in the cells: carotenoids (carotenes and xanthophylls) in the plastids, and flavonoids (anthocyanins and flavones) in the cell sap of vacuoles. These pigments may be dispersed, granular, or crystalline and are found in greater amounts in the outer layers of the diaspore than in the inner ones. The color which usually changes with ripening may be the result of chlorophyll destruction, unmasking other pigments present, or of changes of other pigments or a combination of these factors (Spencer, 1965). The carotene content of red pepper, for example, is more than 30 times greater than that of green pepper. Lycopene (a carotenoid) increases tenfold in tomatoes during ripening. In *Citrus* fruits the flavonoid content increases and the chlorophylls disappear. In oranges the amount of carotenoids also increases, whereas in grapefruits and lemons it decreases. Thomson (1969) observed, with the aid of the electron microscope, the changes occurring in chloroplasts of the epicarp of oranges during ripening and their transfer into chromoplasts. Some varieties of apples obtain their yellow color from carotenes, whereas others become red from anthocyanins. According to Ulbrich (1928), in *Berberis* and *Crataegus* the red color of the fruits results from the presence of carotenes and anthocyanins, whereas dark blue and black colors of fruits of *Ampelopsis, Parthenocissus,* and *Vaccinium* among others are the result of the presence of anthocyanins.

White color is caused by the presence of flavones which may give an ivory white shade or is the result of physical properties. In *Viscum album,* for example, the outer layer of the fruit is translucent, but, in all the other cells, oil drops are present. Refraction of light by these drops causes the white fruit color. In *Symphoricarpos racemosa* the white color results from the many air-filled intercellular spaces (Schoenichen, 1924).

The diaspores may also possess additional special devices enabling them to be seen more easily by animals. This is achieved, for example, by hanging funiculi or by torsion of valves which may be colored from within. In species of *Gahnia* (Cyperaceae), stamen filaments of which the bases remain connected to the fruit base after abscission from the receptacle stretch considerably, thereby exposing the fruits (Fig. 5) (Benl, 1937).

2. ODOR

Upon ripening many diaspores acquire a specific odor which attracts different animals. Sometimes it may be very repulsive for some animals while attracting others. In many fruits the odor is due to essential oils. They may be present in structures typical for the species, as, for example, the oil cavities in *Citrus* and *Psidium* fruits (Fig. 6). Very little is known about adaptations of tissues containing odor-producing substances. Many seeds or fruits have secretory tissues which are also present in vegetative organs, as, for example, the gum ducts in *Pistacia* fruits and the oil ducts of Umbelliferae. It is, however, not clear whether these features assist in seed dispersal or repel animals from eating the fruit.

3. STORAGE MATERIALS

The reserve materials of the diaspores are mainly carbohydrates, such as sugar in grapevines, starch in *Musa* and *Artocarpus* (Haberlandt, 1918), or lipids in *Olea europaea, Persea americana,* and *Durio zibethinus,* or proteins in addition to some other material, e.g. in *Musa, Persea americana,* and *Ficus carica.*

4. SIZE

Fleshy tissue consisting of large, thin-walled parenchyma cells is usually well developed. In many fruits, all or most of these parenchymatous cells contain the nutritive materials, whereas in others only special cells in the fleshy tissue have these materials. In *Prunus nivalis*

FIG. 5. A fruit of *Gahnia aspera* exposed as a result of the stretching of the stamen filaments, the bases of which remain connected to the fruit base while their upper portions
are gripped by the pales. (Adapted from Benl, 1937.)

(Staritsky, 1970), for example, the succulent sugar-containing cells are
arranged around hard clusters of stone cells (Fig. 7). The fleshy diaspores
very often achieve considerable size. The increase in volume is due partly
to an increase in the number of cells in early stages of development,
mainly up to anthesis, and partly to an increase in cell size which occurs
mostly after anthesis (Bain and Robertson, 1951; Nitsch, 1952; Sinnott,
1939; Tukey and Young, 1942; Reeve, 1954a,b). The relative amount of
each of the two factors, cell division and cell enlargement, is, however,
characteristic of the species and sometimes of the horticultural variety
(Havis, 1943). If there is a pronounced change in shape as the ovary enlarges to form the mature fruit, this is usually due to localized cell divi-

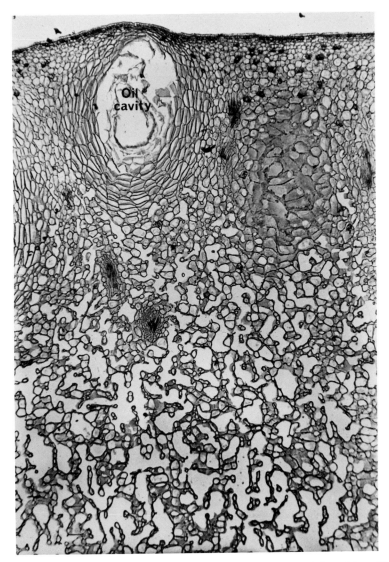

FIG. 6. A portion of the pericarp of *Citrus*. Magnification: × 54. (Adapted from Fahn, 1967.)

sions (Gustafson, 1961). An expansion of intercellular spaces may also contribute to increase in size of the fruit. In apples, the fleshy parenchyma contains 20–30% intercellular spaces ranging up to 2 mm in length (Reeve, 1954a).

An impregnable layer of very hard cells of the endocarp or the seed coat

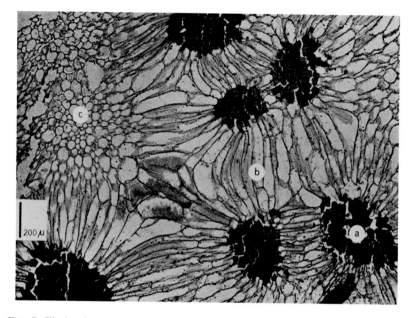

FIG. 7. Fleshy tissue of a ripe pear showing sclereid clusters (a) and elongated paren-chyma cells between two adjacent clusters of sclereids (b). In region c, the elongated parenchyma cells between the clusters are sectioned transversely. (From Staritsky, 1970.)

protects the embryo and endosperm against digestion by animals. Hence, passage through animal intestines not only does not harm the embryo and endosperm, but may even aid germination. It either softens the hard-celled layer, so that penetration of water after the seed has been ex-truded becomes possible, or it removes the soft tissue which very often contains substances inhibitory to germination.

DEVELOPMENT OF FLESHY DIASPORES

The fleshy edible part can develop from different organs of the plant. The following examples are given according to the organ giving rise to the fleshy tissue.

a. PEDUNCLE, BRACTS, AND FRUITS. i. *Ananas* — several flowers grow together to form a multiple fruit.

b. RECEPTACLE. i. *Fragaria* — the diaspore is an aggregate fruit in which the individual fruitlets are achenes, i.e., each is a single-seeded fruit formed by one carpel. The edible portion is composed largely of the receptacle. The principal tissues involved in its development are the cortex and pith. The number of vascular bundles increases with the size of fruit. Cell division in the pith is reduced considerably before that in the cortex and both stop gradually until almost the time of anthesis (Havis, 1943).

ii. *Ficus*—the edible receptacle encloses the achenes.

c. HYPANTHIUM MAINLY, i.e., the basal portion of the perianth and the stamens are fused and adnated to the carpels (an inferior ovary). The receptacle participates in only a small portion of the basal part of the fruit. i. *Pyrus malus*—the outer parenchyma of the fruit develops from the hypanthium. The epidermis is covered by a cuticle. The subepidermal tissue which develops from the outer portion of the hypanthium consists of a several-layered, thick-walled collenchyma tissue the cells of which are tangentially elongated. Intercellular spaces develop in this tissue only shortly before fruit maturation, and they are best developed in the more internal ground parenchyma. In still deeper layers the cells are more-or-less oval and their longer axis usually is radially oriented. The part of the fruit that develops from the ovary is formed by the five folded, but unfused, carpels. Five dorsal carpel bundles are found on the outer side of the locules and ten ventral bundles in the center. The dorsal and ventral bundles are interconnected by branches. The ovary wall develops into a parenchymatous exocarp and a cartilaginous endocarp which lines the locules. The endocarp consists of elongated sclereids with very thick walls which almost completely obliterate the cell lumen (Fig. 8) (MacArthur and Wetmore, 1939, 1941; MacDaniels, 1940; Smith, 1940, 1950; Tukey and Young, 1942).

Other examples of this type are *Cydonia oblonga* and *Pyrus communis*.

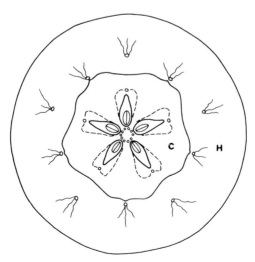

FIG. 8. Schematic drawing of a cross section of an apple fruit showing the fleshy tissue produced by the carpel (C) and hypanthium (H). (Adapted from MacDaniels, 1940.)

d. CARPELS MAINLY. i. *Rubus* (Reeve, 1954a,b) — an aggregate fruit, i.e., the gynoecium is apocarpous, but in the course of ripening the individual fruits of a flower aggregate to form a single fruit. The fleshy tissue of the drupelets develops from the mesocarp by cell elongation. Cells of the endocarp develop into sclereids. The drupelets of the ripe berry cling together by means of hair entanglement. The hairs arise from some of the epidermal cells which elongate to form unicellular, thin-walled trichomes.

ii. *Lycopersicum esculentum* — the fruit consists of a pericarp and placental tissue. The exocarp consists of an epidermis and three or four layers of collenchyma cells. The mesocarp consists of a thick layer of large thin-walled cells which enclose many intercellular spaces. During fruit ripening, some of the cells of the inner and central portion of the carpels may disintegrate. With development of the ovules, after pollination, the parenchymatous tissue of the placenta grows around the funiculi until it completely encloses the developing seeds. The cells of this tissue are thin-walled and they form a homogeneous tissue; they do not fuse with the pericarp but they adhere to it as well as to the seeds. At first this parenchymatous tissue is firm, but as the fruit ripens the cell walls become thinner and the cells are partly destroyed (Fig. 9) (Hayward, 1938).

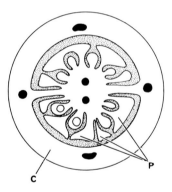

FIG. 9. Schematic drawing of a cross section of a developing tomato fruit showing the enlargement of the parenchymatous tissue of the placenta (P). This tissue forms the juicy tissue of the fruit. C — carpel wall. (Adapted from Hayward, 1938.)

iii. *Citrus* — the fruit develops from a syncarpous gynoecium with axile placentation. The whole pericarp becomes fleshy: the exocarp (flavedo) consists of small, dense collenchyma cells which contain chromoplasts and in which essential oil cavities are present; the mesocarp (albedo) consists of loosely connected, colorless cells with many air spaces; the endocarp is relatively thin and consists of very elongated, thick-walled cells which form a compact tissue. Stalked, spindle-shaped juice vesicles, which fill the locules when the fruit ripens, develop from the cells of the inner epidermis and subepidermal layers. Each juice vesicle is covered externally by a layer of elongated cells which enclose very large, extremely thin-walled juice cells (Figs. 10 and 11) (Schoenichen, 1924; Ford, 1942; Scott and Baker, 1947; Bonerji, 1954; Hartl, 1957).

iv. *Viscum album* — endocarp and mesocarp fleshy, part of mesocarp slimy.

FIG. 10. Juice vesicle from a *Citrus* fruit. (Adapted from Fahn, 1967.)

e. Seed coat. i. *Ephedra* and *Gnetum* species — the outer layer of the seed coat is composed of fleshy parenchymatous tissue; the middle layer is mainly sclerenchymatous (Rodin and Kapil, 1969).

ii. *Punica granatum* — the juicy outer layer develops from the epidermis by elongation of its cells. The sap of these cells develops turgor pressure which preserves the characteristic external shape of these seeds (Fig. 12).

Except for those fruits which are eaten for their flesh, there are many small, hard fruits which are eaten by grazing animals, e.g., *Urtica, Atriplex, Chenopodium, Polygonum, Rumex* and many Gramineae. These pass through the animal intestines with no apparent change.

B. *Epizoochory*

Diaspores can be carried on animals in three ways: (*1*) by hooklike spines which help them to cling to fur, feathers, or garments; (*2*) by sticky

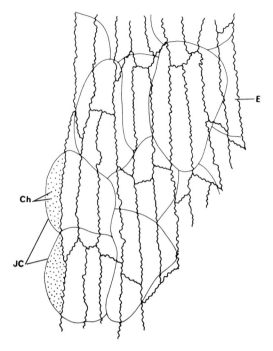

F<small>IG</small>. 11. Portion of a *Citrus* vesicle as seen under the microscope. E—epidermal cell; JC—juice cells; Ch—chromoplasts. (Adapted from Schoenichen, 1924.)

substances present on the diaspore which help them to adhere to birds or other animals; (*3*) by falling into the mud and then clinging along with the mud to the feet of animals. The first two types of diaspores are adaptive and possess appropriate devices with which they can be carried by the animal. The third is passive and demands no special means for dispersal. It has, therefore, no importance for dispersal from the anatomical point of view, although it is a very common method of dispersal.

The hooklike spines by which the diaspore clings to animals may be uni- or multicellular. They can be either trichomes or emergences or may even constitute a whole organ. The simplest type is that of an epidermal cell that develops into a hard, thick-walled, hooklike trichome, e.g., on fruits of *Galium* species, on fruits of *Asperula odorata,* on bracts which surround the fruit of *Parietaria officinalis* and on glumes, paleas, or spikelet axes of many Gramineae (Schoenichen, 1924; von Guttenberg, 1926).

Epidermal cells surrounding the widened base of such a trichome may often be specifically arranged. They may be in a rosette or they may form an elevated base under the trichome, as in the pales of *Lappago racemosa* (Haberlandt, 1918). When inner cells also take part in the construction of

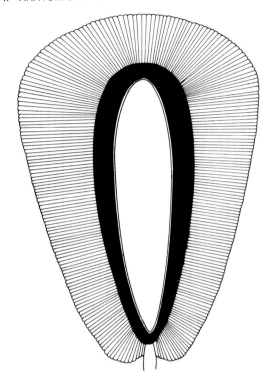

FIG. 12. Schematic drawing of a longitudinal section of a seed of *Punica granatum* showing the radially elongated cells of the outer epidermis of the testa; these cells form the fleshy part of the seed. (Adapted from Eames and MacDaniels, 1947.)

such a base, an emergence is formed. The emergence can terminate with one hooklike cell as, for example, the spines of the fruits of *Ranunculus arvensis* (Fig. 13) and *Caucalis daucoides* (Umbelliferae) (Fig. 14). More than one hooked cell can be situated on one emergence as, for example, on the fruits of *Acaena, Lappula echinata* (von Guttenberg, 1926), and on mericarps of different Umbelliferae (Heywood, 1968).

In *Echinaria capitata* the awn itself is bent in a hooklike form. In *Geum* species the style is persistent and is curved below the stigma. The hook thus formed becomes sharp and hard after the stigma and upper portion of the style are detached (Figs. 15 and 16) (von Guttenberg, 1926).

Hooklike spines rarely appear on seeds. The size of spines, their number and mutual arrangement, the direction of the hook, and the combination of different types of spines on the same disaspore determine the efficiency of this device for dispersal. Carlquist (1966), for example,

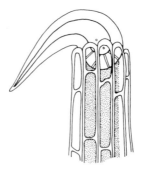

FIG. 13. Distal portion of a spine of a nutlet of *Ranunculus arvensis* showing the pointed, curved cell at its apex by which it becomes attached to the dispersal agents. (Adapted from Fahn, 1967.)

FIG. 14. A scanning electron micrograph of a part of a mericarp surface of *Caucalis platycarpos* showing prominent spines. Magnification: × 75. (From Heywood, 1968.)

examined diaspores of the genus *Bidens* which probably are dispersed by sea birds. He found that different species have different adaptabilities regarding this agent of dispersal (Fig. 17).

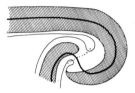

FIG. 15. Schematic drawing of a style of *Geum hirsutum* at the place of its curvature. The sclerified tissue is hatched; the separation line is dotted; the vascular bundle of the style is represented by a thick line. (Adapted from von Guttenberg, 1926.)

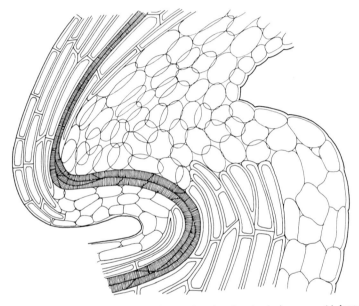

FIG. 16. Detailed drawing of part of Fig. 15 showing the abscission zone. (Adapted from von Guttenberg, 1926.)

It should be mentioned that the hairs or bristles of many seeds that serve as flying apparatus may serve also as an epizoochorous apparatus and cling to feathers or fur of animals, since there are hooks upon the hairs and bristles. Hair or bristles, however, can also serve to protect fruits against animals. Sometimes, the same spines that protect the unripe fruit may assist in its dispersal when ripe, as in some *Medicago* species (McLean and Ivimey-Cook, 1964).

Mucilage can function in seed dispersal in several different or even antagonistic ways, i.e., it can either cause a diaspore to adhere to animals and be further dispersed (epizoochory) or it can cause it to cling to the ground so that further dispersal is prevented (antitelechory; see Sections VI and VII).

FIG. 17. A cypsela of *Bidens pilosa*. (Adapted from Carlquist, 1966.)

The viscid substance of the diaspore can be exuded by glandular hairs or emergences. This is the case in *Psoralea glandulosa* where the sticky hairs are on the legume, in *Siegesbeckia orientalis* (Compositae) where they are on the involucral bracts, in *Verbena officinalis, Salvia glutinosa,* and *Plumbago* where they are on the persistent calyx (Fig. 18), or in species of *Cerastium* and in *Saxifraga tridactylites* where they are scattered all over the plant. The viscid substance, however, may constitute

FIG. 18. Portion of a calyx of *Plumbago capensis* showing emergences secreting viscid substances (V). Magnification: × 34.

a part of the cell wall and become viscous only upon wetting. Cells of which the walls possess layers that become slimy may be situated in different regions of the diaspore which is usually the seed or the fruit. Cells with a mucilaginous layer occur, for example, in the mesocarp of Loranthaceae fruits. This layer is either outside the region of the corolla bundles (Loranthoideae) or inside (Viscoideae) (Danser, 1931). In the fruit (pseudocarp) of *Viscum album* the walls of these cells are constructed of an outer layer of cellulose mucilage and an inner layer of pectic mucilage (see Section VII). In *Loranthus europaeus* the walls are composed of a pectic mucilage layer only (Tomann, 1906; see von Tubeuf, 1923). As already mentioned, these fruits are either swallowed by birds and then the seed, surrounded by mucilage, is excreted usually on a tree (endozoochory) or the seed sticks to the bird's beak when it bites or sucks at the fruit and is wiped off later onto a new host.

Here, as in all other means of dispersal, adhesion to animals may be a secondary device following a primary one. It may follow, for example, autochory as in *Ecballium elaterium,* some *Oxalis* species, and some Acanthaceae (Ulbrich, 1928; van der Pijl, 1969).

C. Synzoochory

Synzoochores are diaspores containing reserve substances which are collected and stored by animals. Of the different types of synzoochores, the ones that are interesting from the anatomical point of view are those in which the fleshy tissue which attracts animals is localized in a specific region of the diaspore and is easily detached from the remainder which is hard and inedible. Such a tissue when adapted to dispersal by ants (*Myrmecochory*) is termed *elaiosome*. When the diaspore is a seed the elaiosome is formed by an aril. An aril may arise from several different parts in the seed such as the base of the ovule, the integuments at the micropylar region, and the funiculus. Ants collect the diaspores because of the edible elaiosome. Most of the diaspores are carried by the ants to their nest and there the elaiosomes are gnawed off. The elaiosome content is sucked from some of the diaspores or the elaiosome is bitten loose on the way to the nest (Berg, 1954). The rest of the seed, or the whole seed in case of the bigger diaspore, is unharmed due to a hard and often smooth testa. The elaiosome is usually white or yellow, differing in color from the rest of the diaspore which is darker in general. It contains drops of oily substances. According to Bresinsky (1963) the substance that attracts the ants is probably the unsaturated fatty acid, ricinolic acid. The attracting ability of the elaiosome is not equally strong in all plants (cf. Berg, 1959). Ants that are attracted to the elaiosome consume the storage material which may be protein, lipoid, starch, and vitamins (van der Pijl,

1969). The elaiosome often is separated from the rest of the seed by a special, thick-walled or crystal-containing tissue (Bresinsky, 1963).

As mentioned, the origin of elaiosomes may be from different organs or tissues occurring in the seed or fruit. According to Sernander (1906) the growth of these elaiosomes occurs usually through cell enlargement rather than cell multiplication. There are exceptions, however, in which the appendage originates from a specially formed meristem (Berg, 1958, 1966; Bresinsky, 1963).

In the simplest cases the whole or part of the seed coat is fleshy, and the edible material is in a diffuse form in this tissue (Sernander, 1906, 1927, *fide* Nordhagen, 1932; Bresinsky, 1963). In species of *Cyclamen,* described by Nordhagen (1932), which may serve as an example for this type, the whole exotesta provides the attractive substances for ants and serves as nourishing tissue. It is composed of hairlike cells filled with oil which also impregnates their walls. According to Sernander there is oil in the endosperm too, which is extruded from its cells to the exotesta. Underneath the exotesta there is a layer of crystal-containing cells. Other examples for this type are species of *Puschkinia* and *Ornithogalum, Allium ursinum,* and *Myrmecodia* (van der Pijl, 1969).

In most cases, however, the elaiosome is a real appendage. In seeds it is very often a *caruncle,* i.e., an outgrowth of the micropyle. In *Chelidonium majus* (Szemes, 1943), the elaiosome is composed mainly of two types of cells—basal, small cells, and outer very elongated cells. In both types of cells, but more so in the outer ones, lipids and small amounts of starch are present (Fig. 19). Inner cells also contribute to construction

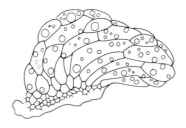

FIG. 19. An elaiosome of *Chelidonium majus* seed. (Adapted from Szemes, 1943.)

of the base of the elaiosome but they lack storage substances. Other plants having this type of elaiosome include *Scilla bifolia, Galanthus nivalis,* species of *Euphorbia, Viola, Primula, Anemone,* and *Ranunculus* (van der Pijl, 1969).

Elaiosomes that develop as *strophioles,* i.e., outgrowths of the funiculus and/or the seed itself are found, for example, in *Stylophorum diphyllum*

(Nordhagen, 1959), *Trillium* (Berg, 1958), and *Scoliopus* (Berg, 1959). In these samples the strophiole is produced by the raphe. In *Stylophorum diphyllum* (Papaveraceae) the elaiosome forms a crest (Figs. 20 and 21)

FIG. 20. A seed of *Stylophorum diphyllum* with a crest-shaped elaiosome. (Adapted from Nordhagen, 1959.)

FIG. 21. Enlarged portion of Fig. 20 showing the seed surface and crest. (Adapted from Nordhagen, 1959.)

consisting of three to five rows of long protuberances, each protuberance consisting of a large terminal cell and a smaller basal one both containing many drops of lipid (Nordhagen, 1959). In *Trillium* (Liliaceae) the elaiosome arises at the time of fertilization from the upper part of the raphe (Berg, 1958). Except for its surface layer, it is of subepidermal origin. This is an example where an elaiosome, formed by cell divisions, becomes a continuous ring-shaped meristematic zone surrounding the vascular strand (Fig. 22). The cells are filled with oil drops and also great quantities of starch grains. In *Scoliopus* (Liliaceae) the elaiosome is produced by the raphe (Berg, 1959) and it forms a ventral protuberance along the seed which extends from near the micropylar almost across the chalazal end. It is formed by considerable enlargement of the epidermal cells which reach a length of 800 μ. The large epidermal cells and the smaller ones inner to these contain oily drops.

Occasionally more than one tissue is involved in development of the elaiosome as in *Dendromecon,* a tree of the Papaveraceae (Berg, 1966). The seed has a necklacelike appendage of the hilum which develops mostly from the micropylar portion of the outer integument but partly also from the adjacent part of the raphe (Figs. 23 to 25).

Of different origin is the elaiosome of *Pedicularis silvatica* (Scrophulariaceae) which develops from the micropylar endosperm haustorium (Berg,

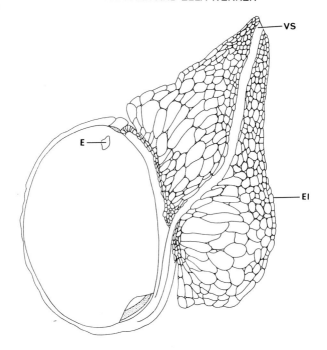

FIG. 22. Longitudinal section of a developing seed of *Trillium ovatum*. El – elaiosome; VS – vascular strand of the funicle; E – developing embryo. (Adapted from Berg, 1958.)

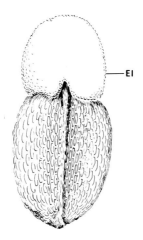

FIG. 23. Seed of *Dendromecon rigida*. El – elaiosome. (Adapted from Berg, 1966.)

1954). Its development occurs from the inner part of the seed. It contains no inner walls and represents a single gigantic cell, thus being unique among the other described elaiosomes. This elaiosome cell is filled with a

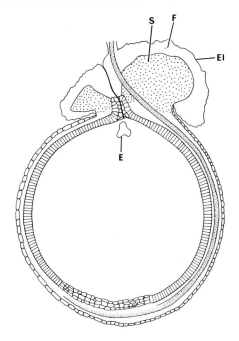

FIG. 24. Schematic drawing of a longitudinal section of the seed of *Dendromecon rigida*. Elaiosome (El) showing two zones one containing fat cells (F) and the other starch cells (S). E—embryo. (Adapted from Berg, 1966.)

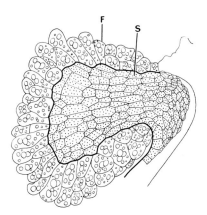

FIG. 25. Part of an elaiosome of *Dendromecon rigida*. F—fat cells; S—starch cells. (Adapted from Berg, 1966.)

branched cellulose network indicating the course of the cytoplasm strands. It contains starch and a considerable amount of lipids most of which are stored near the cell surface. The elaiosome is enveloped by more-or-less collapsed integument cells usually containing starch and traces of lipids. Elaiosomes that develop from endosperm are found in two other genera of the Scrophulariaceae—*Melampyrum* and *Lathraea* (Bresinsky, 1963).

Elaiosomes also develop on larger diaspores than seeds. In some *Anemone* species and in *Lamium* and *Ballota,* the elaiosome appears on the basal part of the fruit just above the receptacle. It consists of the exocarp, rarely also of the mesocarp (Figs. 26 and 27). In the Boragina-

Fɪɢ. 26. Longitudinal section of the lower part of *Ballota nigra* nutlet. Dotted cells are of the elaiosome. (Adapted from Bresinsky, 1963.)

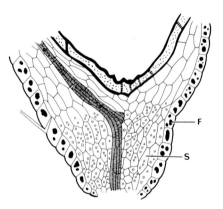

Fɪɢ. 27. Longitudinal section of the lower part, the elaiosome, of an *Anemone nemorosa* achene. F—fat cells; S—starch cells. (Adapted from Bresinsky, 1963.)

ceae and in some genera of Labiatae (e.g., *Rosmarinus* and *Ajuga*) the elaiosome develops in the region of the receptacle, apparently from the basal wall of the carpels (Fig. 28) (Bresinsky, 1963). In *Melica* (Gramineae) the elaiosome develops on the base of the spikelet. In species of *Centaurea* and some other Compositae genera, elaiosomes develop at the base of the cypselae (fruits). Here the development of the elaiosome is accompanied by loss of the pappus (van der Pijl, 1969).

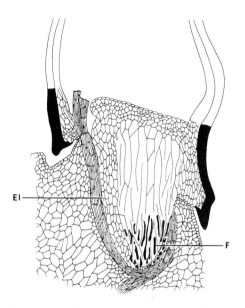

FIG. 28. Longitudinal section of the lower part of a nutlet of *Symphytum*. El — elaiosome; F — fat. (Adapted from Bresinsky, 1963.)

Shortly after seed ripening, most elaiosomes shrink, collapse, and lose most of their attractiveness to ants. It is important, therefore, that quick desiccation be prevented or that seeds can be reached by ants soon after their ripening. This is accomplished in several ways. Many species have very weak supporting tissue in their stems which results in the plant being prostrate. The peduncle may either fall over and, when the fruit opens, the seeds fall straight onto the ground (*Trillium;* Berg, 1958) or some-times may even roll (*Cyclamen;* Nordhagen, 1932). Thus both mecha-nisms bring the seeds to the ground protecting them from desiccation.

Myrmecochory may follow dispersal by other ways, e.g., autochorous explosive apparatus in *Impatiens* (see Section VII), *Cyclanthera,* and *Acanthus.* Certain *Polygala* species with elaiosomes are at first dispersed by wind (Sernander, 1906, see Nordhagen, 1959). These combinations probably derive, according to Sernander, from elaiosomes being a most advanced and recent type of dispersal device. They may however also indicate a transition from zoochory to other ways of dispersal, as myr-mecochory may have developed in many cases from other forms of zoochory.

The majority of myrmecochores are mesophytic plants of temperate forests. Nevertheless, a xerophytic group exists in the Mediterranean countries and the Californian chaparral (Nordhagen, 1959). There are also some hydrophytes, such as *Pedicularis silvatica* (Berg, 1954), in which myrmecochory occurs.

V. Anemochory

Wind is the most active of all agents in dispersal of seeds. Dispersal by wind can be achieved by active transport of diaspores by wind or by slowing down the falling of diaspores by air resistance and thus enabling them to move further away from the mother plant. Aerodynamics of anemochorous diaspores, which depends on the relationship between surface area and weight, is discussed in Dingler's book which appeared in 1889. Various anatomical and morphological adaptations of diaspores to the anemochorous mode of dispersal as well as other adaptations which expose these diaspores to the wind, exist in the plant kingdom. The means of exposing diaspores to the wind may involve development of an ab-scission zone at the base of a diaspore or opening of fruits. Some spe-cialized devices for exposure are found in many epiphytic orchids. In the fruits of these plants there are elaterlike hygroscopic hairs which gradually push the seed out of the fruit (Fig. 29) (van der Pijl, 1969). The capitulum of the Compositae may hide or expose fruits by movement of involucre bracts in accordance with the amount of moisture (see Section VII). In these two examples there is a combination of autochory and anemochory in successive stages.

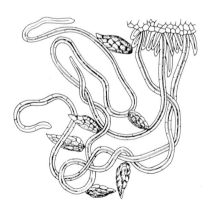

FIG. 29. Seeds with placental elaters of *Anguloa roezli*. (Adapted from van der Pijl, 1969.)

The tissues which enable the diaspores to remain a relatively long period in the air and be carried by wind may be of various origins, e.g., the testa, the fruit coat, calyx, bracts, hypsophylls, and even the whole plant. Here too, the principle of transfer of function can be observed in allied taxa. In some *Ricotia* species (Cruciferae) the pod is dehiscent and may contain winged seeds and in others it may be indehiscent and winged. In *Sterculia alata* the seed is winged, and in the allied *Tarrietia* the one-seeded pod is similarly winged; in *Rheum* (Polygonaceae) the fruit is winged along three angles, but in *Rumex* the wings are produced by the persistent sepals (Corner, 1958). In the Dioscoraceae there are genera in which there is a transfer of the wing from the seed to the capsule which then becomes a samara (Burkill, 1961).

There are many variable structures adapted for anemochory. The classification of the types of plants according to these structures, as given by van der Pijl (1969) is followed here.

A. *Meteoranemochores (Flyers)*

1. DUST DIASPORES

Small dimensions and light weight of seeds provide for ready dispersal by air currents. This may be achieved by an undeveloped embryo and a small amount of reserve material. The seeds having these qualities are the dust diaspores. Such seeds appear, for example, in *Gypsophila, Pyrola, Sempervivum, Sedum, Campanula, Rhododendron, Digitalis, Thymus,* and the Orobanchaceae. The weight of many of these individual seeds does not exceed 0.003–0.004 mg (Ulbrich, 1928) and, in *Orobanche ionantha* (Fig. 30), seed weight is only about 0.0001 mg (von Guttenberg, 1926). These seeds may reach great distances and considerable heights.

2. BALLOONS

In addition to minute seed size, the seed or part of it or a larger diaspore

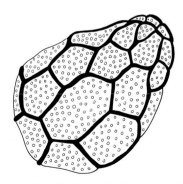

FIG. 30. Seed of *Orobanche*. Magnification: × 210. (Adapted from Schoenichen, 1924.)

may be surrounded by a balloonlike sack of which the walls are one or a few layers thick. They may also be surrounded by very large empty cells or by loose tissue with very large intercellular spaces all of which are filled with air. A combination of these features can also occur. This causes the specific weight of the diaspore to diminish and the surface area (*Angriffsfläche* as called by von Guttenberg) to increase.

The cells of the outer layer usually are partly thickened so that collapse of the cells with drying of the seed is prevented. An example of the loose testa formed by the epidermis of the outer integument which surrounds the embryo is found in the Orchidaceae (Fig. 31). Here the anticlinal cell

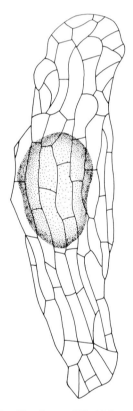

FIG. 31. Seed of *Orchis*. Magnification: × 210. (Adapted from Schoenichen, 1924.)

walls usually are thicker than the periclinal walls. On the latter, discontinuous thickenings may appear (Clifford and Smith, 1969). A loose testa is also found in the Saxifragaceae, Burmanniaceae, Sarraceniaceae, Droseraceae, Nepenthaceae, Pyrolaceae, and Ericaceae. The loose enveloping of the outer layer of cells may occur only at a few regions. In the fruit of *Astrantia major,* for example, five vesiclelike cavities are present (Schoenichen, 1924).

Large balloonlike legumes are found in some Leguminosae, e.g., *Colutea, Crotalaria, Cicer,* and *Medicago.* An inflated calyx carrying a small fruit is found in some species of *Trifolium, Astragalus, Salvia, Physalis,* etc. An inflated corolla persisting after fruit ripening may also occur in some species of *Trifolium* and in *Callipeltis cucullaria.* Balloonlike structures formed by hypsophylls are found, for example, in species of *Ostrya.* The same can be formed by glumes or paleas as in *Briza* and *Melica.*

Enlargement of cells which become filled with air occurs, for example, in the epidermis of seeds of *Cuscuta.* Of the same type are *Nemophila* and *Codon* of the Hydrophyllaceae (von Guttenberg, 1926).

A loose tissue with large air-filled intercellular spaces is found in the pericarp of *Atriplex inflata,* some species of *Chenopodium,* and in *Valerianella.* A similar kind of tissue is found in the bracts that form a part of the diaspore in species of *Atriplex* and *Suaeda* (von Guttenberg, 1926).

3. PLUMED DIASPORES

These are diaspores equipped with trichomes. The trichomes are often one-celled, although many-celled and branched trichomes of intricate form and origin may also occur. In some cases only one hair may be present. In species of *Aeschynanthus (Trichosporum),* Gesneriaceae, the seed at the chalazal end is elongated into an emergence bearing a single hair. The seed of these species is 1–2 mm long whereas the hair is up to 3 cm long (von Guttenberg, 1926). In *Dichrotrichum, Agalmyla, Laiphaimos* (Fig. 32), two hairs are formed by elongation of the integument into a hair on one side and the long funiculus forming a hair on the opposite side (von Guttenberg, 1926; Ulbrich, 1928; van der Pijl, 1969).

Hairs may cover a diaspore over its entire surface. This is the case, for example, in seeds of *Gossypium* and some *Ipomoea* species. Hairs of *Gossypium* are one-celled outgrowths of the epidermis of the seed coat. The hairs that grow on the narrower side of the seed are a few millimeters long, whereas most hairs, those which grow on the wide sides of the seed, are 2–4 cm or more long and 12–42 μ wide. The walls of the hair cells are unlignified but are covered with a cuticle that prevents moistening.

The hair may be located in specialized regions and, hence, be of different origin. In Asclepiadeceae and Apocynaceae, a tuft develops from the outer integument at the micropylar side. According to van der Pijl (1969), it is apparently a split wing, whereas in the Bromeliaceae (*Tillandsia*) the tuft may have developed from an aril. In Compositae there

FIG. 32. Seed of *Leiphaimos azurens* (Gentianaceae). (Adapted from Ulbrich, 1928.)

is a calyx pappus the hair of which may develop into intricate "feathers" or bristles (Figs. 33 and 34). A tuft of hairs which are outgrowths of the funiculus (and the placenta) appears in *Salix* and *Populus*. A persistent style which bears hair is present on the nutlets in some Ranunculaceae and Rosaceae. A crown of hairs on the narrow margins of the flat seeds appears in *Heliosperma quadrifidum* (von Guttenberg, 1926). All of these examples represent some of the variations of hair appearing on diaspores.

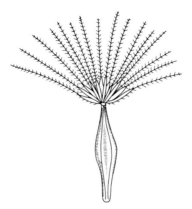

FIG. 33. Cypsela of a Compositae showing the feathery pappus. (Adapted from Fahn, 1967).

FIG. 34. Part of a pappus ray of *Taraxacum officinale*. (Adapted from Schoenichen, 1924.)

In many species the hairs are capable of movement, for example, they often spread when dried and converge when wet. Such hairs appear in *Salix* and *Populus,* in many Apocynaceae, Asclepiadaceae, Tamaricaceae, etc. The hygroscopic mechanism is described in Section VII,B.

Hair may sometimes take part in other methods of dispersal, as in epizoochory or hydrochory. In seeds of *Catopsis nutans* (Bromeliaceae), each hair is sharp and hooked at its tip. The hairs serve as a device for flying, but they also enable the seed to anchor on smooth and thin branches. The same is true for *Phygmatidium,* an epiphyte of the Orchidaceae (Müller, 1895, see Haberlandt, 1918).

4. Winged Diaspores

The shape of the wings of a diaspore, either one-sided or wholly surrounding the diaspore, flattened in one plane or curved, single or several in number, determines how it is carried by the wind. The diaspore can glide or be dynamically propelled (Fig. 35). The origin of wings also

Fig. 35. Fruit of *Ailanthus altissima.*

varies. The wing-bearing diaspores usually are large and can be 15 cm or more in length.

The wing is anatomically built so that it possesses a large surface area, a relatively small weight, and is firm. These features are acquired in different ways, depending on the organ from which they are formed.

When the wing is a part of the seed, it develops from the outer integument. It, therefore, often has no vascular bundles unless those of the funicle are included. The wing may be one- or multilayered. Its strength is derived from the various types of wall thickenings. The wing may consist of, or include, a layer of tubelike cells with some of the walls thickened or with annular, spiral, or reticulate wall thickenings, e.g., in some Scrophulariaceae, Bignoniaceae (Fig. 36), and Rubiaceae. In *Zanonia macrocarpa* of the Cucurbitaceae the large wing has a particular type of cell arrangement which provides both strength and large air-filled spaces (Fig. 37) (Haberlandt, 1918). All wings are covered by a cuticle which helps to avoid moistening.

The anatomy of wings of larger diaspores is in most cases more complicated, depending on the organ from which they are formed. Wings are usually strengthened by strands of sclerenchymatous and vascular bundles. A wing of simple structure is that which is attached to a pine seed. It is built, in its thinner portions, of the upper, thick-walled epidermis of the cone scales; at its thicker portions, it also includes some thick-walled subepidermal layers (Haberlandt, 1918). Wings may develop as emergences of the pericarp, peduncle, calyx, bracts, and other types of leaves. The outer fruit of *Calendula officinalis* is an example of wings that develop from the wall of a false fruit (Fig. 38). *Salsola kali* (Stevens, 1943) and *Holmskioldia* exhibit a winged calyx. A modified calyx is also found in *Molucella* and a modified involucel in *Scabiosa.* A modified bract bearing a peltate wing is found, e.g., in *Lecocarpus* (Compositae) (Carlquist, 1966).

FIG. 36. Portion of a seed wing of *Stenolobium alatum* showing cells with wall thickenings (wholemount). Magnification: × 140.

Here, too, the principle of transference of function can be observed. In Dioscoraceae, for example, most genera have winged seeds. In two genera, *Avetra* and *Rajania,* however, there is a transfer of the wing to the capsule wall (Burkill, 1961).

B. Chamaechores (Rollers)

In plants with this type of dispersion the diaspore consists of whole plants, of the whole epigeal part of plants, or of smaller portions to which the seeds are still attached. The diaspore, when dried, becomes globular

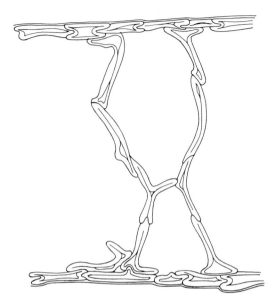

Fig. 37. Cross section of thicker part of the seed wing of *Zanonia macrocarpa*. (Adapted from Haberlandt, 1918.)

in shape and so it can be rolled by the wind. This manner of dispersal is common in steppes, deserts, and prairies (Ridley, 1930).

From the anatomical point of view the plant portions that roll must become weakly attached to the remainder of the plant when dry. The fruits or seeds must also be weakly attached so that they may be detached from the plant while rolling. In the branches there must be different capacity for shrinkage between their inner and outer sides. *Gundelia tournefortii, Salsola kali, Prangos goniocarpa, Ferula biverticillata,* and *Allium schuberti* are examples of tumbleweeds.

The seeds or larger diaspores of chamaechores, which are detached from the plant only upon rolling, may themselves be equipped with some dispersal device, such as the winged diaspores of *Salsola kali* (Stevens, 1943).

C. Parasite–Host Diaspores

Coordination between parasite and host in seed dispersal may occur, an example of which is given by Atsatt (1965). The self-incompatible annual herb *Orthocarpus densiflorus* (Scrophulariaceae) is an obligate parasite of other flowering plants. One of these plants is *Hypochoeris glabra* of the Compositae, with which it came in contact only less than a hundred years ago. The seed of *Orthocarpus* is surrounded by a rigid

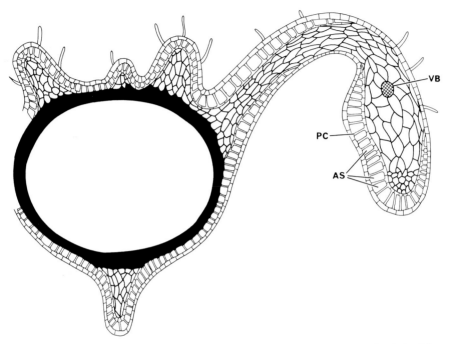

FIG. 38. Cross section of a cypsela from the periphery of the capitulum in *Calendula officinalis*. VB—vascular bundle; AS—air spaces; PC—palisadelike cell. (Adapted from Schoenichen, 1924.)

net formed from the outer integument and is about 1 mm in diameter. The portion of the seed that contains the embryo is only about 0.5 mm in length and is completely free within the loose-fitting net. The diaspores of *Hypochoeris* are cypselae and are of two kinds. The outer whorl produces cypselae with a pappus of barbed bristles and a central mass of hairs; the pappus of the inner whorl is less open and lacks the fine central hairs. The seeds of both plants mature at approximately the same time. Falling *Orthocarpus* seeds often become attached to the pappus of *Hypochoeris* (Fig. 39), and either the net of the *Orthocarpus* is pierced by a pappus bristle (Fig. 40) or the seed becomes entangled by the fine hairs of the outer cypsellae. Thus, both diaspores are dispersed together, assuring further parasite–host interrelations.

Among the anemochorous plants, there are different types of dispersal units, which are typical for different habitats. Dust diaspores and rollers (tumbleweeds), for example, are found mainly on open plains. Except for rollers, the winged diaspores usually are relatively large. They are typical to trees and climbers. Different structural adaptations of diaspores which are dispersed by the same agent can be found even in closely related

Fig. 39. *Orthocarpus* seeds attached to the pappus of *Hypochoeris*. (Courtesy of R. R. Atsatt, University of California.)

taxa. In Compositae, for example, are found both plumed and winged diaspores which are dispersed by wind.

VI. Hydrochory

Water plays a role in seed dispersal both indirectly, serving as a trigger for opening mechanisms, movements of tissues associated with the diaspores, and adhesion of diaspores and other organs of different land plants, and directly upon diaspores of land, shore, and marsh plants. The first type is based on the moistening by water and it will be described under autochory.

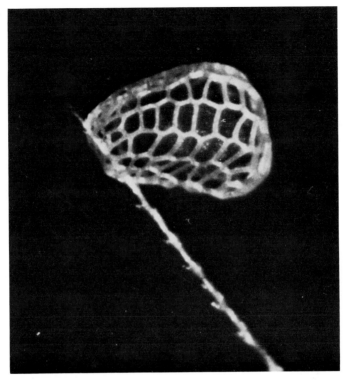

Fig. 40. The net of a seed of *Orthocarpus* pierced by a pappus bristle. (Courtesy of R. R. Atsatt, University of California.)

Two types of direct action of water on dispersal can be distinguished: (*a*) direct action of rain drops which hit on certain parts of the plant; and (*b*) flow of water in which the diaspore is carried.

The direct action of rain drops or, as it is called, the springboard mechanism (Brodie, 1955) can be regarded, with the same justification, either as hydrochorous or as autochorous. This mechanism was described by Brodie (1951, 1955) for *Salvia lyrata, Ocimum basilicum, Sagina decumbens,* and *Oenothera rosea.* In *Salvia lyrata* the calyx tube and pedicel are built so that when drops of rain hit on the distal part of the upper lip of the calyx, the springy pedicel is depressed (Fig. 41). When it returns to its normal position the nutlets are violently ejected (Fig. 42). The calyx tube does not lose its resiliency upon being wetted. A further effect of water is on the seeds. These extrude mucilage as soon as they are wetted. This last phenomenon is explained in Section VII,D.

Diaspores that are directly dispersed by water currents may either stay afloat for a long time or be submerged. Those that stay afloat have a type

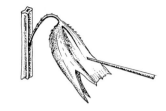

FIG. 41. Depression of calyx tube of *Salvia lyrata*. (Adapted from Brodie, 1955.)

FIG. 42. Expulsion of nutlet in *Salvia lyrata*. (Adapted from Brodie, 1955.)

of structure that helps to diminish their weight and increase their surface area. This structure is usually an air-filled floating tissue and is found mainly in fruits and seeds. The floating tissue may have various forms:

1. It is spongy with many intercellular spaces filled with air, e.g., fruits of *Cerbera odollam, Laguncularia racemosa, Nipa fruticans* (Schimper, 1891), *Potamogeton natans* (Schenk, 1885, see von Guttenberg, 1926).

2. Air is enclosed in the lumina of the cells and intercellular spaces may be absent or nearly so. This tissue of air-filled cells comprises different layers of the diaspores of different species. In *Cocos nucifera* the mesocarp consists of lignified air-filled cells elongated in the transverse direction and penetrated by many fiber strands, whereas the endocarp is very hard and the exocarp is leathery. In *Cycas circinnalis* the floating tissue is the inner layer of the seed coat, and in *Pancratium maritimum* (Fig. 43) it is the whole seed coat.

3. A large air-filled cavity within an external envelope surrounding an inner core. In *Nymphaea* an aril encloses the seed as an air-filled sac which serves as a swim bladder.

4. A combination of more than one type of floating tissue also occurs, e.g., *Limnanthemum nymphaeoides* has both a large air space within the seed coat and the epidermal cells around the margins of the flat seed are elongated and filled with air.

The cell walls of the floating tissue must prevent penetration of water. This is accomplished by impregnation of the walls by suberin, lignin, cutin, or other substances. An inner layer protecting the embryo usually is present.

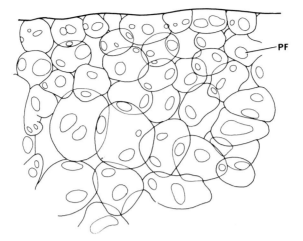

F I G. 43. Cross section of the outer part of the testa of *Pancratium maritimum* showing dead empty cells with large areas of primary pit fields (PF).

Enlargement of volume and a decrease in weight may also be obtained by mucilage as suggested by Gutterman (1971) for *Blepharis persica*. In this species the seeds of those plants that grow in wadi beds are dispersed in a hydrochorous manner by the water of floods (see also Section VII).

VII. Autochory

Autochorous plants are those equipped with an autonomous mechanism involved in seed dispersal. Some mechanisms provide for dispersal to relatively great distances (*telechory*), whereas others prevent dispersal (*antitelechory*). Still others may promote telechory in some plants and antitelechory in others.

Some autochorous mechanisms carry out autonomous functions only as a result of some external trigger such as moisture or pressure.

A. *Dispersal by Turgor Mechanism*

A solely telechorous mechanism of dispersal is the *turgor mechanism*. Living cells take part in its action. The principle of this mechanism is based on the fact that elastic tissue is stretched in a certain direction by another tissue with a very high turgor pressure, or the tissue with high turgor pressure is itself stretched in a certain direction as a result of being set against a tissue that offers a resistance. In both cases, great tension develops. When tension exceeds a critical value, a barrier is removed by opening of the fruit along an abscission tissue which reaches final stages of development at the same time. As a result, the stretched tissue suddenly contracts and the seeds are ejected with great force. The direction of stretching of the elastic tissue and the form and location of the abscission tissue determine the way the seeds are ejected when tension is released.

The turgor mechanism is found in some fleshy fruits as in *Ecballium elaterium, Impatiens parviflora, Cardamine impatiens, Lathraea clandestina, Dorstenia contrajerva, Oxalis acetosella,* and *Biophytum.*

In *Ecballium elaterium,* the seed dispersal mechanism is as follows. The fruit is ellipsoidal and is attached to a long stalk, which is bent downward at an acute angle (Fig. 44). The pericarp (which develops from

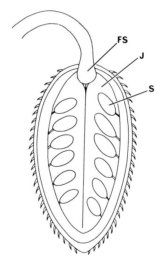

FIG. 44. Schematic drawing of a longitudinal section of a fruit of *Ecballium elaterium.* S—seed; J—juicy tissue; FS—flower stalk. (Adapted from von Guttenberg, 1926.)

an inferior ovary) is fleshy and its outer portion consists of an epidermis and chloroplast-containing parenchyma in which vascular bundles are embedded. Further inward, the pericarp is white and consists of elliptical cells the thick walls of which are pitted and rich in pectic substances. The long axis of these cells is at a right angle to the longitudinal axis of the fruit, and there are large intercellular spaces between the cells (Fig. 45). Still further inward is the tissue enveloping the seeds which consists of large, vesiclelike, extremely thin-walled cells, between which there are no spaces. These cells have a very thin layer of cytoplasm, and their cell

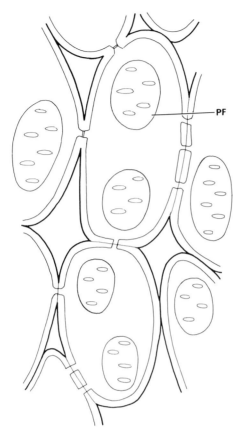

FIG. 45. A few cells from the inner white portion of the pericarp of *Ecballium elaterium*. PF—areas with primary pit fields. (Adapted from von Guttenberg, 1926.)

sap contains the glucoside, elaterinidin. This substance is present in such large amounts that in the ripe fruit the osmotic pressure of the sap is about 27 atms. As a result of the turgor pressure of the elaterinidin-containing cells, the elastic cells of the white portion of the pericarp expand mainly in the width of the fruit. Abscission tissue develops as the fruit matures around the tissue which is continuous with the stalk within the pericarp (Fig. 44). When the pressure which develops in the inner juicy tissue surrounding the seed exceeds that of the force keeping the cells of the separation layer together, then the fruit becomes detached from the stalk. Simultaneously the pericarp, especially its white portion, contracts and the fruit contents—the large juicy cells together with the seeds—are ejected with great force through the hole produced by detachment of the "inserted stalk." The amount of contraction of the pericarp in a transverse direction was recorded as 17%, and in a longitudinal direction 11% (von Guttenberg, 1926).

In *Impatiens parviflora* the fruit is a fleshy capsule in which the septa are extremely delicate. It is cylindrical but somewhat swollen in the upper portion in which the seeds develop (Fig. 46). This upper part of

FIG. 46. Longitudinally sectioned closed fruit of *Impatiens*. (Adapted from von Guttenberg, 1926.)

the fruit remains inactive so far as the opening mechanism is concerned. In the lower portion, however, tension is developed between the outer tissue which has an expansion potential and the inner tissue which offers resistance. When the fruit is mature, the abscission tissue between the carpels ruptures and each valve abruptly curls inward and, as a result, the seeds are expelled (Fig. 47). The expansion tissue is located below the outer epidermis of thick-walled cells. This tissue consists of radially elongated parenchyma cells and lacks intercellular spaces (Fig. 48). The cells have a high sugar content when the fruit ripens and the osmotic pressure in their cell sap may reach 25–26 atms. This pressure would re-

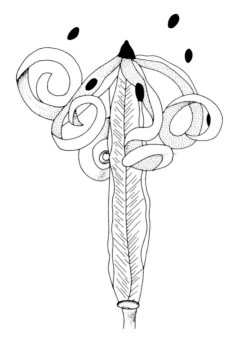

Fɪɢ. 47. An *Impatiens* fruit the valves of which have curved inward and in doing so have ejected the seeds. (Adapted from von Guttenberg, 1926.)

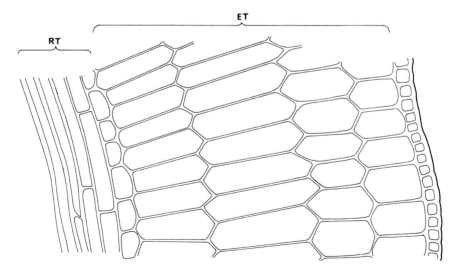

Fɪɢ. 48. Portion of a longitudinal section of the pericarp of *Impatiens* showing the tissue active in opening the fruit. ET — expansion tissue; RT — resistance tissue. (Adapted from von Guttenberg, 1926.)

sult in the rounding of the cells were it not for the resistance offered by the inner portion of the pericarp consisting of two or three layers of thick-walled cells the longitudinal axes of which are parallel to that of the fruit (Fig. 48). These cells elongate by 10% as a result of turgor pressure in the outer tissue, and they contract again to the same extent with opening of the fruit. The outer tissue elongates parallel to the longitudinal axis of the fruit by about 32% with the opening of the fruit (von Guttenberg, 1926).

A similar case to that of *Impatiens* is *Cardamine impatiens* (Overbeck, 1925), but, whereas in *Impatiens* the direction of rolling of the valves is inward, in *Cardamine* it is outward.

B. Dispersal by Imbibition Mechanism

Whereas the turgor mechanism is present only in telechory, the other two autochorous mechanisms, the *imbibition* and *cohesion mechanisms,* act both in telechory and antitelechory. In contrast to the turgor mechanism the site of action of these two is in the walls of dead cells.

The imbibition mechanism, also termed *shrinkage* or *swelling mechanism,* is based on antagonistic action of the walls belonging to cells of two antagonistic groups or of different walls of the same cell (Fig. 49). When a cell wall loses or imbibes water, it shrinks or swells, respectively, in a direction perpendicular to that of the cellulose microfibrils. Therefore, if the angle of the cellulose microfibrils in different cells varies or the direc-

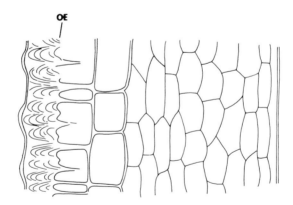

FIG. 49. Longitudinal section through a tooth of a capsule of *Vaccaria pyramidata.* The outer epidermis (OE) has very thick outer walls in which the lamellae are oriented almost perpendicularly to the surface of the teeth. These walls, when wet, are capable of extensive swelling in the longitudinal direction of the teeth, whereas the inner walls of the outer epidermal cells, walls of the cells below this epidermis, and walls of the inner epidermal cells form the resistance tissue. (Adapted from Fahn, 1967.)

tion of the cells themselves varies, their reaction toward moisture will be different. The walls of the cells taking part in this mechanism are usually thick. The movement resulting here is either bending or torsion.

The imbibition mechanism is often responsible for dehiscence of fruits. When the tension developed in the drying tissue overcomes the forces which keep together the cells of an abscission tissue the fruit opens.

Another way in which this mechanism functions is by spreading or folding of plant organs or their parts, upon wetting, thus exposing or hiding the diaspores. Spreading when dried and infolding when wet is termed *xerochasy* and when the opposite occurs, it is termed *hygrochasy*. In the first case telechory results because the diaspore in dry weather can easily be further dispersed by other means such as wind and animals. In the second case antitelechory results. This kind of dispersal is common in arid zones. Dissemination here does not take place soon after fruit ripening, but is delayed until the rainy season when conditions are favorable for germination. Thus the diaspores have little chance to be transported to localities where microecological conditions may be unfavorable.

Telechory resulting from imbibition mechanism can be found in many plants. In *Vicia, Lupinus* (Fig. 50), and some other genera of the Legumi-

FIG. 50. Oblique section of a pod valve of *Lupinus hirsutus*. OE—outer epidermis. (Adapted from Fahn and Zohary, 1955.)

nosae (Fahn and Zohary, 1955), the sclerenchyma cells of the endocarp are oriented at an angle of about 45° to the longitudinal axis of the legume, whereas the elongated, thick-walled outer epidermal or epidermal and hypodermal cells forming the exocarp are oriented at a similar angle but in the opposite direction. In the valves of these legumes the microfibrillar orientation relative to the cell axis is the same in both the endo- and exocarp. However, as the cell axes in these two strata of the pericarp are themselves differently oriented, tension develops during the drying out of the valves. This tension results in twisting of the valves after the

forces that keep the cells together in the mature abscission zone are overcome. The legume then dehisces explosively, the valves contort, and the
seeds are expelled. There are many variations in the structure connected
with this type of opening mechanism. In the legumes of *Wisteria sinensis*
(Monsi, 1943) and *Lupinus angustifolius* (Fahn and Zohary, 1955), for
example, all the sclerenchymatous cells of the endocarp have a uniform
orientation, but they are divided into two zones in which the orientation of
the cellulose microfibrils differs. These zones, therefore, also differ in the
direction of greatest shrinkage upon drying. There are also legumes in
which the endocarp sclerenchyma consists of two layers that differ in
their cell orientation, e.g., *Astragalus fruticosus*, *Astragalus hamosus*
(Fig. 51), and *Hedysarum pallens*. In all these species the same result,

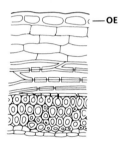

Fɪɢ. 51. Portion of a cross section of a pod valve of *Astragalus hamosus*. OE—outer
epidermis. (Adapted from Fahn and Zohary, 1955.)

with the aid of the same mechanism, is achieved. The anatomical structure, however, by which this mechanism functions varies in the different
species.

   The imbibition mechanism can also aid in dispersal of diaspores which
may creep or leap on the ground. Bristles, involucre bracts, calyxes,
awns, or other rigid organs of the diaspores which are capable of hygroscopic movements can cause the whole diaspore to move on the ground.
This type of movement is found in Compositae, such as in species of
Centaurea, in which the plume is relatively small for anemochory, in
Dipsaceae, such as *Scabiosa* and *Knautia* and in species of *Trifolium.*
Creeping is achieved by outward and inward movements of the bristles or
bracts equipped with short hairs or teeth by which they adhere to the
ground. In the Gramineae, for example in *Avena,* the lower parts of the
awns exert torsion movements as a result of changes in moisture. These
movements cause pressure of one awn of the spikelet upon the other, thus
turning the whole diaspore over and exerting a leap. The layers of
sclerenchymatous tissue responsible for the torsion occupy most of the
awn's tissue (Figs. 52 and 53) (Ulbrich, 1928).

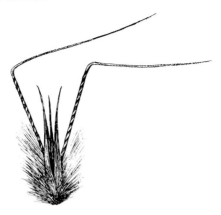

FIG. 52. Spikelet of *Avena*.

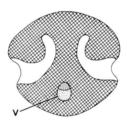

FIG. 53. Schematic drawing of a cross section of an awn of *Avena*. Sclerified tissue cross-hatched; V – vascular bundle.

Hairs that function as flying organs of diaspores can also be equipped with an imbibition apparatus, e.g., *Tamarix* (Figs. 54 and 55), *Anemone pulsatilla* (Fig. 56), *Strophantus, Dryas octopetala, Clematis,* and others. The hairs are close together when wet and spread apart when dry, thus being effective for wind dispersal.

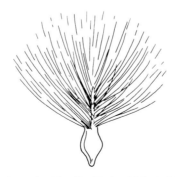

FIG. 54. A *Tamarix* seed with tuft of hairs. (Adapted from Fahn, 1967.)

FIG. 55. The base of a single hair of the tuft of *Tamarix* seed showing the characteristic structure of the abaxial wall involved in hair movement. (Adapted from Fahn, 1967.)

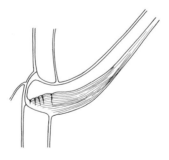

FIG. 56. Basal part of hair of the nutlet of *Anemone pulsatilla*. (Adapted from Haberlandt, 1918.)

The opening of scales of ovulate pine cones is also caused by an imbibition mechanism (Harlow *et al.,* 1964; Allen and Wardrop, 1964). The scale in some species is constructed of two major zones. The adaxial side is constructed principally of strands of vascular tracheids extending from the cone axis, and the abaxial side consists of relatively short rectangular thick-walled cells. Orientation of the cellulose microfibrils in the short rectangular cells is at right angles to the long axis of the scales and this causes shrinkage of this zone upon drying along the cells and opening of the scales (Fig. 57). The opening of the cone causes separation of the

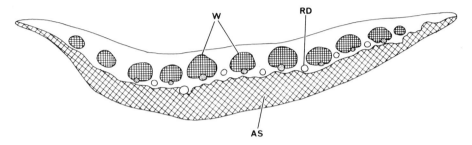

FIG. 57. Schematic drawing of a cross section of a pine cone scale. AS—abaxial sclerified tissue; W—adaxial strands of wood tissue; RD—resin ducts. (Drawn on the base of a photograph in Harlow *et al.*, 1964.)

adaxial epidermis of the scale which forms the wings of the seeds and is followed by dispersal of the exposed seeds by wind (see Section V).

The ejaculation of *Geranium* seeds is another example of the imbibition mechanism. The five one-seeded carpels form mericarps with long, beaklike prolongations originating from the outer part of the style. They are firmly united to one another and to the central axis formed by the inner parts of the style at the tip. The wall of each beaklike prolongation contains a heavy strand of fibers. In the outer fibers the cellulose is oriented transversally, in the inner ones it is oblique or longitudinally oriented. Upon drying the outer strand of fibers contracts longitudinally much more than do the inner fibers so that it eventually causes more shrinkage on the outer side than on the inner side resulting in the tearing of the mericarps from the central axis. The beaklike prolongation now released, coils upward violently, thus ejecting the seeds from the ovary cells (Figs. 58 and 59).

A combination of a hygrochastic mechanism with an additional purely mechanical one is found in the fruit of *Blepharis persica* (Gutterman *et al.*, 1967). Here, in the dry condition the sepals enclose the fruit. Upon wetting, the sepals diverge exposing the capsule. This allows the tip of the capsule to be wetted and consequently it is weakened. The tension in the septum due to differential shrinkage of its tissues overcomes the restraining force of the tip and the capsule explodes along an abscission

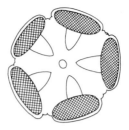

FIG. 58. Schematic drawing of a cross section of the beaklike part of the fruit of *Geranium sanguineum*. The sclerified strand of each mericarp is cross-hatched. (Adapted from von Guttenberg, 1926.)

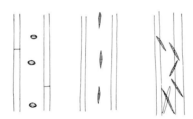

FIG. 59. Portions of the sclerified cells of the beaklike appendage of a mericarp of *Geranium sanguineum*. On the extreme left is a portion of an abaxial sclerified cell, and the other two are portions of adaxial cells. (Adapted from von Guttenberg, 1926.)

zone. In addition to sudden dehiscence of the fruit, the springy hooklike seed stalks act as ballistic apparati for dispersal of the seeds (Figs. 60 and 61).

The imbibition mechanism is responsible for different types of anti-telechory. One of these is found in *trypanocarpy* or *boring fruits*. It is based on the fact that a long appendage on the upper side of the diaspore is divided into two parts, the lower part is capable of spiral twisting by a hygroscopic mechanism whereas the upper part is always straight and at right angle to the lower one. This structure is similar to that described above for *Avena*. This apparatus is found, for example, in some species of *Erodium* where the appendage is the "beak" of the mericarp and in some Gramineae, such as *Stipa barbata, Aristida raddiana, Andropogon hirtus,* and others, where the appendage is the awn (Zohary, 1937). Changes in moisture cause the spiral portion to twist and untwist as a result of the imbibition mechanism, while the upper portion is held in place by pressure against the ground. The movement of the spiral portion causes the penetration of the diaspore, which is pointed and stiff at its base, into the soil. Hairs present on the diaspore are directed upward so that they prevent return of the diaspore to the soil surface with the untwisting of the appendage.

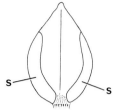

FIG. 60. A schematic drawing of a closed capsule of *Blepharis persica*. S—sepals. (Adapted from Gutterman *et al.*, 1969.)

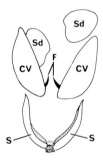

FIG. 61. A schematic drawing of a just opened fruit of *Blepharis persica* after moistening, showing the ejaculation of the seeds (Sd). CV—capsule valves; S—sepals; F—funiculi. (Adapted from Gutterman *et al.*, 1969.)

Antitelechory based on the imbibition mechanism can also be found in some oriental annual species of *Alyssum* (Zohary and Fahn, 1941). After fruit ripening the pedicel of these species is bent upward thus becoming appressed to the axis. When moistened by the first rain the pedicel spreads horizontally thereby exposing the seeds with their mucilaginous epidermis to the rain drops. The movement of the pedicel is due to the existence of two kinds of sclerenchymatous cells. On the upper side of the pedicel these cells consist of thick-walled fibers. The transverse orientation of the cellulose microfibrils of the thick walls is indicated by the transverse arrangement of pits. On the lower side of the pedicel the cells are considerably thinner with diagonally oriented pits and microfibrils arranged in a steep helix. Cells with the transversally oriented microfibrils are the active elements because of their capacity for longitudinal shrinkage whereas those with diagonal orientation constitute the resistance layer.

In *Salvia horminum* (Verschaffelt, 1890; Fahn, 1947) the hygrochastic movements are composed of three combined movements: (*a*) deflexing and horizontal straightening of the fruiting calyx, (*b*) widening of the calyx tube, and (*c*) opening of the calyx teeth (Fig. 62A–E). The nutlets of *Salvia* spp. (Haberlandt, 1918; Hedge, 1970) and species of other genera

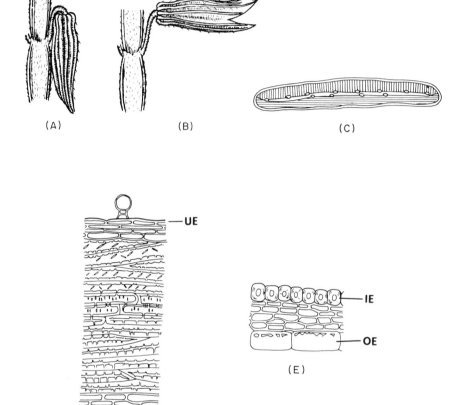

(A)   (B)   (C)

(E)

(D)

Fɪɢ. 62. The calyx and pedicel of *Salvia horminum* after fruit ripening. (A) Dry. (B) Moistened. (C) Schematic cross section of the pedicel showing the position of the active tissue (horizontal lines) and resistance tissue (vertical lines). (D) Portion of a median longitudinal section of the pedicel. Fibers with pits elongated in a steep diagonal direction are from the resistance tissue; fibers with the transverse pits are from the active tissue. UE—adaxial epidermis. (E) Cross section of the calyx tube. IE—inner epidermis which expands equally in all directions is responsible for the closing and opening of the calyx. OE—outer epidermis which together with the ribs of the calyx tube represent the resistance tissue. In both the cellulose microfibrils are oriented lengthwise to the calyx. (Adapted from Fahn, 1947.)

of the Labiatae (Mehra and Atal, 1961) have a mucilaginous outer layer which helps them to cling to the soil when moistened (see below). This increases the antitelechorous effect.

An imbibition mechanism is also responsible for movements of the involucre bracts of some Compositae growing in arid zones. When the fruits are ripe the bracts converge and cover the receptacle and cypselae almost entirely. When moistened the bracts bend outward, exposing the fruits for dispersal. In *Anvillea garcini,* for example (Fahn, 1947), a sclerenchymatous tissue of the bracts effects these movements (Fig. 63).

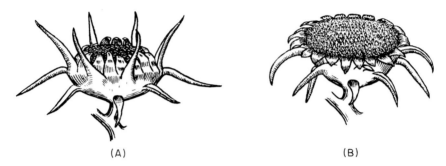

(A)   (B)

FIG. 63. Ripe capitulum of *Anvillea garcini.* (A) Dry. (B) Moistened. (Adapted from Fahn, 1947.)

C. Dispersal by Cohesion Mechanism

Whereas imbibition mechanism is based on the water content of the cell walls, the cohesion mechanism is based mainly on the water content of the cell lumen. The walls of the cells which are active in this mechanism usually are thin. When water present in the cells begins to evaporate, the volume of the cells decreases by curving of the walls inward or by their wrinkling which is the result of cohesion of water molecules and their adhesion to cell walls. A group of such cells may draw along with it a certain organ in which the cell walls are thicker or it may cause the rupture of an abscission tissue.

The location of the cohesion tissue in relation to the organ of which the movement is caused by this tissue determines whether telechory or anti-telechory is involved.

Examples for cohesion mechanism resulting in telechory are found in many Compositae (e.g., *Senecio, Tragopogon,* and *Geropogon*). Here the cohesion tissue occurs on the abaxial side of the bract base (Fig. 64). In some Umbelliferae (e.g., *Ammi visnaga*) this tissue is located between the bases of the rays of the umbel (Figs. 65 and 66) and serves for anti-telechory.

In many plants a combination of imbibition and cohesion mechanisms causes antitelechorous seed dispersal. In *Plantago cretica* (Zohary and Fahn, 1941) the flowers and their bracts, after ripening of the fruits, be-

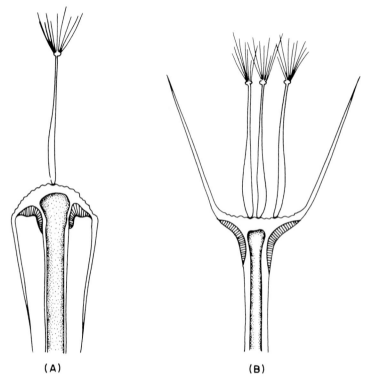

(A)   **(B)**

FIG. 64. Ripe capitulum of *Geropogon* showing action of cohesion tissue (hatched areas). (A) Dry. (B) Moistened. (Adapted from Fahn, 1967.)

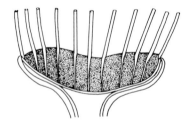

FIG. 65. Schematic drawing of a longitudinal section through the base of the compound umbel of *Ammi visnaga* showing the bases of the partial umbels to be embedded in cohesion tissue (stippled). (Adapted from von Guttenberg, 1926.)

come closely appressed to the peduncle with the sepals converging closely. The thick long hairy scapes curve outward and downard like a watch spring so that the whole plant has the appearance of a spherical body densely beset with long white hairs (Fig. 67A). With the first winter

 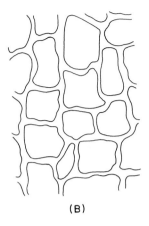

(A)   (B)

FIG. 66. Portions of sections through the cohesion tissue perpendicular to the rays of a compound umbel of *Ammi visnaga*. (A) Dry. (B) Moistened. (Adapted from von Guttenberg, 1926.)

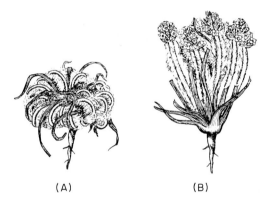

(A)   (B)

FIG. 67. *Plantago cretica* plants. (A) Dry. (B) Moistened. (Adapted from Zohary and Fahn, 1941.)

rain, the scapes turn back upward, the pedicels and bracts spread, and the sepals open again by diverging from one another. During this season the seeds exposed to the wind and rain are washed out from the capsules (Fig. 67B). The movement of the scape is controlled by an imbibition mechanism. There are two kinds of sclerenchymatous cells occupying the greater part of the scape (Fig. 68A)—cells of the lower side have narrow lumina, remarkably thick walls and transversely oriented pits (Fig. 68C), whereas cells of the upper side have relatively thinner cell walls, wider lumina, and pits running diagonally (Fig. 68B). This difference in structure of the cell wall and the different orientation of the

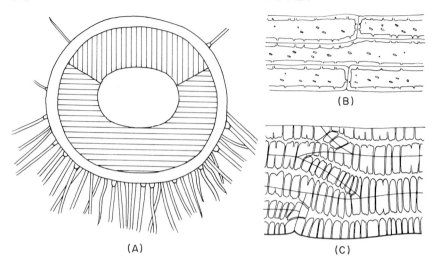

(A)   (C)

Fig. 68. Scape of *Plantago cretica*. (A) Schematic drawing of a cross section showing the active tissue (horizontal lines) on the upper and resistance tissue (vertical lines) on the lower side of the scape. (B) Portion of a longitudinal section of the resistance tissue. (C) Portion of a longitudinal section of the active tissue. (Adapted from Zohary and Fahn, 1941.)

cellulose micellae enable the scape movements. The cells in the lower side of the scape constitute the active elements whereas those of the upper side serve as a resistance layer (Fig. 68A). In the sepals the movement is determined by a cohesion mechanism which is confined to a region of parenchymatous cells present at the base of the inner side of the sepals. Similarly, the bract as well as the very short pedicel possess a paren- chymatic region that acts by cohesion mechanism. The cohesion tissue of the pedicel is situated on the upper side of the pedicel and passes directly into the cohesion tissue of the inner side of the sepals in those places not interrupted by the midrib of the latter (Fig. 69). This tissue when

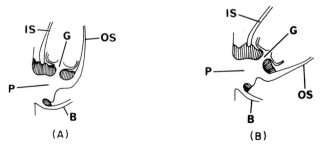

(A)   (B)

Fig. 69. Schematic longitudinal sections of flowers of *Plantago cretica* after ripening of fruit. (A) Dry. (B) Moistened. Cohesion tissue hatched. B—bract; OS—outer sepals; IS— inner sepals; G—gynophore; P—pedicel. (Adapted from Zohary and Fahn, 1941.)

saturated with water expands considerably in the direction of the calyx and forces the latter as a whole to spread horizontally or nearly so. In *Plantago cretica* an additional antitelechorous mechanism operates. The seeds when wetted become fixed to the soil by mucilage (see Section VII,D,1 below).

The well-known Jericho rose *(Anastatica hierochuntica)* also spreads its fruit-bearing branches when wetted with the aid of the imbibition mechanism (Steinbrinck and Schinz, 1908).

D. Some Specific Antitelechorous Mechanisms

In addition to imbibition and cohesion mechanisms which function both in telechory and antitelechory, there are a few autochorous mechanisms which are specific for antitelechory. The ones that are interesting from the anatomical point of view are adherence by mucilage or *myxospermy,* as defined by Zohary (1937), and *geocarpy.*

1. ADHERENCE BY MUCILAGE

Many seeds, nutlets or one-seeded fruits when moistened after their release, adhere to the soil surface by their mucilage. This feature has the advantage that the adhering diaspore cannot be carried further away to unfavorable localities by wind or rain (Klebs, 1885, see Mühlethaler, 1950; Zohary, 1937). Mucilage is common in Cruciferae, Labiatae, Compositae, Plantaginaceae, and other families (Zohary, 1962). Because of uneven distribution of the mucilaginous hairs, adherence to the soil may in some cases be oriented so that the radicle end always touches the soil, as is the case in *Blepharis persica* (Gutterman *et al.,* 1967).

The functions of mucilage, however, can be many. Mucilage may cause adherence to animals (Section IV,A) or reduction of the specific weight of diaspores in water (Section VI). Mucilage may have other advantages not connected with dispersal. It may take part in regulation of germination by (*a*) preventing desiccation of the germinating seed and (*b*) in case of excess of water when the seed becomes entirely covered by mucilage, hindering the passage of oxygen and preventing germination. This latter mechanism is suggested by Gutterman *et al.* (1967) for seeds of *Blepharis persica.*

In spite of the variability in functions of mucilage, its structure has some characteristics common to all mucilaginous diaspores. Mucilage is the pectinous matrix of cell wall layers which swell considerably upon wetting. According to Frey-Wyssling (1959) this capacity of the wall matrix to swell so markedly indicates that such cell walls consist of considerable amounts of unesterified galacturonic acid with an especially large capacity for hydration. The slime membranes are laid down as such from the start and only rarely develop by metamorphosis from already existing secondary walls (Mühlethaler, 1950).

The slime may either contain no cellulose or it may contain cellulose microfibrils. The first type of slime was reported in seeds of *Linum usitatissimum* (Fig. 70) and of *Plantago* species. Here, the mucilage con-

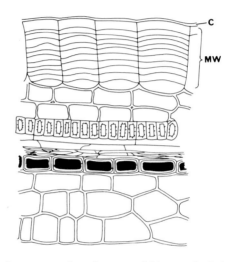

FIG. 70. Portion of a cross section of a testa of *Linum usitatissimum* showing the mucilaginous outer wall (MW) of the epidermis. C — cuticle. (Adapted from Hayward, 1938.)

stitutes the secondary wall. The second type of slime formed upon wetting has more structural strength than pure slime. This prevents the slime from being washed away. This type was found in the epidermal cells of seeds of Cruciferae, such as *Lepidium sativum*. In this case the cellulose mucilage constitutes the primary wall (Mühlethaler, 1950; Frey-Wyssling, 1959). Upon wetting the slimy wall absorbs water and swells. If a layer of cuticle is present on the outer side of the cell it is lifted and/or fissured by the slime.

There are cases in which an inner secondary wall is formed on the side walls. The matrix between the cellulotic helical bands of this wall swells forcibly when wetted so that the cells elongate considerably outward. The mucilaginous content is then discharged to the outside and the helical strand of the secondary wall is torn and unwound. Such is the case in epidermal cells or epidermal hairs of seeds of *Ruellia* species (Haberlandt, 1918) and of *Cobaea scandens* (Frey, 1928). A similar case is found in the tip cells of the multicellular hairs of *Blepharis persica* (Gutterman et al., 1967; Witztum et al., 1969) in which annular or helical secondary wall thickenings are present. The mucilage of the mucilaginous primary wall when examined with the aid of the electron microscope was found to be composed of cellulose microfibrils in a uronic gel (Fig. 71).

FIG. 71. Electron micrograph of the mucilage of the seed of *Blepharis persica*. Magnification: × 48,300. (Courtesy of Witztum *et al.,* 1969.)

A different mucilaginous cell structure is found in the epidermis of seeds of *Cuphea viscosissima* (Lythraceae). From the outer wall of each cell, a threadlike outgrowth protrudes into the cell (Fig. 72). This outgrowth consists of a wall with folds forming a dense helix and is filled with a substance that swells when moistened. Upon wetting of the seed a small round section of the outer wall above the thickening opens on one side like a lid, and the inner substance of the threadlike outgrowth swells. Later, the outgrowth is pushed out like a finger of a glove and acquires a

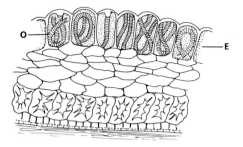

FIG. 72. Cross section of the testa of *Cuphea* showing the epidermis cells (E) with the threadlike outgrowths (O), the walls of which exhibit helically arranged folds. (Adapted from Correns, 1892.)

hairlike structure. The inversion of the thread is a result of the swelling of a substance present in the lumen of the epidermal cell (Correns, 1892).

The presence or absence of slime in the epidermal cells of seeds may have an inter- or intraspecific taxonomic significance as is the case in the Cruciferae (Vaughan *et al.,* 1963; Vaughan, 1968).

2. GEOCARPY

Plants the fruits of which are produced beneath the soil surface are called geocarpic plants. These plants thus have antitelechorous seed dispersal. Geocarpic plants interesting from the anatomical point of view are those defined by Zohary (1937) as *hysterogeocarpic,* i.e., those in which the fertilized ovary penetrates into the soil by means of a long gynophore or peduncle. These differ from *protogeocarpic* plants in which both the flowers and fruits are subterranean. Examples of hysterogeocarpic plants are *Faktorovskya aschersoniana, Trifolium subterraneum, Callitriche pedunculata,* and *Arachis hypogaea.*

The transfer of the ovary from its original aerial position to its final hypogeal one is accomplished through growth and positive geotropism of the gynophore. In *Arachis hypogaea,* as described by Jacobs (1947), the gynophore develops from the upper part of the short ovarian stalk by an intercalary meristem which becomes active after pollination. The shape of the gynophore meristem is that of a cylinder with a thick plate across the center. The cylinder is composed of meristematic tissue in the cortex which extends from 1 to more than 6 mm from the ovary tip, and of a diaphragm-shaped pit meristem 2.6–3.7 mm from the tip. The meristem in a longitudinal cross section is therefore H-shaped. The tracheary elements in this zone have only helical thickenings and no pitted or reticulate elements. Sieve tube elements in the intercalary meristem are also continuous with those above and below the meristem. Fruit development does not occur until the ovary is underground and elongation has stopped.

A special structure that protects the ovules during penetration into the soil has already developed in the unopened floral bud. A group of epidermal cells at one side of the base of the single style becomes radially elongated. Shortly after fertilization, these cells push aside the base of the style and form a pointed, hard, protecting cap. In addition the remainder of the epidermal cells near the tip of the ovary become lignified in their outer walls and increase in radial diameter.

VIII. Conclusion

The principal gross morphological devices serving seed dispersal are limited in number. Hooks and sticky substances occur in epizoochory; food-containing tissues occur in endozoochory and myrmechochory; structures reduced in weight and with relatively enlarged surface area occur in anemochory and hydrochory. There is, however, a great variability in shape, size, and origin as well as in the many diverse anatomical structures of these devices. In autochory, only the microscopical and submicroscopical structures are involved in the dispersal mechanism.

Although generally each mode of dispersal has its special type of morphological and anatomical adaptation, the same device may sometimes serve different modes of dispersal. Such is the case with mucilage which may prevent the diaspore from dispersal to great distances by sticking to the ground in some cases and may adhere to animals and be dispersed by them in other cases. In some plants it may happen that one and the same diaspore is dispersed by two different mechanisms, one following the other.

Seed dispersal always occurs together with abscission. The location of the abscission tissue determines whether the diaspore will be a seed, a fruit, or even a whole shoot.

References

Allen, R., and Wardrop, A. B. (1964). The opening and shedding mechanism of the female cones of *Pinus radiata. Aust. J. Bot.* **12,** 125.

Ashri, A., and Ladijinski, G. (1964). Anatomical effects of the capsule dehiscence alleles in sesame. *Crop Sci.* **4,** 136.

Atsatt, R. R. (1965). Angiosperm parasite and host: Coordinated. *Science* **149,** 1389.

Bain, J. M., and Robertson, R. N. (1951). The physiology of growth of apple fruits. I. Cell size, cell number and fruit development. *Aust. J. Sci. Res., Ser. B.* **4,** 75.

Baum, H. (1948). Vergleichend morphologische, anatomische und entwicklungsgeschichtliche Untersuchungen über die *Astragalus*-Frucht. *Ann. Naturhist. Mus. Wien* **56,** 246.

Benl, G. (1937). Eigenartige Verbreitungseinrichtungen bei der Cyperaceengattung *Gahnia. Flora (Jena)* **31,** 369.

Berg, R. Y. (1954). Development and dispersal of the seed of *Pedicularis silvatica. Nytt. Mag. Bot.* **2**, 1.

Berg, R. Y. (1958). Seed dispersal, morphology and phylogeny of *Trillium. Nor. Vid. Akad. Oslo, Mat.-Nat. Kl.* No. 1.

Berg, R. Y. (1959). Seed dispersal, morphology and taxonomic position of *Scoliopus,* Liliaceae. *Skr. Nor. Vid. Akad. Oslo, Mat.-Nat. Kl.* No. 4.

Berg, R. Y. (1966). Seed dispersal of *Dendromecon:* Its ecologic, evolutionary and Taxonomic significance. *Amer. J. Bot.* **53**, 61.

Bonerji, I. (1954). Morphological and cytological studies on *Citrus grandis* Osbeck. *Phytomorphology* **4**, 390.

Bresinsky, A. (1963). Bau, Entwicklungsgeschichte und Inhaltsstoffe der Elaiosomen. *Bibl. Bot.* No. 126.

Brodie, H. J. (1951). The splash cup dispersal mechanism in plants. *Can. J. Bot.* **29**, 224.

Brodie, H. J. (1955). Springboard dispersal operated by rain. *Can. J. Bot.* **33**, 156.

Burkill, I. H. (1961). The organography and the evolution of Dioscoreaceae, the family of the Yams. *J. Linn. Soc. London, Bot.* **56**, 319.

Carlquist, S. (1966). The biota of long-distance dispersal. II. Loss of dispersibility in Pacific Compositae. *Evolution* **20**, 30.

Clifford, H. T., and Smith, W. K. (1969). Seed morphology and classification of Orchidaceae. *Phytomorphology* **19**, 133.

Cooper, W. C., Rasmussen, G. K., Rogers, B. J., Reece, P. C., and Henri, W. H. (1968). Control of abscission in agricultural crops and its physiological basis. *Plant Physiol.* **43**, 1560.

Corner, E. J. H. (1958). Transference of function. *J. Linn. Soc. London, Bot.* **56**, 33.

Correns, C. (1892). Ueber die Epidermis der Samen von *Cuphea viscosissima. Ber. Deut. Bot. Ges.* **10**, 143.

Danser, B. H. (1931). The Loranthaceae of the Netherlands Indies. *Bull. Jard. Bot. Buitenz.* **11**, 233.

Dingler, H. (1889). "Die Bewegung der pflanzlichen Flugorgane." Ackermann, Munich.

Eames, A. J., and MacDaniels, L. H. (1947). "An Introduction to Plant Anatomy," 2nd ed. McGraw-Hill, New York.

Fahn, A. (1947). Physico-anatomical investigations in the dispersal apparatus of some fruits. *Palestine J. Bot., Jerusalem Ser.* **4**, 36.

Fahn, A. (1967). "Plant Anatomy." Pergamon, Oxford.

Fahn, A., and Zohary, M. (1955). On the pericarpial structure of the legumen, its evolution and relation to dehiscence. *Phytomorphology* **5**, 99.

Ford, E. S. (1942). Anatomy and histology of the Eureka lemon. *Bot. Gaz.* **104**, 288.

Frank, R. (1963). Anatomy and morphology of the rachis in selected species of the tribe Hordeae (Gramineae). M.Sc. Thesis, Hebrew University of Jerusalem (in Hebrew).

Frey, A. (1928). Das Wesen der Chlorzinkjodreaktion und das Problem des Faserdichroismus. *Jahrb. Wiss. Bot.* **67**, 597.

Frey-Wyssling, A. (1959). "Die pflanzliche Zellwand." Springer-Verlag, Berlin and New York.

Gustafson, F. C. (1961). Development of fruits. *In* "Handbuch der Pflanzenphysiologie" (W. Ruhland, ed.), Vol. 14, p. 951. Springer-Verlag, Berlin and New York.

Gutterman, Y. (1971). Personal communication.

Gutterman, Y., Witztum, A., and Evenari, M. (1967). Seed dispersal and germination in *Blepharis persica* (Burm.) Kuntze. *Isr. J. Bot.* **16**, 213.

Gutterman, Y., Witztum, A., and Evenari, M. (1969). Rain clocks in *Blepharis persica. Mada* **13**, 276 (in Hebrew).

Haberlandt, G. (1918). "Physiologische Pflanzenanatomie." Engelmann, Leipzig.

Harlow, W. M., Côté, W. A., Jr., and Day, A. C. (1964). The opening mechanism of pine cone scales. *J. Forest.* **62,** 538.

Hartl, D. (1957). Struktur und Herkunft des Endocarps der Rutaceen. *Beitr. Biol. Pflanz.* **34,** 35.

Havis, A. L. (1943). Developmental analysis of the strawberry fruit. *Amer. J. Bot.* **30,** 311.

Hayward, H. E. (1938). "The Structure of Economic Plants." Macmillan, New York.

Hedge, I. C. (1970). Observations on the mucilage of *Salvia* fruits. *Notes Roy. Bot. Gard. Edinburgh* **30,** 79.

Heywood, V. H. (1968). Scanning electron microscopy and microcharacters in the fruits of the Umbelliferae-Caucalideae. *Proc. Linn. Soc. London, Bot.* **179,** 287.

Holden, D. J. (1956). Factors in dehiscence of the flax fruit. *Bot. Gaz.* **117,** 295.

Jacobs, W. P. (1947). The development of the gynophore of the peanut plant, *Arachis hypogaea.* I. The distribution of mitoses, the region of greatest elongation and the maintenance of vascular continuity in the intercalary meristem. *Amer. J. Bot.* **34,** 361.

Jensen, T. E., and Valdovinos, J. G. (1967). Fine structure of abscission zones. I. Abscission zones of the pedicels of tobacco and tomato flowers at anthesis. *Planta* **77,** 298.

Klebs, G. (1885). Beiträge zur Morphologie und Biologie der Keimung. *Unters. Bot. Inst. Tubingen* **1,** 581 (cited by Mühlethaler, 1950).

MacArthur, M., and Wetmore, R. H. (1939). Developmental studies in the apple fruit in the varieties McIntosh Red and Wagener. I. Vascular anatomy. *J. Pomol. Hort. Sci.* **17,** 218.

MacArthur, M., and Wetmore, R. H. (1941). Developmental studies of the apple fruit in the varieties McIntosh and Wagener. II. An analysis of development. *Can. J. Res., Sect. C* **19,** 371.

MacDaniels, L. H. (1940). The morphology of the apple and other pome fruits. *Cornell Univ., Agr. Exp. Sta., Mem.* **230.**

McCown, M. (1943). Anatomical and chemical aspects of abscission of fruits of the apple. *Bot. Gaz.* **105,** 212.

McLean, R. C., and Ivimey-Cook, W. R. (1964). "Textbook of Theoretical Botany," Vol. 2. Longmans, Green, New York.

Markgraf, F. (1925). Das Abbruchsgewebe der Frucht von *Aegilops triaristata* Willd. *Ber. Deut. Bot. Ges.* **43,** 117.

Mehra, P. N., and Atal, C. K. (1961). Studies in the mucilage yielding seeds. I. Nutlet structure and mucilage formation in *Lallemantia royleana* Benth., *Ocimum basilicum* Linn. and *O. canum* Sims. *Res. Bull. Panjab Univ.* **12,** 169–182.

Monsi, M. (1943). Untersuchungen über den Mechanismus der Schleuderbewegung der Sojabohnen-Hülse. *Jap. J. Bot.* **12,** 437.

Mühldorf, A. (1926). Über den Ablösungsmodus der Gallen von ihren Wirtspflanzen nebst einer kritischen Übersicht der Trennungserscheinungen im Pflanzenreiche. *Beih. Bot. Zentralbl., Abt. 1* **42,** 1.

Mühlethaler, K. (1950). The structure of plant slimes. *Exp. Cell Res.* **1,** 341.

Müller, F. (1895). Orchideen von unsicherer Stellung. *Ber. Deut. Bot. Ges.* **13** (cited by Haberlandt, 1918).

Nitsch, J. P. (1952). Plant hormones in the development of fruits. *Quart. Rev. Biol.* **27,** 33.

Nordhagen, R. (1932). Über die Einrollung der Fruchtstiele bei der Gattung *Cyclamen* und ihre biologische Bedeutung. *Beih. Bot. Zentralbl.* **49 E,** 359.

Nordhagen, R. (1959). Remarks on some new or little known myrmecochorous plants from North America and East Asia. *Bull. Res. Counc. Isr., Sect. D* **7,** 184.

Overbeck, F. (1925). Über den Mechanismus der Samenabschleuderung von *Cardamine impatiens. Ber. Deut. Bot. Ges.* **43,** 469.

Pfeiffer, H. (1928). Die pflanzlichen Trennungsgewebe. *In* "Handbuch der Pflanzenanatomie" (K. Linsbauer, ed.), Vol. 5, Sect. 1, Part 2, Borntraeger, Berlin.

Reeve, R. M. (1954a). Fruit histogenesis in *Rubus strigosus*. I. Outer epidermis, parenchyma, and receptacle. *Amer. J. Bot.* **41**, 152.

Reeve, R. M. (1954b). Fruit histogenesis in *Rubus strigosus*. II. Endocarp tissues. *Amer. J. Bot.* **41**, 173.

Reinders-Gouwentak, C. A., and Bing, F. (1948). Action de l'acide α-naphtylacétique contre la chute des fleurs et des fruits de la tomate et son influence sur la couche séparatrice des pedicelles. *Proc. Kon. Ned. Akad. Wetensch.* **51**, 248.

Rethke, R. (1946). The anatomy of circumscissile dehiscence. *Amer. J. Bot.* **33**, 677.

Ridley, H. N. (1930). "The Dispersal of Plants Throughout the World." Reeve, Ashford.

Rodin, R. J., and Kapil, R. N. (1969). Comparative anatomy of the seed coats of *Gnetum* and their probable evolution. *Amer. J. Bot.* **56**, 420.

Schenk, H. (1885). Die Biologie der Wassergewächse. *Verh. Naturw. Ver. Preuss* **42**, (cited by von Guttenberg, 1926).

Schimper, A. F. W. (1891). "Die indo-malayische Strandflora." G. Fischer, Jena.

Schoenichen, W. (1924). "Biologie der Bluetenpflanzen." T. Fisher, Freiburg.

Scott, F. M., and Baker, K. C. (1947). Anatomy of Washington navel orange rind in relation to water spot. *Bot. Gaz.* **108**, 459.

Sernander, R. (1901). "Den skandinaviska vegetationens spridningsbiologi." Lundequistska Borkhandeln Uppsala (cited by Nordhagen, 1959).

Sernander, R. (1906). Entwurf einer Monographie der europäischen Myrmekochoren. Uppsala. *Sv. Vet. Ak. Handl.* **41**, No. 7 (cited by Nordhagen, 1932).

Sernander, R. (1927). Zur Morphologie und Biologie der Diasporen. *Nova Acta Regiae Soc. Sci. Upsal.* [4] Volumen extra ordinerm editum (cited by Nordhagen, 1932).

Sinnott, E. W. (1939). A developmental analysis of the relation between cell size and fruit size in cucurbits. *Amer. J. Bot.* **26**, 179.

Smith, W. H. (1940). The histological structure of the flesh of the apple in relation to growth and senescence. *J. Pomol. Hort. Sci.* **18**, 249.

Smith, W. H. (1950). Cell-multiplication and cell-enlargement in the development of the flesh of the apple fruit. *Ann. Bot. (London)* [N.S.] **14**, 23.

Spencer, M. (1965). Fruit ripening. *In* "Plant Biochemistry" (J. Bonner and J. E. Varner, eds.), 2nd ed., p. 793. Academic Press, New York.

Staritsky, G. (1970). "The Morphogenesis of the Inflorescence, Flower and Fruit of *Pyrus nivalis* Jacquin var. *orientalis* Terpó." Veenman & Zonen, Wageningen.

Steinbrinck, C., and Schinz, H. (1908). Ueber die anatomische Ursache der hygrochastischen Bewegungen der sog. Jerichorosen usw. *Flora (Jena)* **98**, 471.

Stevens, O. A. (1943). Russian thistle. Life history and growth. *N. Dak. Agr. Ext. Bull.* **326**.

Stösser, R. (1967). Über die Ausbildung des Trenngewebes und seine kausale Beziehung zu Fruchtfall und Fruchtentwicklung bei Süsskirschen. *Angew. Bot.* **41**, 194.

Stösser, R. (1970). Die Induktion eines Trenngewebes bei Fruchten von *Prunus avium* L. durch 2-chloräthylphosphone. *Planta* **90**, 299.

Stösser, R., Rasmussen, H. P., and Bukovac, M. J. (1969a). A histological study of abscission layer formation in cherry fruits during maturation. *J. Amer. Soc. Hort. Sci.* **94**, 239.

Stösser, R., Rasmussen, H. P., and Bukovac, M. J. (1969b). Histochemical changes in the developing abscission layer in fruits of *Prunus cerasus* L. *Planta* **86**, 151.

Szemes, G. (1943). Zur Entwicklung des Elaiosoms von *Chelidonium majus*. *Wien. Bot. Z.* **92**, 215.

Thomson, W. W. (1969). Ultrastructural studies on the epicarp of ripening oranges. *Proc. Int. Citrus Symp., 1st*, Vol. 3, p. 1163.

Tomann, G. (1906). Vergleichende Untersuchungen über die Beschaffenheit des Fruchtschleimes von *Viscum album* und *Loranthus europaeus* und dessen biologische Bedeutung. *Sitzungsber. Akad. Wiss. Wien, Math.-Naturwiss. Kl., Abt. 1* **115** (cited by von Tubeuf.

Tukey, H. B., and Young, J. O. (1942). Gross morphology and histology of developing fruit of the apple. *Bot. Gaz.* **104**, 3.

Ulbrich, E. (1928). "Biologie der Früchte und Samen (Karpobiologie)." Springer-Verlag, Berlin and New York.

van der Pijl, L. (1969). "Principles of Dispersal in Higher Plants." Springer-Verlag, Berlin and New York.

Vaughan, J. G. (1968). Seed anatomy and taxonomy. *Proc. Linn. Soc. London* **179**, 251.

Vaughan, J. G., Hemingway, J. S., and Schofield, H. J. (1963). Contributions to a study of variation in *Brassica juncea* Coss. & Czern. *J. Linn. Soc. London, Bot.* **58**, 435.

Verschaffelt, J. (1890). De Verspreiding der Zaden by *Brunella vulgaris, B. grandiflora, Salvia horminum*, en *S. lanceolata. Bot. Jahrb.* **2**, 148.

von Guttenberg, H. (1926). Die Bewegungsgewebe. *In* "Handbuch der Pflanzenanatomie" (K. Linsbauer, ed.), Vol. 5, Sect. 1, Part 2, Borntraeger, Berlin.

von Tubeuf, K. F. (1923). "Monographie der Mistel." Oldenburg-Verlag, Munich.

Witztum, A., Gutterman, Y., and Evenari, M. (1969). Integumentary mucilage as an oxygen barrier during germination of *Blepharis persica* (Burm.) Kuntze. *Bot. Gaz.* **130**, 238.

Zohary, M. (1937). Die verbreitungsökologischen Verhältnisse der Pflanzen Palästinas. I. Die antitelechorischen Erscheinungen. *Beih. Bot. Zentralbl., Abt. 1* **56**, 1.

Zohary, M. (1948). Follicular dehiscence in Cruciferae. *Lloydia* **11**, 226.

Zohary, M. (1962). "Plant Life of Palestine (Israel and Jordan)." Ronald Press, New York.

Zohary, M., and Fahn, A. (1941). Anatomical-carpological observations in some hygrochastic plants of the oriental flora. *Palestine J. Bot., Jerusalem Ser.* **2**, 125.

Zohary, M., and Fahn, A. (1950). On the heterocarpy of *Aethionema. Pal. Jour. Bot.* **5**, 28–31.

5

SEED GERMINATION AND MORPHOGENESIS

Graeme P. Berlyn

I. Introduction

The seed is a packet of energy, some of which is in the form of information; it is the state of minimum entropy in the life cycle of angiosperms and gymnosperms. The seed contains a quiescent embryo and the coacervate complex of raw materials needed for cell assembly. All of the information required for germination, growth, morphogenesis, and procreation of the species is contained within this highly ordered structure. Furthermore, the instructions are precisely programmed to produce a specific four-dimensional pattern of development under the set of environments to which the species is adapted by the process of evolution.

Germination in seed plants may be defined as the sequential series of morphogenetic events that result in the transformation of an embryo into a seedling. This is a truly remarkable process, involving cell division, cell expansion, and the formation of plant organs, i.e., leaves, stems, and roots. Throughout the unfolding of the four-dimensional pattern of germination a high degree of order pertains and the whole system is dependent on a series of complicated but well-integrated chemical and physical transformations. It is a half-closed system, i.e., it is initiated when the quiescent embryo is reactivated, but the terminal end of the system is open because the point where germination ends and seedling growth commences is undefined (cf. Mayer and Poljakoff-Mayber, 1963). This chapter will take a broad view of germination, extending it to include all of the nonrepetitive morphogenesis in the establishment of the seedling sporophyte. Thus the process of germination may be subdivided in the following series of events (cf. Torrey, 1967): (*1*) imbibition—the physical absorption of water; (*2*) hydration and activation; (*3*) cell division and cell extension; (*4*) protrusion—the physical emergence of the embryo from the seed; and (*5*) completion of nonrepetitive morphogenesis—the establishment of the primary plant body. The major concern in this chapter will be with events *3, 4,* and especially *5.*

In this chapter, first an overview of germination and morphogenesis will be presented, which will be followed by a detailed discussion of the morphogenetic events involved in the germination of two well-studied genera—the angiosperm *Zea* L. and the gymnosperm *Pinus* L.

II. Overview of Germination

A. Seed Structure

Seed structure has a basically similar plan in seed plants. The embryo may be located in different positions in the seeds of various species, and, with the possible exception of the Orchidaceae (Mayer and Poljakoff-Mayber, 1963), it is provided with an energy source. The energy source in the case of gymnosperms is the haploid tissue of the female gametophyte, but the situation is more complicated in the angiosperms where the endosperm, perisperm, or even the cotyledons of the embryo itself may be the main energy source. At seed maturity the embryo may be actually embedded in the nutritive tissue (e.g., *Zea, Pinus, Ricinus,* and *Triticum*) or the nutritive tissue may have been entirely absorbed into the embryo (*Pisum, Phaseolus*). The embryo may entirely fill the seed (e.g., Rosaceae, Cruciferae, Fagaceae, and Juglandaceae) or it may occupy only a small portion of the seed volume (e.g., Ranunculaceae, *Zea*).

The mature embryo is a polarized structure having a root pole and a shoot pole. It consists of an epicotyl (plumule) and a radicle, connected by a hypocotyl; however, the integrated nature of the entity is manifested by the continuity of the primordial tissue systems: protoderm, procambium, and ground meristem. Even if xylary differentiation is at first discontinuous, it is nonetheless based on the prior continuity of the all-pervasive organization of the procambial tissue system.

B. *Protrusion and Elevation*

The mode of germination can be separated into two general types— epigeal (a.k.a. epigeous) and hypogeal (a.k.a. hypogeous). In epigeal germination the cotyledons are, of course, elevated above the soil surface, whereas in hypogeal germination they remain below the soil surface. In castor bean and pine, germination is epigeal, and the cotyledons absorb nutritive material from the haploid female gametophyte in the case of pine and from the triploid endosperm in the case of *Ricinus*. The form of germination may vary within a genus, the classic example being *Phaseolus vulgaris* (French bean) and *Phaseolus multiflorus* (scarlet runner bean). *Phaseolus vulgaris* is epigeal and nonendospermic, whereas *P. multiflorus* is hypogeal and nonendospermic (Avebury, 1892; Skene, 1947; Priestly, 1964; Mayer and Poljakoff-Mayber, 1963). Voluminous information on seed and germination type may be found in Avebury (1892).

Protrusion of the embryo from the seed is one of the most dramatic and crucial steps in the germination process. It takes place against the resistance of the seed coat, or, in the case of indehiscent fruits, against the fruit wall. Seldom is the rupture of the seed envelope accomplished by imbibitional swelling alone. Germination is affected by various environmental factors including oxygen, temperature, water, and light (see Chapter 1, Volume II of this treatise) as well as by complex changes occurring in the envelope of the seed as in the embryo, e.g., mechanical strength of the seed coat may be reduced by over 40% in some species (Skene, 1947). This complexity in the protrusion process obviously has selective advantages in reduction of exposure to insects and fungi, minimizing mechanical damage to the embryo, and permitting the mechanism of seed coat dormancy to exist.

In the great majority of seed plants, protrusion is effected by growth forces evoked by the reactivation of the dormant embryo. There is no absolute rule that specifies which part of the embryo will actually penetrate the seed coat, but the organ of protrusion is usually characteristic of at least the species. Most commonly the organ of protrusion is the radicle, but in some species it is the cotyledons and in others the hypocotyl. In *Marah oreganus* (cf. Schlising, 1969) the minute radicle and epicotyl are carried far out of the seed by expansion of the bases of the two thick fleshy cotyledons. The cotyledon bases are fused and form a hollow tube as they elongate and at the tip of the tube is the embryonic axis. After several months of subterranean elongation, the expansion phase is terminated and the epicotyl grows up the hollow tube, reaching the soil surface by early March in California. Schlising (1969) considered this fused hypogeal germination an adaptation that aids in successful seedling establishment in seasonably arid regions such as "Mediterranean" California. In many forms (e.g., plums, walnut, hazel, Cyperaceae) the seed coat splits along predifferentiated lines of weakness. In other cases, such as *Cocos, Potamogeton,* and *Tradescantia,* protrusion is accomplished by extrusion of preformed plugs of tissue (Skene, 1947). In the coconut there are three "eyes" at the base of the indurate endocarp, i.e., a plug for each of the three carpels. There is only one mature embryo, however, and it displaces only the plug corresponding to the functional carpel.

Subsequent to protrusion there is the mechanical problem of elevating the plumule to the soil surface without incurring damage to the delicate embryonic leaves and the shoot meristem. The unique solution for the case of *Marah oreganus* has been described above. In grasses the problem was resolved by the evolution of the bullet-shaped coleoptile which forces its way to the soil surface where it photoblastically opens to expose the plumule. In other species the hypocotyl or epicotyl usually forms a hook which allows the meristem to be dragged to the surface; the hook opens on exposure to light and turns the shoot apex upward.

C.   *Cell Division and Cell Expansion*

A question that often arises is whether cell division or cell elongation is the initial process of germination (Evenari *et al.,* 1957; Haber and Luippold, 1960b; Mayer and Poljakoff-Mayber, 1963; Wareing, 1969). Under most conditions this is an extremely difficult question to answer. Evenari *et al.* (1957) concluded that in lettuce seeds both cell division and cell expansion were taking place at 12 to 14 hours after sowing at 25°C. Mayer and Poljakoff-Mayber (1963) cite Goo (1952) as showing that in *Pinus thunbergii* Parl. cell division occurs before cell extension; however,

Goo never directly answered this question. Goo (1952) sampled embryos during the three phases of water absorption: (*1*) initial sharp rise, (*2*) slow rise or plateau, and (*3*) fast rise. Goo concluded that, although he had thought phase *1* was only due to diffusion of water (Goo, 1951), he now felt that the cell division in phase *2* was caused by physiological processes originating in phase *1*. In my own work on *Pinus lambertiana* the two processes were visually observed to be simultaneous. However, incorporation of tritiated thymidine took place within 3 hours after explantation, whereas elongation was observed at approx. 48 hours. However, it is clear that in *Zea mays* (Toole, 1924; Picklum, 1953; Sass, 1955), barley (Caldecott and Smith, 1952), and *Vicia faba* (Wolff, 1954), cell extension precedes cell division. Nevertheless, the work of Pollock and Olney (1959) suggests that mitosis precedes cell extension in after-ripening *Prunus cerasus* L. seeds. Thus it can be concluded that, although cell division and cell extension are covariate processes during germination, their onset may not be simultaneous.

It is possible to germinate a number of cereal grains after they have been exposed to high doses of ionizing radiation, viz., 500–1000 krad (Schwartz and Bay, 1956; Moutschen *et al.,* 1956; Sicard and Schwartz, 1959; Haber and Luippold, 1960a; Haber *et al.,* 1961; Haber and Foard, 1964; Haber, 1968). These plants will germinate and form small seedlings that can grow for a maximum of 3 weeks; however, they do not undergo any mitoses nor is there any deoxyribonucleic acid (DNA) synthesis. The high dose of radiation is critical in the induction of these "gamma plantlets"* because mitosis must be completely suppressed so that gross genic imbalances are not manifested, i.e. distribution of genes on fragments does not critically impair DNA as long as mitosis is not present. In gamma plantlets of wheat, Haber (1968) reports that the cells can synthesize ribonucleic acid (RNA) and that this RNA is transportable to the cytoplasm where it functions in protein synthesis. The plantlets are able to fix CO_2 and increase in dry matter. They also demonstrate normal functions such as migration of trichoblast nuclei into root hairs and nuclear disintegration in maturing vessel and sieve elements. However, no new leaf primordia are formed nor are guard cells differentiated in the preexisting embryonic leaves, except in the apical 2–3 mm of the first leaf (Foard and Haber, 1961). Cell size is usually 10 times greater than normal and the cells are conspicuously vacuolated. Thus, in this very special circumstance, germination can proceed without concurrent cell division. Nevertheless, special factors are required to effect this result, viz., (*1*) dry seed at time of irradiation, (*2*) a very high dose of radia-

---

*Specifically this term has been used for wheat seedlings, but it seems appropriate for any species expressing this phenomenon of germination without cell division after irradiation.

tion, (*3*) a highly differentiated embryo, and (*4*) a relatively large volume of nutritive tissue. Note that the radiation dose must far exceed the evolutionary experience of the organism. It is doubtful if the teratological results stemming from these unusual experimental conditions provide insights into the normal situation because in the natural environment germination is always attended by the dual processes of cell division and cell extension. However, these data do show that the two processes are separable. This separation could also be induced in *Lactuca sativa* seeds with appropriate temperature, radiation, hormonal, and osmoticum treatments (Haber and Luippold, 1960b). But under normal germination conditions, as indicated previously (Evenari *et al.*, 1957), cell division and cell extension are virtually simultaneous in this species.

D. *Vascular Differentiation*

This is probably the topic that has been of most interest to students of morphogenesis during the past decade and, according to Esau (1965b), the differentiation of the primary vascular system of the plant is one of the most fundamental processes in plant development. Esau (1965b) states "The role of the vascular tissues in the life of the plant cannot be properly understood unless they are studied developmentally in relation to the ontogeny of the whole plant; and conversely, the development of the vascular plant cannot be discussed comprehensively without reference to the differentiation of the vascular system."

Vascularization of the germinating seedling is especially critical in establishing the autotrophic phase of the life cycle of green plants. I emphatically agree with Esau (1965b) that differentiation of the vascular system must be studied in relation to development of the whole seedling. Vascular differentiation will, therefore, be discussed in the context of the two model species, *Zea* and *Pinus*. References to other species will be brought in where relevant, but the main intent is to present an integrated and unified concept that will, in the final analysis, provide deeper insights into the general problem of germination and morphogenesis.

III. Germination of *Zea mays* L.

Maize is perhaps the best studied organism in the plant world. It is especially well researched with respect to classic genetics (cf. Rhoades, 1955; Sprague, 1955) and, more recently, with respect to biochemical genetics (cf. O. E. Nelson, 1967). The literature on the physiology, anatomy, and morphology of maize is also voluminous, and some preliminary integrated genetic and morphological studies have been recorded (e.g., Gelinas *et al.,* 1969; Mericle, 1950). Thus from almost every standpoint *Zea mays* is a prime experimental vehicle for morphogenetic work. The anatomy of germination has been reviewed by Avery (1930), Hayward (1938), Kiesselbach (1949), Picklum (1953), and Sass (1955). The present review begins with the basic developmental anatomy of germination and then considers the modern experimental approach that has contributed greatly to our knowledge of the germination process over the past decade. Much of this recent work has used *Zea mays* as the experimental material and, thus, by summarizing the work with *Zea* we can get a good grasp of the current status of germination and morphogenesis as a whole.

A. *Anatomy of the Dormant Embryo*

The mature caryopsis of maize consists of an embryo embedded in a starchy endosperm and covered by a multilayered pericarp (Fig. 1). The pericarp may be conceptually separated into an outer portion (exocarp) consisting of compacted thick-walled pitted cells, and a thin-walled inner portion that evidences considerable distortion (Kiesselbach and Walker, 1954; Randolph, 1936; Sass, 1955; Winton and Winton, 1932). Sass (1955) states that the integuments are completely obliterated in the mature seed and do not form a true seed coat over the caryopsis; however, there may be some variation here in the extent of obliteration in different genetic stocks and individuals, particularly with respect to the micropylar end of the seed (Hayward, 1938; Weatherwax, 1923). After fertilization the nucellus is crushed by the proliferating endosperm, and at maturity only a thick-walled cuticular nucellar membrane may be distinguished (Esau, 1965a; Randolph, 1936; Sass, 1955). Normally the endosperm is triploid, being the fusion product of two haploid polar nuclei and one of the haploid sperm nuclei. The outermost layer of cells of the endosperm is cytologically distinct and constitutes the aleurone layer. These cells have the small globoid-containing aleurone grains typical of cereals. The globoids consist of virtually insoluble calcium and magnesium salts of phytic acid (inositol hexaphosphoric acid) which are, nonetheless, mobilized during the germination process (Frey-Wyssling and Mühlethaler, 1965). The single membrane-bounded aleurone grains also contain proteins and appear to be of universal importance in germinating systems; currently there is a good deal of interest in their function (Morton and Raison, 1963; Newcomb, 1967; Opik, 1966). In *Zea,* as in other grasses, the activation of the endosperm is thought to be controlled by an endosperm-mobilizing hormone (most likely a gibberellin) originating in the scutellum (Esau, 1965a; Naylor and Simpson, 1961; Paleg *et al.,* 1962), but the aleurone layer has also been implicated in this process (Esau,

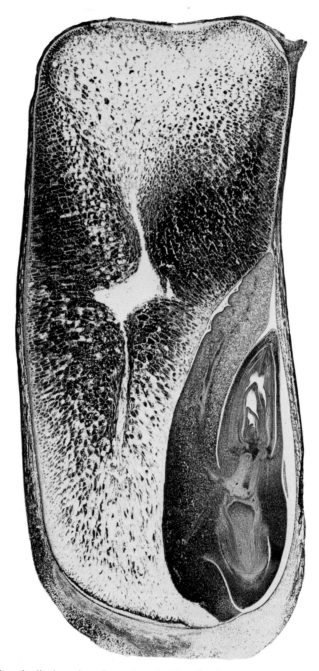

FIG. 1. Longitudinal section of corn kernel. Magnification: × 20. (From Sass, 1955.)

1965a). Mobilization is definitely an enzyme-mediated process with Q_{10} values ranging from 2.6 to 8.7 in barley (Paleg, 1961); however, the the process may not be entirely universal (Lam, 1968).

As shown in Fig. 1, the maize embryo is embedded in the endosperm at the base of the caryopsis and is appressed to one side below the silk scar. The primary root is directed toward the pedicel. The mature embryo consists of a scutellum (cotyledon), an epicotyl with five embryonic leaves enclosed in a coleoptile, a hypocotyl, and a primary root meristem encased in the coleorhiza. There are also three "seminal" roots, two posterior, and one anterior. There is still controversy over the true morphological nature of a number of these structures. For example, the scutellum has been an object of debate at least since the time of Celakovsky (1897). Even the Latin origin of the name is disputed (Esau, 1965a; Fahn, 1967). Although the diminutive of the Latin word for shield *(scutum)* was used to mean a tray in the time of Vespasian, the use of the term with respect to the structure in question is probably related to the stem word — the scutellum resembles a small shield in form (provided one is imbued with the proper amount of anatomical imagination).

Celakovsky (1897) termed the axial area between the scutellum and the coleoptile the mesocotyl, interpreting the scutellum as homologous to a foliage leaf. The lateral positioning of the epicotyl with respect to the scutellum poses a problem as to the evolutionary origin of the distal organization, and some workers (e.g., Jacques-Felix, 1958; Worsdell, 1916) have considered the scutellum as the true embryonic axis and the extant epicotyl as a derivative of a lateral appendage. Avery (1930) and Hayward (1938) interpreted the scutellum as the cotyledon, and the coleoptile as homologous with a foliage leaf, being the second leaf of the plant. Hayward (1938) concluded that this interpretation obviates the need for the term mesocotyl and argued that the putative mesocotyl region is more correctly designated the first internode. On the basis of embryological studies, Weatherwax (1920) also concluded that the scutellum is homologous to a foliage leaf and is thus a lateral organ. The most commonly accepted ontological theory appears to be that expressed by Guignard (1961) which states that the scutellum is a cotyledon, the coleoptile is the first leaf, and the axial region between the scutellum and the coleoptile (the mesocotyl of Celakovsky) is the scutellar internode. Nevertheless, the foliar interpretation of the coleoptile is not unequivocal, and it has been interpreted as an excrescence from the scutellum — a scutellar sheath (Worsdell, 1916). *Zea mays* lacks an epiblast but in those grasses where it occurs it is regarded as a vestige of a second cotyledon. It does not seem necessary to call the scutellum (cotyledon) the "first leaf of the plant," and, similarly, it is doubtful if the coleoptile deserves this designation for, although the coleoptile may have evolved from a foliage leaf more recently than did the scutellum, at this date it is greatly transmogrified. Embryologically the coleoptile arises from scutellar tissue and not from the stem apex. Operational terminology is preferable at least until the requisite evolutionary data are available for devising precise phylogenetic designations.

B. Initiation of the Germination Process

During imbibition, water uptake occurs primarily through the pericarp despite the presence of the fractured pedicel which superficially would appear to offer less resistance to water movement. The caryopsis swells rapidly and, according to Sass (1955), the first cells to reactivate are those of the primary root meristem and the coleorhiza. Stein and Quastler (1963) using radioautographic techniques were able to show that the tissues most closely achieving maturation at the conclusion of the embryogenic phase are the first tissues to resume DNA synthesis. These are the tissues of the root cap, coleorhiza, and the scutellar node region. The nuclei of the dormant shoot and root meristems were found to contain 4 times the DNA content of the haploid sperm nucleus. The root area was the first to reactivate (50 hours) and the shoot apex was the last, not being activated until some 70 hours after initial soaking.

The organ of protrusion is the coleorhiza. It bursts through the pericarp and, in turn, is penetrated by the reactivated radicle. Under favorable conditions, protrusion follows the onset of water sorption by ca. 20 to 30 hours (Kiesselbach, 1949; Sass, 1955). The germination process is extremely rapid and follows a rather fixed program. Lateral root primordia are observable within 48 hours after protrusion. The scutellum remains embedded in the endosperm in this typical hypogeous germination and the plumule itself emerges in a secondary protrusion. Actually the scutellum bears most of the load in this second pericarp fracture, but the plumule-coleoptile complex itself probably makes some contribution to the fracture process. The coleoptile and its contained plumule are elevated toward the soil surface by elongation of the first internode (scutellar node). In turn this elongation is due to the activity of an intercalary meristem located just below the second or coleoptile node. This is a unique phenomenon in that in all succeeding internodes the intercalary meristem is located near the base of the node but not below it.

C. Anatomical Development of the Germinating Shoot

When the base of the coleoptile reaches the soil–air interface the edges of the coleoptile spread apart and expose the plumule. At embryonic

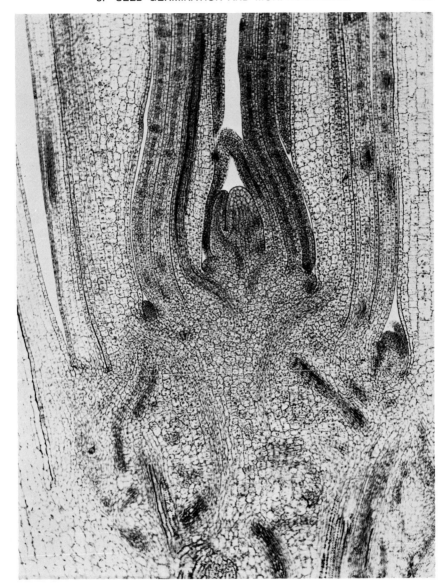

FIG. 2. Longitudinal section of germinating (1-week) shoot apex of *Zea mays*. Magnification: × 140. (From Sass, 1955.)

maturity the plumule consist of five leaves (see Figs. 1 and 2). The speed of development at this stage of germination is truly amazing. A very small maize seedling will contain all of the primordia for the organs it will have in its lifetime. After reactivation, about ten additional leaf primordia complement the original five, and, subsequently, the apex enters a transition phase leading to or marking floral induction (see Clowes, 1961).

Leaf formation in *Zea mays* has been described in detail by Mericle (1950), Ledin (1954), and Sharman (1942, 1945, 1947). The first evidence of primordia formation is found in the tunica. This is rapidly followed by mitotic activity in the corpus leading to the formation of a crescent-shaped anlage. The primordium folds over at the tip, and the base of the crescent enlarges marginally into a ring which forms the adnate base of the inchoate leaf. The blade expands through the activity of a marginal meristem. In his excellent monograph on apical meristems, Clowes (1961) summarizes this stage of maize leaf development in terms of plastochrons. A *plastochron* is a measure of time in developmental units, i.e., the time period between the formation of successive leaf primordia (or whorls or pairs of leaf primordia). The encircling of the base requires two plastochrons, and at the end of this period the marginal meristems are evident. The leaf base overlaps during the third plastochron and blade expansion terminates in the fourth plastochron during which time axillary buds are also formed. The ligule is formed from the adaxial epidermis during the fifth plastochron, and apical growth ceases in the sixth plastochron. Vascularization of the leaf is similar to that of the stem, and Fig. 3 shows the structure of the corn leaf, displaying the typical graminaceous pattern. The large epidermal cells are termed bulliform cells and function as motor cells in the curling and uncurling of the leaf in response to water stress.

Picklum (1953) reported that during germination plastochrons are rather constant in absolute time, averaging about 60 hours. This condition is said to persist until the full complement of leaves is attained. However, Abbe and his colleagues (Abbe *et al.,* 1951; Abbe and Stein, 1954) found that, under the environmental conditions and the maize strains used in their study, the duration of the plastochron decreased from 5 days at the onset of germination to 12 hours at the termination of leaf production.

The scutellum is composed primarily of typical parenchymatous tissue except for the crenate abaxial surface which has a uniseriate dermal covering of epithelial cells interrupted by numerous glands. This portion of the scutellum is in contact with the endosperm, suggesting a possible haustorial function for the specialized dermal layer. Prior to germination the scutellar parenchyma is uninuclear, but during the germination process the nuclei become lobed and some of these lobes separate from the parent nucleus giving rise to micronuclei. The function, fate, and metabolic significance of these cells are not known.

According to Sass (1955) the stem apex of maize has a classic tunica–corpus organization (Schmidt, 1924). There is little evidence of combined

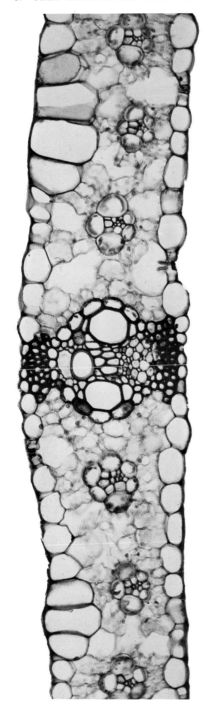

FIG. 3. Cross section of *Zea* leaf. Magnification: × 375.

tunica–corpus cytohistological organization which pertains in many angiosperm stem apices, or at least it is doubtful if the combinatorial concept would have been formulated from observations based solely on *Zea mays*. The tunica in maize is essentially uniseriate (Clowes, 1961), and divisions in this layer are almost exclusively anticlinal. The uppermost layer of the corpus resembles the tunica structurally and operationally; the divisions in this layer vary from predominantly anticlinal in less active apices to mixed periclinal–anticlinal in highly active apices. The bulk of the corpus exhibits more-or-less random planes of cell division. Steffensen (1968) has made a detailed reconstruction of cell development in the shoot apex of maize using radiation to produce aberrations that serve to mark cellular descent in two strains heterozygous for an albino factor. He concluded that a quite small number of cells, located in the domed apex, are operational as initials and give rise to the bulk of the vegetative tissues of the corn plant.

During germination, procambium can be observed in the seedling axis at approximately the level of the second youngest leaf primordium and, according to Sass (1955), the strands differentiate from the ground meristem which is, in turn, derived from the corpus of the apex. Visually the strand can be traced acropetally to a single vertical column of cells which usually terminates in the apex at the locus of this leaf primordium. Often the earliest evidence of the primordium and of the tip of the strand appear simultaneously. Steeves and Sussex (1970) present the notion that there exists a transition region between the cells of the promeristem and the procambium. The transition region is termed the provascular tissue and is morphologically and biochemically distinct. The concept is based upon experimental evidence originally provided by the work of Wardlaw (1946, 1950) and, more recently, by McArthur (1967) using the leaf puncture approach. The experiments have been described in detail by Steeves and Sussex (1970). The basic procedure is to isolate the apical meristem from the influence of previously formed leaves by making surgical cuts around the meristem. The new leaf primordia are then destroyed (punctured) as they are formed. After several successive leaf primordia are destroyed the apices are fixed and examined histologically. Working with *Geum chiloense*, McArthur (1967) found tissue subjacent to the promeristem to contain a provascular cylinder lacking leaf gaps, traces, and with no mature xylem and phloem. Steeves and Sussex (1970) hypothesize that this type of provascular tissue exists in the intact shoot apex and that vascular differentiation and the development of leaf traces and gaps are determined by the leaves themselves. The hypothesis is not apodictic and further cytohistochemical studies are needed to clarify the pattern of differentiation here. However, it has long been recognized that the densely cytoplasmic prismatic-shaped procambial cells appear to be encased in a specialized tissue often termed the residual meristem (Esau, 1943; Sass, 1955). Ontogenetically the procambium is said to be derived from the residual meristem and that portion of the residual meristem that does not develop into procambium eventually differentiates as interfascicular parenchyma (Sass, 1955).

As the vascular bundles differentiate the first vascular element to mature is a protophloem sieve tube (Fig. 4), but the first protoxylem is

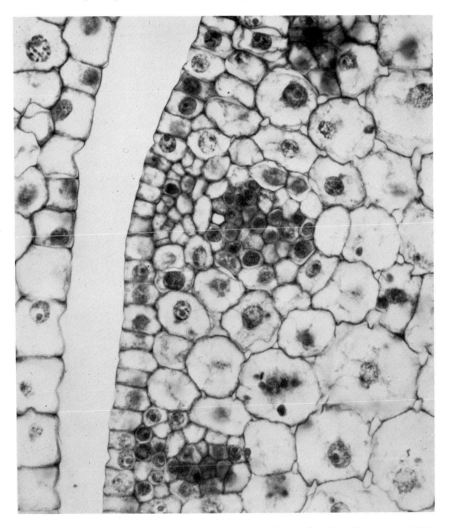

FIG. 4. Procambial strand with one protophloem sieve tube. Magnification: × 1300. (From Sass, 1955.)

evident soon thereafter (Fig. 5). The first protophloem sieve elements develop without companion cells but, nevertheless, at maturity appear to be functionally enucleate even if nuclear remnants are present at the base of the cell. In the area between the protophloem and protoxylem a cambi-

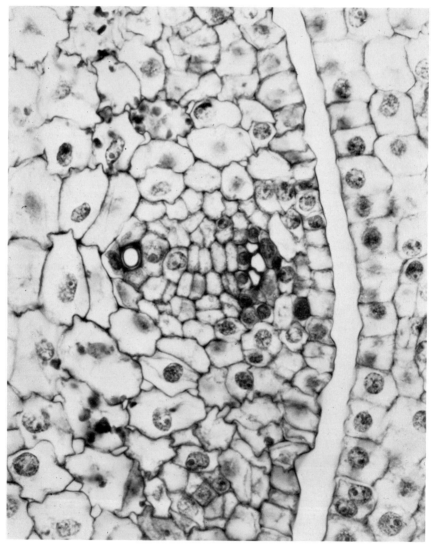

FIG. 5. Developing vascular bundle with three differentiated sieve tube elements, cambiform region, and one mature protoxylem element. Magnification: × 1250. (From Sass, 1955.)

form layer develops, and this meristematic region apparently gives rise to the metaphloem elements. These latter structures are provided with companion cells that are formed through longitudinal divisions of the incipient metaphloem sieve elements. The protophloem is eventually crushed and obliterated during bundle ontogeny. The peripheral cells undergo a sclerification process and form the bundle sheath, whereas the hypodermal zone of the stem also becomes remarkably sclerified and is often termed the *rind* (Fig. 6).

The first protoxylem cells to differentiate in the stem are vessel elements possessing the annular type of secondary thickenings. Following the maturation of the first vessel, an additional three or more columns of

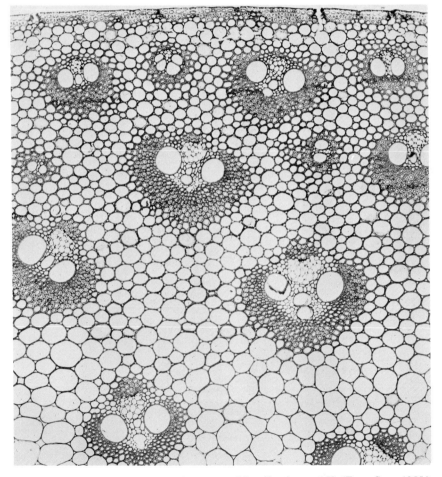

FIG. 6. Dissected siphonostele of *Zea* stem. Magnification: × 160. (From Sass, 1955.)

cells develop as vessels in centrifugal sequence. The first two columns all have annular thickenings, but the third and subsequent columns can have annular, helical, or both types of thickenings. The two large metaxylem vessels develop on the edges of the cambiform layer and are pitted (Fig. 6). The xylem derivatives of the cambiform layer are quite unusual. They are rather narrow, prismatic, metaxylem cells with imperforate but pitted end walls, somewhat similar to the vascular tracheids found in some woody species such as the Ulmaceae. As the stem matures the first-formed protoxylem elements are ruptured, giving rise to the protoxylem lacuna typical of the mature vascular bundle (Fig. 6).

D. Anatomical Development of the Root

Turning to the root (Fig. 7), it is apparent that the root meristem is well organized before germination. Sass (1955) describes the root meristem of *Zea mays* in terms of the classic histogen theory of Hanstein (1868) with some modifications. The histogens of the *Zea* root are the calyptrogen, dermatogen–periblem initials, and stelar initials. This, of course, is not the only concept of maize root meristem organization and differing points of view will be considered later.

According to the histogen theory the calyptrogen is a saucer-shaped group of initials that gives rise to the ground meristem of the root cap. The derivatives of the calyptrogen divide one or more times before maturing as the spongy parenchyma of the root cap. The dermatogen–periblem (D–P) initials are present as a one-cell thick layer about 3 to 5 cells wide. According to Sass these cells are identified by activity and not by structural distinctions. In the center of the D–P zone, division planes are exclusively anticlinal or nearly so, whereas the division planes at the margins are tangential to the domed apex. The outermost layer which divides anticlinally differentiates as epidermis while the inner derivatives of the D–P initials give rise to the cortex. Sass (1955) states that the inner zone of the periblem is particularly active mitotically, resulting in a stratified zone which he describes as evidencing cambiform activity. As this activity ceases the innermost cell layer forms the endodermis and marks the furthest centripetal excursion of the cortex.

The stelar initials give rise to the stele almost directly without much of an intercalated plerome region. The columns of metaxylem can be traced to the second or third cell behind the initials (Figs. 7 and 8). The pericycle differentiates as a single layer of cells abutting the endodermis on the centrifugal side and the phloem sieve tubes on the centripetal side (Fig. 9). Typically, the vascular elements of the root develop centripetally, e.g., the first protophloem sieve tubes develop in contact with the pericycle. As in the stem these sieve elements lack companion cells. The meta-

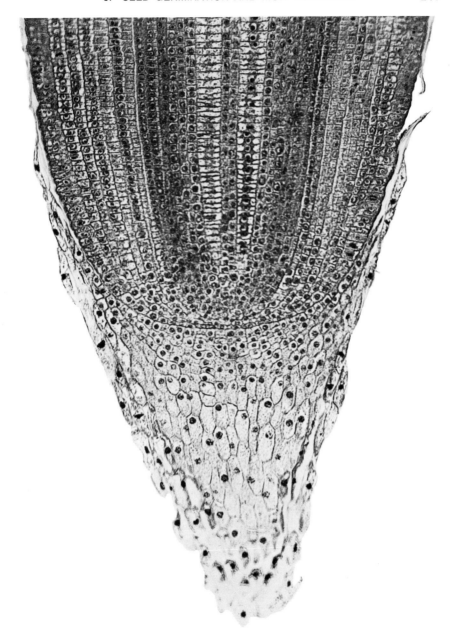

FIG. 7. Longitudinal section of *Zea* root tip. Magnification: × 220. (From Sass, 1955.)

phloem is provided with companion cells and follows in centripetal succession from the radial files of two or more protophloem elements. Un-

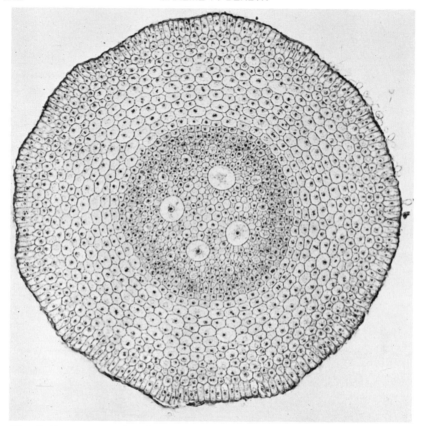

FIG. 8. Transverse section of *Zea* root meristem in the zone of differentiation. Magnification: × 360. (From Sass, 1955.)

like the stem the protophloem of the root is not crushed to any significant degree but most of the phloem is of the metaphloem type. The mature phloem alternates with the xylem arcs (Fig. 10).

The protoxylem elements differentiate much later than the protophloem, and their first conspicuous differentiation even lags behind that of the much larger metaxylem vessels although the protoxylem eventually matures before the metaxylem. The phases of wall thickening and lignification follow this same pattern of differentiation in these two cell types and are essentially homologous processes. Sass (1955) reported that between the xylem and phloem there exists a curious zone of small-diameter, highly lignified cells of unknown function and significance. Maize has a central parenchymatous pith as do monocots in general, and the surface of the root is protected by a thick sheath of hypodermal sclerenchyma (Fig. 10).

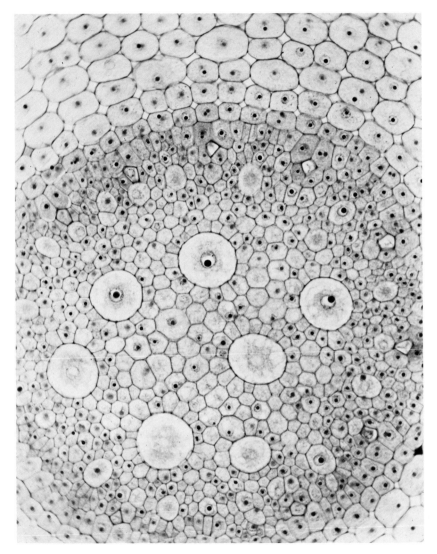

Fɪɢ. 9. Transverse section of *Zea* root metistem showing cytohistological detail of stele–cortex interface. Magnification: × 600. (From Sass, 1955).

Lateral root primordia originate in the pericycle opposite xylem areas. Emergence through the sheath occurs at a considerable distance in back of the root initial area; the histogen organization of the lateral root meristem is evident early in lateral root ontogeny, i.e., before the root has grown halfway through the cortex of the parent root.

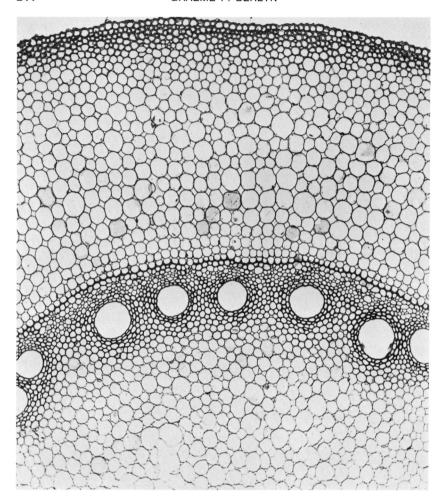

FIG. 10. Transverse section of mature tissue of maize root showing tissue cylinders from epidermis to pith. Magnification: × 140. (From Sass, 1955.)

The above discussion and description of root meristem organization in *Zea mays* are based on the classic research of Sass and others over many years. However, more recently, Clowes (1954, 1956, 1959, 1961, 1963) has made a series of intensive investigations of the root meristem of *Zea mays* which have led to conclusions that are at considerable variance with the previous studies. Some of his observations even conflict with long accepted data on roots in general and maize roots in particular. Clowes (1961) states, "To some extent, serious work on this subject was delayed by the acceptance of the histogen theory of Hanstein . . . . All that the histogen theory tells us about meristems is that the initiating cells of the various tissue regions are separate and all that the subsequent investigations show is that the epidermis can be part of the cortex, or part of the cap, or independent."

Clowes (1961, 1969) has considerably more appreciation of the Körper-Kappe theory of Schuepp (1917) which he feels should have set anatomists straight about root meristems a long time ago. However, Schuepp himself regarded this theory as complementary to that of Hanstein (Schuepp, 1926). This nonexclusivity is, of course, acknowledged by Clowes (1961) but apparently does not perturb his interpretation of the situation. The essential feature of the Schuepp theory is that "T" divisions in the *Kappe* (cap) are 180° out of phase with those of the *Körper* (body). Normally in the Körper the cross bar of the T is near the tip of the root (distal), whereas in the Kappe it is proximal (Fig. 11). For

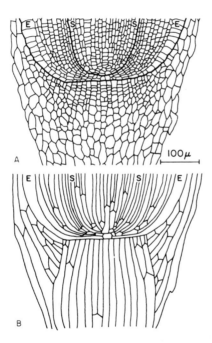

FIG. 11. Longitudinal section of maize root (A) and its Körper-Kappe pattern. (E) Epidermis; (S) outer layer of stele. (From Clowes, 1961.)

most roots the root cap is only part of the Kappe but for *Zea mays,* the root cap is the only tissue belonging to the Kappe, i.e., the root cap and the Kappe are identical (synomymous) structures in the maize root. The Körper-Kappe theory permitted Clowes to conceive the root meristem as a series of layers and apparently provided a stimulus for development of the currently popular quiescent center theory.

Clowes (1961) describes the existence of a columella in the root cap of corn (Fig. 12) which is formed as a result of the absence of longitudinal

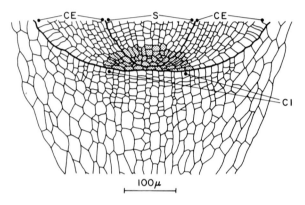

FIG. 12. Longitudinal section of maize root showing the quiescent center (shaded). (CI) Cap initials; (S) stele; (CE) cortex-epidermis complex. (From Clowes, 1963.)

divisions in the central part of the root cap. The exclusively transverse cell division planes in this region result in columns of cells in the center of the cap. It should be noted that the existence of this columella abrogates the pattern of T division planes. As a matter of fact, Clowes (1961) recognizes three regions in the root apices of maize and of these only the peripheral region has the properties usually attributed to the Kappe. The third region is a transition zone which is intermediate in division orientation. Clowes also states that the root cap has its own initials but, needless to say, they are not referred to as the calyptrogen.

It was while studying the roots of germinating maize seedlings that Clowes (1954, 1956) first formulated his quiescent center theory. Geometric considerations led him to this idea which has since been supported by a considerable body of cytochemical evidence. The geometric considerations are as follows. The columella is formed because the central cells of the root cap do not divide longitudinally. Just above the cells of the columella (proximal) is a layer of cells belonging to what Clowes calls the cortex–epidermis complex (even though they belong to two different tissue systems, the ground and dermal systems, respectively). Clowes states that there is no evidence that these cells slip in relation to the subjacent columella cells and, therefore, concludes that these cells cannot divide longitudinally. In the center of the root, the cells of the cortex–epidermis complex (D–P initials of Sass) are only one cell high. These cells are squeezed in between the columella and the stele, and Clowes states (as does Sass) that there is no exchange of cells between the stele and these central cells of the cortex–epidermis complex or between these central cells and the columella. Thus, Clowes concludes that these cells, the putative dermatogen–periblem initials, do not divide at all; they are nonmeristematic (Fig. 12).

A number of possibilities can be considered here. First of all, Clowes's own data (1963) cast doubt on the nonmeristematic statement with regard to the putative D–P initials. He determined a mitotic cycle time of 174 hours for cells of the quiescent center and, although this represents a low rate of mitotic activity, it is not zero. It shows that not only can these cells divide, but they do divide. Sass (1955) claims that in the center of the axis the D–P initials divide anticlinally (longitudinally), but on the periphery they divide tangentially (with respect to the root apex). This would require some gliding growth, and, although such growth is not established in this instance, it is a common phenomenon in plants, e.g., in vessel ontogeny. Clowes's argument of "no evidence" is not sufficient to eliminate the possibility that gliding growth occurs and the geometric argument by itself is not unequivocal. Recently Clowes (1969) has modified his 1961 declaration of total nonmeristematicity of the quiescent center, acknowledging that "it has always been known that some mitosis does take place there." However, this latter statement may refer to only a part of the quiescent center, not necessarily the central cells of the cortex–epidermis complex.

Sass (1955), however, used a different approach. He began his observations with the fertilized egg and followed the organism through embryogeny before investigating the cytohistology of germination. Clowes notes that the quiescent center in maize does not form until a few days after protrusion; and after the inception of the quiescent center the cellular pattern in the root apex does not change (i.e., remains as it was when all zones were active). It would thus be easily subject to misinterpretation. Nevertheless, Sass says that the D–P initials are identified by high mitotic activity, and Clowes's (1963) fine structure and cytochemical evidence do not support this contention.

Clowes (1963) found that the cells of the quiescent center have less RNA, protein, and DNA. They possess smaller nuclei (90 compared with 150 μ^3 in cap initials) and smaller nucleoli (4.6 as opposed to 12.3 μ^3) than the other cells of the root meristem. They also contain smaller Golgi bodies, fewer mitochondria per cell, and less endoplasmic reticulum. It must be noted that these data are generalized for the entire quiescent center which may include as many as 500 cells and, therefore, the data are not specific for the D–P initials alone. Clowes (1961) accepts the notion that the stelar initials are distinct from those of the columella, and he also recorded that differentiation proceeds rapidly and proximally from the stelar initials. In order to reconcile this operational duality, Clowes (1961, 1963) concludes that the number of stelar initial cells would have to be large.

Although Clowes was the first to propose the quiescent center theory for roots, it has been known for a long time that the cells at the tip of the meristem divide very slowly (Baldovinos De La Pena, 1953). Respiration in the tip itself has been shown to be quite low as compared to that in more proximal regions (Goddard and Meeuse, 1950). Baldovinos De La Pena noted that in the first millimeter of the maize root the rate of cell division was extremely low. The second millimeter back of the root caps was termed a transition zone between cell division and cell enlargement and was characterized by both rapid cell division and cell enlargement with the proximal portion of the region showing only cell enlargement. Protein synthesis in this region kept pace with cell enlargement, but in the fourth millimeter protein content declined although rapid cell expansion was occurring. The fifth and sixth millimeters marked the zone of differentiation. Baldovinos De La Pena's work is partially corroborated by Cook (1959) who also found mitosis confined to the first 2 mm from the tip. However, Cook designated the region from 2 to 8 mm as the zone of expansion and assigned the region greater than 8 mm behind the tip as the zone of maturation. Fresh weight per cell increased from 14 in the meristem to 88 ng at 8 mm. Dry weight per cell was 2.2 in the meristem and 8.3 ng at 8 mm, whereas protein increased uniformly from 0.68 in the meristem to 1.01 ng/cell at 16 mm. Transaminase content was low in the meristem, increased at a decelerating rate to obtain a maximum at 10 mm, and thereafter declined. Mertz (1961) found a somewhat similar pattern for distribution of ascorbic acid oxidase. In this case 70% of the activity was associated with the cell wall fraction, and Mertz surmised that the correlation of bound oxidase activity and cell elongation indicates that the enzyme may function primarily in cell wall growth and not in respiration as is commonly assumed.

Clowes (1958) has studied the changes in cell volume of differentiating cells of germinating maize root apices (Fig. 13). At 300 μ from the tip of the root, meristematic cells are of a relatively uniform volume of ca. 800 μ^3. From 300 to 1000 μ, there is only a slight increase in volume but above 1000 μ the increase is rapid. At 3000 μ the epidermal cells attain a volume of $2.3 \times 10^4 \mu^3$ (29-fold increment), whereas the inner cortical cells are $3.6 \times 10^4 \mu^3$ (45-fold increment) and the central stelar cells reach a volume of $1.2 \times 10^5 \mu^3$ (150-fold increment). The cross-sectional area of the root undergoes a 5-fold increase as one proceeds proximally from the tip to 1600 μ. The cortex and epidermis also increase from the tip for 2000 μ (5-fold for the cortex and 2-fold for the epidermis). These data are supported and corroborated by the work of Erickson and Sax (1956) and Jensen and Kavaljian (1958), among others.

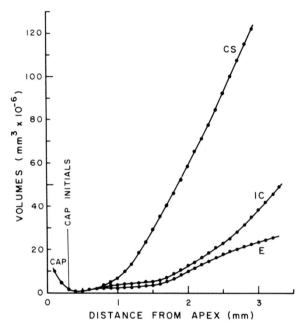

FIG. 13. Changes in the volumes of cells at increasing distance from the root apex of *Zea mays*. (CS) Central stele; (IC) inner cortex; (E) epidermis. (From Clowes, 1961.)

Twenty years ago, Swift (1950) used microspectrophotometric instrumentation to disclose that in maize root tips the amount of DNA per nucleus may be 2, 4, 8, 16, or 32C, where C stands for the amount of DNA in the haploid sperm nucleus. The data revealed that DNA content was a function of distance from the apex—the greater the distance the greater the DNA per nucleus. Most of the meristematic cells had either the 4 or 8C level of DNA (see, also, Stein and Quastler, 1963). Vessel elements generally had the most DNA per nucleus (up to 32C), and Swift concluded that, during the differentiation after the last division of the xylem mother cells, there is periodic doubling of the DNA. This, of course, does not imply that the duplicate DNA, which could be due to polyteny and/or polyploidy, is a cause of differentiation per se; more probably it is a result—a termination. Swift's work has a basis in earlier studies (cf. Huskins, 1947) and has since been verified for *Zea mays* and for a number of other species (D'Amato, 1952; List, 1963). Stein and

Quastler (1963) reported the existence of polyploidy during germination in maize and concluded that the mature cells of the coleoptile typically are tetraploid. However, Partanen (1959) notes that in many species of plants there is no polyploidy and concludes that there is little or no correlation between differentiation and polyploidy, a view shared by Steeves and Sussex (1970). In addition there are polyploid plants that grow and develop more rapidly than the diploid representative of the species, and polyploid poplars are capable of forming adventitious roots. Thus polyploidy by itself cannot be designated as an absolute seal with respect to growth and differentiation.

Clowes (1961, 1963) recognized that, although the quiescent center theory was gaining acceptance, there were still some unanswered questions raised by the earlier work of Brumfield (1942). In these studies the root meristems of *Vicia* and *Crepis* had been irradiated with sublethal doses of X-rays. With this procedure Brumfield was able to identify mutant cells by factors such as chromosome morphology and increased size due to polyploidy. These specific aberrations could be used to trace cell lineages, i.e., all cells with a specific aberration were assumed to come from a single cell that was directly mutated by the X-rays. After irradiation the roots were allowed to grow several centimeters before harvesting (3 weeks time). This growth period was thought to be sufficient to eliminate from the meristematic region all cells formed by the meristem before irradiation. Brumfield found that in many roots the treatment resulted in sectorial chimeras which included only cells showing a specific (and identifiable) aberration. Furthermore these chimeras were discretely contained in about one-third of the cross-sectional area of the root despite the fact that such an area includes portions of the cap, stele, and cortex. Brumfield concluded that there are about three common initials for the whole root; however, he did not locate these initials cytologically.

In 1963, Clowes reported the results of experiments designed to duplicate and extend Brumfield's work. Clowes used germinating maize seedlings for these studies and found that the cup meristem cells (those on the edge of the quiescent center) became themselves quiescent soon after irradiation with X-rays. The mitotic cycle in these cells changed from 12 to 134 hours; concomitantly the mitotic cycle time for cells of the quiescent center changed from 174 to 62 hours. Cytologically Clowes observed that after the sublethal irradiation a smallish group of cells in the quiescent center became mitotically active and produced cells that eventually reconstituted the root meristem whereupon the original pattern of mitotic frequency (cup meristem plus quiescent center) was reestablished.

These radiobiological results of Clowes are logical when considered in the light of the 1906 law of Bergonié and Tribondeau (1959) and the work of Sparrow and Miksche (1961). According to Bergonié and Tribondeau, X-rays are more effective on cells that have high actual or potential mitotic activity and this would apply to the cup meristem of the root. The cells of the quiescent center are predominantly in the G_1 (pre DNA synthesis) phase of their cell cycle and have, on the average, smaller nuclear volume than cells of the cup meristem; therefore, according to the nuclear volume theory of Sparrow and Miksche the cells of the quiescent center should be less radiosensitive, as indeed they are. This view is supported by the fact that following irradiation many fewer micronuclei are evident in the quiescent center than in other regions of the root meristems. In summary the quiescent center can be regarded as a pool of diploid cells in the steady-state condition which under normal conditions divides very slowly. Although mitotically inactive these cells maintain their meristematic potency and under suitable stimuli such as sublethal X-irradiation, they can be induced to divide. Thus Clowes concluded that Brumfield's results apply only to irradiated meristems. On the basis of surgical experiments, Clowes (1961) concluded that the meristems of "normal" roots contain a rather large number of initials; however, these results may apply only to surgically treated tissue. There does remain, then, the possibility of periodic activity in the quiescent center of normal roots.

IV. Germination of *Pinus*

There has been a great deal of interest in the developmental anatomy of pines for over a hundred years, and the literature is both rich and voluminous. Unlike the situation for *Zea,* there is little coordinate genetic data, but the genus *Pinus* has been used extensively in tissue culture and experimental studies. As a gymnosperm it presents a marked contrast to maize in phylogeny, anatomy, morphology, and physiology. For these reasons and because of my own familiarity with the genus, *Pinus* has been selected for detailed discussion with respect to germination and morphogenesis. The following account will emphasize work done in my laboratory on *Pinus lambertiana;* however, information from other species and laboratories will also be included (Baron, 1966; C. L. Brown and Gifford, 1958; Carpenter, 1941; Fosket and Miksche, 1966; Goo, 1952, 1956; Mirov, 1967; Sacher, 1954, 1956; Smith, 1958; Smith and Silen, 1963; Spurr, 1949, 1950; Stanley, 1958; Stone, 1957; Tepper, 1962, 1964; Unger, 1954).

A. *Dormant Embryo*

1. MATURE SEED AND POLYEMBRYONY

The fully developed mature embryo of pine lies in the corrosion cavity of the massive fleshy female gametophytic tissue (1N) which, in turn, is surrounded by a seed coat that is derived from the central stony layer of the integument. Most of the outer fleshy layer of the integument is removed at maturity except for a membranous wing attached to the adaxial (amicropylar) end of the seed. The inner fleshy layer of the integument is crushed and pressed into a papery membrane that lines the seed coat and wraps around the gametophyte and remnants of the nucellus. The nucellus is reduced to a cap of tissue surrounding the micropylar end of the gametophyte. It often extends about one-fifth the length of the gametophyte. Despite the common occurrence of the multiple fertilization of the several archegonia present in the female gametophytes of *Pinus,* there is reportedly only one embryo per mature seed in 98% of the seeds of the majority of the *Pinus* species (Buchholz, 1946). However, pine exhibits both cleavage and simple polyembryony as a constant feature of embryogenesis (with the exception of species such as *Pinus radiata* that are reported to have only one archegonium per gametophyte in which case only cleavage polyembryony occurs) (Berlyn, 1962, 1967; Berlyn and Passof, 1965). Theoretically, multiple pine embryos could arise from three sources: (*1*) multiple fertilization (simple polyembryony due to the fertilization of more than one archegonium per gametophyte); (*2*) apical tier-derived embryos; and (*3*) rosette embryos. The latter two categories represent cleavage polyembryony. As a consequence, seeds of some species and/or some individuals may show more than one embryo per mature seed; as many as eight embryos (several of which were viable) have been dissected out of a single seed (Berlyn, 1962). The occurrence of prolonged polyembryony in pine is probably a function of climate as well as genetics, but Krugman (1965)* indicates that in some instances it can be a highly heritable trait since certain individual trees show a consistently high percentage of polyembryony year after year despite environmental fluctuations. Johansen (1966)† has made observations on a grove of Mexican pines (a new or little known species) that indicated that all the mature seeds were polyembryonic. Berlyn (1962) reported that a seed lot of *Pinus lambertiana* and *Pinus cembra* exhibited about one-third polyembryonic seeds. In the same study, a lot of *Pinus strobus* seed showed about 7% polyembryony (Table I). Thus, in *P. strobus* there

---

*Dr. Stanley L. Krugman, U.S.F.S. Pacific Southwest Forest and Range Experiment Station, Berkeley, California.

†Dr. Donald A. Johansen, 861 East Columbia Ave., Pomona, California.

TABLE I

Length and Frequency Data from a Random Subsample of Pine
Embryos Excised from Mature Seeds

Measurements	*Pinus* species		
	P. lambertiana	*P. cembra*	*P. strobus*
Mean length of single[a] mature embryos (mm)	8.26	4.55	4.91
95% Confidence interval of true mean of single mature embryos (mm)	7.70–8.82	4.31–4.79	4.74–5.08
Coefficient of variation (%)	16.32	12.64	8.28
Standard deviation	1.348	0.575	0.407
Mean lengths of multiple[b] embryos (mm)			
Embryo 1	5.48	3.33	3.95
Embryo 2	2.01	0.87	0.75
Embryo 3	1.32	0.51	0.25
Embryo 4	0.72	0.21	0.12
Mean summed length of multiple embryos (mm)	8.96	4.28	5.07
Proportion of seeds[c] examined with more than 1 embryo (%)	32.43	30.56	7.41

[a] Sample size for each species is 25 embryos.

[b] Measurements on multiple embryos are on the basis of 41 embryos (total) for *P. lambertiana,* 40 embryos for *P. cembra,* and 8 embryos for *P. strobus.*

[c] Proportions are on the basis of 12 cases of polyembryony out of 37 seeds for *P. lambertiana,* 11 cases out of 36 seeds for *P. cembra,* and 2 cases out of 27 seeds for *P. strobus.*

is usually decisive (and probably early) termination of embryonic selection in favor of a single embryo, whereas in *P. lambertiana* pregerminal embryonic selection is frequently prolonged and may not be terminated decisively in favor of a single embryo.

Questions of viability and origin of the multiple embryos have been considered. Development of two pine seedlings from a single seed has been reported in over twenty species (Toumey, 1923; Jacobs, 1924; Clare and Johnstone, 1931; Gravatt *et al.,* 1940; Johnstone, 1940; M. L. Nelson, 1941; Johansen, 1950; Black, 1960; Berlyn, 1962; Cayford and Waldron, 1965), but the occurrence is very infrequent (e.g., in 1% of 1200 *P. lambertiana* seeds tested by Jacobs, in 1 instance among 1354 *Pinus banksiana* seeds tested by Cayford and Waldron, and in 14 instances among 8464 seeds germinated by Johnstone).

Careful microdissection revealed that polyembryonic seeds often contained more than two embryos (Table I). Furthermore the frequency of polyembryonic seeds was much greater than the frequency of twin seedlings. There are no recorded instances of field germination of more than two embryos. This is probably owing to the fact that upon germination, postgerminal selection occurs, whereupon the largest embryo (or infrequently the larger two embryos) dominates. This, in turn, results from the fact that the larger embryos germinate faster. During initial stages of reactivation the corrosion cavity undergoes considerable enlargement due to hydration and possibly also due to renewed activity of the extracellular enzyme system involved in the genesis of the cavity. The largest embryo sends its radicle down the cavity, and usually this process destroys or isolates subjacent embryos that are not able to reactivate fast enough to extract nutrients from the gametophyte and establish their own germination pattern. The exact mechanism *in vivo* is not known. The mass of gametophyte in a seed of a particular species usually supports a relatively finite amount of pregerminal growth, i.e., the summation of embryo lengths within a seed usually falls within (or close to) the 95% confidence interval including the true mean length of single mature embryos (Table I).

Berlyn (1962) concluded that it was possible to determine the origin of many of the multiple embryos. The rosette tier could be discounted because these embryos were never composed of more than a few hundred cells, had no cotyledons, and they were not viable in a variety of culture media (White's, Knops's, and Hoagland's). At the present level of pine embryo culture technology, it is not possible to culture pine embryos that lack cotyledons. Thus even if the rosette embryos were viable, they would still be too immature for successful embryo culture. In many cases of polyembryony, there was a developmental progression of four embryos attached to a single suspensor system. The largest embryo was always the embryo furthest from the micropyle, and the embryos graded down in size and development, the embryo closest to the micropyle being the smallest. The conclusion was that a single suspensor system containing four embryos represented the cleavage products of the apical initial tier resulting from a single fertilized egg. Polyembryony of other origins was also observed. In one case, eight embryos were present in a single *Pinus lambertiana* seed and, in this instance, there were two separate embryo systems. Three embryos of one system had initiated cotyledons before entering dormancy, and it is likely that these systems resulted from the fertilization of two archegonia. In other cases of polyembryony it is common to find two rather large embryos with the suspensor system or systems totally obliterated. In these situations it is not possible to say anything about the origin of the embryos with respect to cleavage or simple polyembryony. Embryos such as these can very easily be raised to the seedling state by first culturing them on semisolid Knops's solution medium (Berlyn, 1967) and then transferring them to soil (cf. Haddock, 1954).

Embryos of almost every stage from spherical to fully developed may occur in a given seed, reflecting the indeterminate pattern of predormancy developmental competition among embryos. However, only embryos with well-developed cotyledons have been grown to the seedling state, and usually there are no more than two such embryos to a seed. Nevertheless, in a number of instances it has been possible to grow callus from four embryos from a given seed (Berlyn, 1962, 1963). This requires a supplemented medium similar to that of C. L. Brown and Lawrence, (1968).

2. Gross Measurements

De Ferré (1965) has compiled an extensive tabulation of the gross anatomical characteristics of pine embryos with particular reference to cotyledons. The mean number of cotyledons in the genus ranged from 3.80 in *Pinus contorta* to 13.76 in *Pinus coulteri*. Individual variation in cotyledon number and embryo length in *Pinus lambertiana* is presented in Figs. 14 and 15. Mean dormant embryo length was 11.05 mm with a

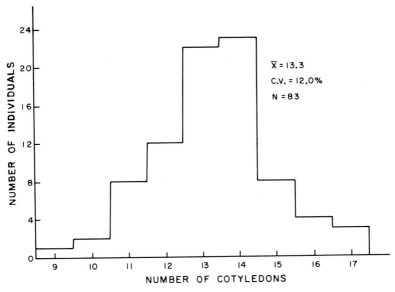

FIG. 14. Step distribution function for cotyledon number for *Pinus lambertiana*. (C.V.) Coefficient of variation. (From Berlyn, 1967.)

coefficient of variation of 9.33%. Average cotyledon number was 13.3 with 14 being the modal value. Cotyledon numbers ranged from 9 to 17 and the coefficient of variation was 12%. Average cotyledon length was 2.8 mm. The average length of the hypocotyl–shoot axis was 3.2 mm, and the root meristem was at least 3 mm below the hypocotyl node in 76%

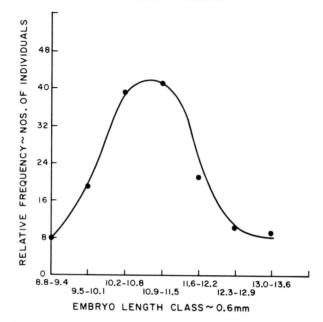

Fig. 15. Frequency distribution function for dormant embryo length in *Pinus lamberti-ana*. (From Berlyn, 1967.)

of the embryos. The juncture zone, the region where the root cap abuts on the hypocotyl, is ca. 0.1 mm above the root initials.

3. ANATOMICAL ORGANIZATION

The essential components of the dormant, mature, sugar pine embryo are cotyledons, epicotyl anlage, hypocotyl, root meristem, massive root cap, and suspensor (Fig. 16). In turn the cotyledons consist of epidermis, mesophyll, procambium, and secretory cells. The cells of the apical meristem are zoned into regions of specialized function and morphology. The hypocotyl consists of epidermal, cortical, secretory, procambial, and pith cells in that order (centripetally), whereas the root cap is composed of the column which is organized as a rib meristem, and a peripheral region.

4. EPICOTYL

The epicotyl of the dormant pine embryo is situated between the over-hanging cotyledons and is a prominent feature in longitudinal section (Fig. 17). It caps the hypocotyl–shoot axis and may be defined in three-space as an elliptical paraboloid. The upper part of the apex is circular in cross section (Fig. 18), but the base is clearly elliptical (Fig. 19). The axes for this sugar pine apex are $320 \times 455\ \mu$ and the height is $150\ \mu$. In

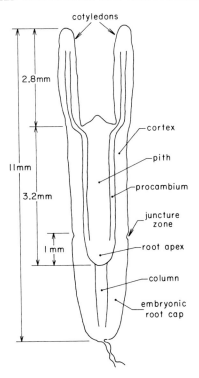

FIG. 16. Diagrammatic representation of mature, dormant, sugar pine embryo. (Redrawn from Greenwood, 1969.)

Pinus ponderosa, Tepper (1962) reports the average cross-sectional axes of four apices was $250 \times 200 \; \mu$.

A detailed view of the dormant sugar pine apex is presented in Fig. 20. It is quite similar to that of ponderosa pine given by Tepper (1962, 1964). The distal cells, the putative apical initials (Esau, 1965a), are located at the uppermost extremity of the apex and show a fairly consistent pattern of anticlinal walls. They have been designated summital cells by Tepper (1962, 1964), a nonspecific term commensurate with our knowledge of their functional role in morphogenesis. Below the summital cells is a very small centrally located zone of cells with unpolarized division planes. This zone represents the future central mother cell zone, and it abuts on the subjacent incipient rib meristem which fans out rapidly into the pith region. Ringing the meristem is the incipient peripheral zone and, although at this stage the zone is not well differentiated by typical histological stains, it has specific optical properties as evidenced by interference microscopy. The outermost layer of this peripheral zone (protoderm) is quite similar to the summital region in cellular characteristics.

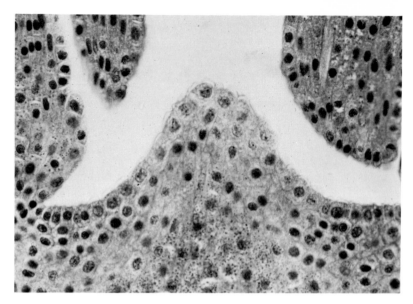

FIG. 17. Stem apical meristem of dormant *Pinus lambertiana* embryo minor axis. Magnification: × 200. (From Berlyn, 1967.)

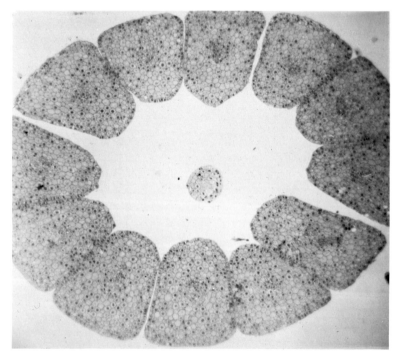

FIG. 18. Cross section of dormant embryonic apex and cotyledons. Magnification: × 60. (From Berlyn, 1967.)

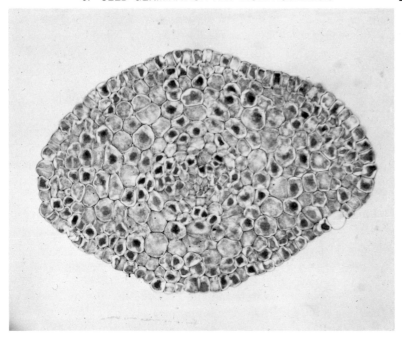

FIG. 19. Cross section of base of dormant embryonic apex. Magnification: × 200. (From Berlyn, 1967.)

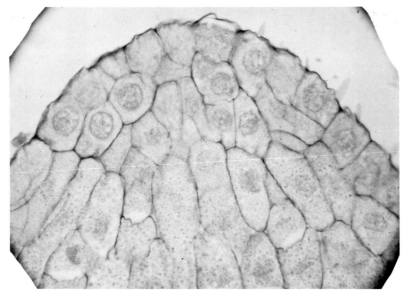

FIG. 20. Stem apex of dormant sugar pine embryo. Magnification: × 400. (From Berlyn, 1967.)

Nuclei of the summital cells are larger, stain less intensely, are more regular in outline, and are provided with more prominent nucleoli in comparison with other cells of the hypocotyl–root axis. Tepper (1962) points out that, although numerous proteinaceous granules (aleurone grains) are present in all cells of the dormant embryo (Figs. 17 and 20), those in the summital cells are less conspicuous. This holds for both *Pinus lambertiana* shown here (Berlyn, 1967) and *Pinus ponderosa* studied by Tepper (1962). The micrographs of these two species are strikingly similar. Note that there are no leaf buttresses or primordia on the dormant embryonic apex. It is in a zero plastochronic condition and exhibits the same general regions described in buds of mature pine trees by Sacher (1954), although the zones are much foreshortened in vertical dimensions. Tepper similarly interprets the dormant embryonic apex of *P. ponderosa* as zonate, although Fosket and Miksche (1966) describe the dormant and even the 5-day apex of *P. lambertiana* as undifferentiated. In my own work on *P. lambertiana,* the dormant apex was considered to be zonate and the zones were even more differentiated in the 5-day apex. However, the environmental conditions used by Fosket and Miksche were different from those employed in my laboratory, and the presence of the zones is to some extent a matter of interpretation. Spurr (1949) did not analyze the dormant embryonic apex of *Pinus strobus,* but its zonal character is evident in his Fig. 22.

Although there are no leaf buttresses present on the dormant embryonic apex, slight excrescences can occasionally be detected well down on the periphery of the meristem (Berlyn, 1967). These excrescences denote the sites of future needle buttresses. A buttress is said to enter the primordium class when its height exceeds the height of its axil (Tepper, 1962).

A cross-sectional view of the dormant stem apical meristem about 60 μ from the top of the apex is shown in Fig. 18. In this sample there are twelve cotyledons, each with a clearly evident procambial strand. A more highly magnified view of the strand is shown in Fig. 21. There is no mature vascular tissue in the entire dormant embryo but in this strand the protophloic procambium can be morphologically separated from the rest of the procambium. The xylic procambium can be located as the procambium most centripetal (adaxial direction) from phloic procambium. The basis for this interpretation lies in observations of the more advanced phases of germination (Berlyn, 1967).

The individual cotyledonary procambial traces (Fig. 18) exist as separate columns for a short distance in the upper portion of the hypocotyl before they coalesce into a procambial ring. Below this insertion point the concentric siphonostelic arrangement of the procambial system is

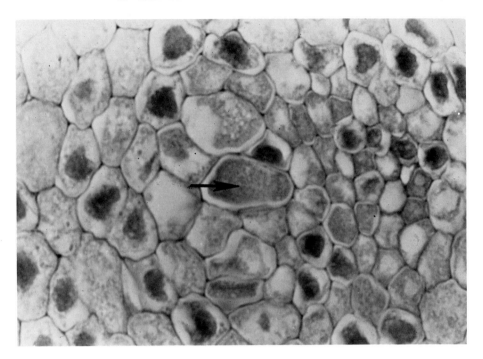

FIG. 21. Cross section of cotyledonary procambial strand in dormant embryo. Protophloic procambium indicated by arrow. Magnification: × 800. (From Berlyn, 1967.)

gradually rearranged into a stele with four to six poles of xylem, indicating a transition from stemlike to rootlike anatomy (Fig. 22). However, a pith is always present in the hypocotyl and may even be retained for 120 or more days in the primary root. The cortex is a hollow cylinder, twelve to fifteen cells thick, and consists of cells with dimensions on the order of 12 μ in length by 18 μ in diameter (Fig. 22). The pith consists of a solid cylinder about fifteen cells in diameter and is composed of cells ca. 36 μ long and 10 μ in cross-sectional diameter (Fig. 22). Bordering the procambium on the cortical side are several layers of perivascular cells which are at least positionally pericycle. These cells are distinguishable from the cortical parenchyma by their smaller dimensions and lack of intercellular spaces (Berlyn, 1967; Greenwood, 1969). Endodermis is not morphologically obvious in the embryonic hypocotyl of sugar pine; nor is there clear evidence for phi layers (Wilcox, 1962) or periderm at this stage. Resin duct cells (incipient epithelial parenchyma) located at the termina of xylic procambial arcs are identified as a circular group of differentially staining, vacuolate cells (Fig. 22). The ducts form schizogenously but only after germination.

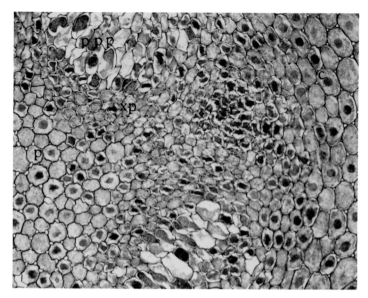

FIG. 22. Cross section of dormant hypocotyl. (ppp) Protophloic procambium; (xp) xylic procambium; (p) pith; (c) cortex; (r) resin duct cells. Magnification: × 200. (From Berlyn, 1967.)

In the hypocotyl–shoot axis, it is again clear that the procambium can be separated into the larger protophloic procambium and the remaining procambium. The delineation of the remaining procambium into meta-phloic procambium, xylic procambium, and residual procambium is not obvious, but by tracing back from the seedling stages the positional re-lationships can be established (Fig. 22). The protophloic procambial cells average about 150 to 250 μ in length* and are approximately 20 μ in diameter. Each cell possesses a single multinucleolar nucleus which is essentially a prolate spheroid with dimensions 40 × 12.5 μ (Figs. 23 and 24). The cytoplasm is multivacuolate and the end walls of these distinc-tive cells are encrusted with a heavily safranophilic material; however, fluorescence microscopy reveals that the substance is not callose (Berlyn, 1966), possibly it is P-protein. Cross sections of the end walls were not obtained in dormant material, presumably because of the massive in-crustation at this stage; however, in 7-day seedlings, it was possible to make preparations of the flattened end walls, and these revealed large simple pits resembling a sieve plate (Fig. 25). The cellulosic filaments are firmly attached to the isotropic matrix. These protophloem sieve cells

*These figures are based on data from several hundred enzyme-macerated cells (Berlyn, 1967).

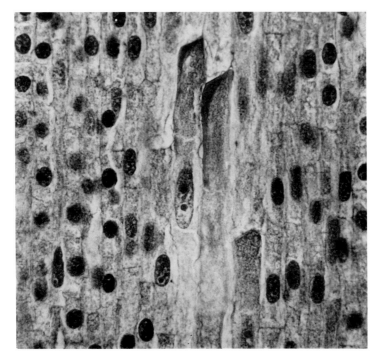

F<small>IG</small>. 23. Longitudinal section of dormant embryonic hypocotyl with protophloic procambium. Magnification: × 400. (From Berlyn, 1967.)

(sometimes referred to as precursory phloem because of their lack of
sieve pitting) are vertically aligned for some distance, and their similarity to angiospermous sieve tubes is apparent (Fig. 23) despite the
heresy of such a comparison. Evert and Alfieri (1965) have also pointed
out this similarity in their observations of secondary coniferous phloem
tissue.

In contrast to the massive protophloic procambial cells the xylic procambial cells are much reduced in dimension, averaging on the order of
70–120 μ long × 10–12 μ in diameter (Fig. 26). The multinucleolar nuclei
are longer and narrower, averaging 35 μ long × 7 μ in diameter. They
are also prolate spheroids and possess a complex internal structure possibly including membranes, plastids, and vacuoles. These xylic nuclei
do not degenerate during germination as rapidly as the phloic nuclei, and
the variations in dimension and structure of the xylic nuclei appear to be
covariate with the pattern of differentiation, particularly with the phases
of wall formation.

The secretory cells (subdermal and intracortical) are still intact in the

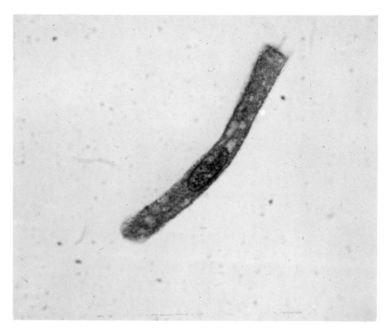

FIG. 24. Maceration of protophloic procambial cell. Magnification: × 400. (From Berlyn, 1967.)

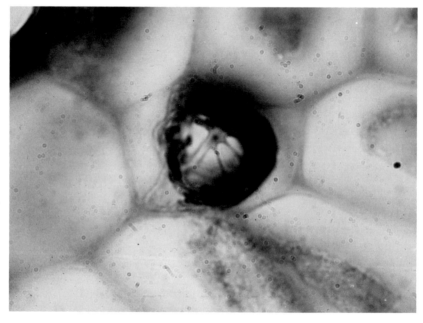

FIG. 25. End wall and pitting in protophloic procambium. Magnification: × 2500. (From Berlyn, 1967.)

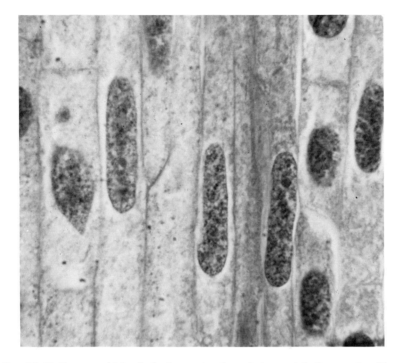

FIG. 26. Xylic procambial cells in dormant embryonic hypocotyl of sugar pine. Magnification: × 1000. (From Berlyn, 1967.)

mature embryo (Fig. 27), but their lightly staining nuclei (Fig. 28) and crystalline packed nucleoli give evidence of senescence [see Kordan (1964) for a report on birefringent materials in nucleoli]. The remarkably large nuclei of these cells are ca. 80 μ long × 13 μ in diameter. Spurr (1949, 1950) speculates that secretory cells may have a translocatory function in embryogeny which is transformed to excretory function in the mature and germinating pine embryo. Shortly after germination, secretory cells fill with resin and other extractives prior to degeneration of the cells. Presumably upon germination the differentiating resin duct cells form a functional resin duct system and thus also assume an excretory role.

5. ROOT MERISTEM AND ROOT CAP

The root initials in mature pine embryos appear to be identical in structure and function within the genus and they appear "in embryo after embryo," as described by Chamberlain (1935, p. 270). Chamberlain indicates that the spherical group of initials are nonspecific in function and that no dermatogen or calyptrogen (Hanstein's histogens) exists within the initial group; however, Spurr (1949) and Berlyn (1967) estab-

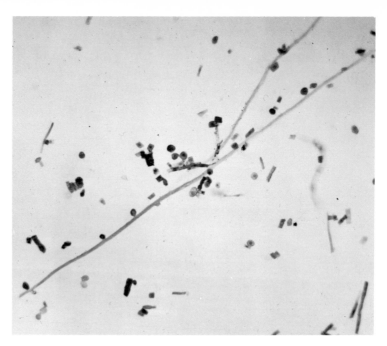

FIG. 27. Secretory cell, enzyme maceration of sugar pine. Magnification: × 60. (From Berlyn, 1967.)

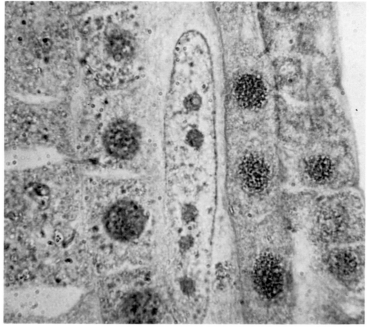

FIG. 28. Secretory cell nucleus in dormant embryonic hypocotyl of sugar pine. Magnification: × 1000. (From Berlyn, 1967.)

lished that at least the column of the root cap is the product of discrete initials. Chouinard (1959) stated that the pine root has a quiescent center, but a comprehensive analytical study of this root meristem is yet to be made although the problem is currently under investigation in my laboratory.

Occasionally the question has arisen as to whether *Pinus* really has a root cap (de Janczewski, 1874; Flahault, 1878; Spurr, 1949). Spurr's conclusion that *Pinus* has a perfectly respectable root cap, even in the embryo, is supported in the present study. However, the root cap that forms upon germination is physically distinct from the embryonic root cap. In excised embryos (Berlyn and Miksche, 1965, Fig. 6), a separate root cap has clearly formed within the massive embryonic root cap which now forms a cellular casing around the root. This casing may keep pace with root growth for at least 120 days in culture. However, instances also occur (Berlyn, 1967) in which the naked root meristem will burst through the hypocotyl and grow out for a short time without a root cap.

B. *Germinating Embryo and Seedling*

1. PHYSIOLOGICAL AND ANATOMICAL CHANGES IN EARLY GERMINATION

Stone (1957), Stanley (1958), Stanley and Conn (1957), and Goo (1956) have studied the physiology of germinating pine embryos with respect to water absorption and respiration. According to Stanley (1958), germination in sugar pine begins when the embryo contains approximately 23% of the seed moisture. This marks the initiation of the first active state of water absorption which is followed by a steady-state condition; this plateau is then followed by the third phase characterized by rapid water absorption. In *Pinus thunbergii,* cell division supposedly begins in phase 2 and becomes rapid in phase 3 (Goo, 1956). Goo also studied the composition of reserve foods in dormant *P. thunbergii* seed and their subsequent utilization during germination (see, also, Katsuta, 1961). The reserve foods consist of ca. 90% fats and proteins. The fats are rapidly oxidized and there is a migration of nitrogenous compounds from the gametophyte into the embryo. In dormancy the gametophyte also contains considerable starch, but by the time of radicle emergence the starch content of the gametophyte is virtually nil. Conversely, the embryo gains in starch content as germination proceeds. Protein synthesis is not evident in the gametophyte, but, as expected, proceeds rapidly in the embryo during the course of germination.

In the typical epigeous germination of pine seedlings, the radicle elongates first; subsequently the cotyledons begin actively to elongate and are gradually elevated by hypocotyl development. As reported for *P. thunbergii* (Goo, 1952), cell division in sugar pine begins almost simultaneously throughout the embryo with a slight lag in the stem and root meristem regions. Measurable cellular growth occurs in the first day of explantation (of decoated, unstratified seeds).

The first evidences of reactivation of the shoot apex are the mobilization of aleurone grains and increased hydration of the apical initials. Mitosis first occurs in the axils of the cotyledons at sites of needle primordia initiation. Upon explantation to agar tubes (Berlyn, 1967), the apices enter the first plastochron after 3 to 4 days. Figure 29 shows a

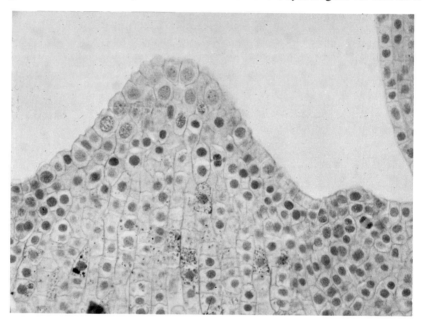

FIG. 29. Longitudinal section of the sugar pine shoot apex 4 days after onset of germination. Magnification: × 200. (From Berlyn, 1967.)

4-day apex with a needle buttress. According to Tepper (1962, 1964) the disappearance of aleurone grains in the apex follows a specific pattern. The grains first disappear from the axils of the cotyledons where the first divisions will occur. However, the grains may be mobilized before division actually commences. The aleurone grains are then mobilized in the apical initials (summital cells) and lastly from the central zone and rib meristem. If one compares Fig. 17 with Fig. 29, the mobilization process can be clearly perceived. The reactivation of the stem apex follows a precise developmental program. The wave of cell division proceeds from the basal peripheral region and moves toward the apical initials. By 8 days in Berlyn's (1967) material and 12 days in Tepper's (1962, 1964) study, the apices were fully reactivated (Fig. 30). At this stage there are several leaf primorida and cytohistological zonation is well-defined. Cells in all regions of the stem apex undergo mitosis. Mitotic figures (late prophase through telophase) are shown in Fig. 31 which is drawn from Tepper's (1962) thesis on *Pinus ponderosa*. The figure is composed of data

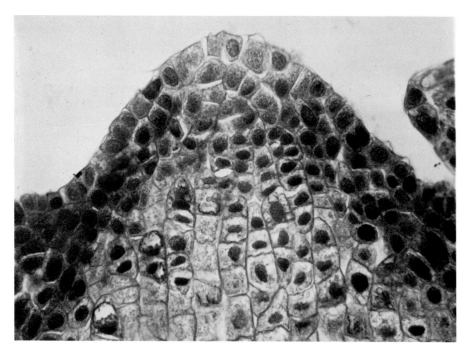

FIG. 30. Longitudinal section of 8-day sugar pine stem apex with clearly defined cyto-histological zonation. Magnification: × 300. (From Berlyn, 1967.)

FIG. 31. Mitotic diagram from the three most median sections of six apices on 12-day-old *Pinus ponderosa* seedlings. (From Tepper, 1962.)

from the three most median sections of six apices from 12-day-old *P. ponderosa* seedlings. The pattern of mitosis indicates that division is most rapid in the peripheral zone and rib meristem and, of course, is most concentrated in developing needle buttresses and primordia.

Chouinard (1959) observed similar patterns in the reactivation of *Pinus banksiana* seedlings as did Unger (1954) in *Pinus resinosa*. Chouinard also claimed to have observed a quiescent center in *P. banksiana,* but Tepper (1962) attributes this to his failure to look at seedlings older than 10 days. I have recently confirmed Teppers conclusions using *Pinus lambertiana.*

2. VASCULAR DIFFERENTIATION AND CYTOHISTOLOGICAL ZONATION

Serial cross sections of 4-day sugar pine embryos reveal a remarkably synchronous pattern of differentiation of the cotyledonary vascular bundles (Figs. 32 and 33). The protophloem matures first and appears to be functionally active in transporting food materials from the gametophyte. In Fig. 33 some of the incipient metaphloem cells are beginning to show thickened, unlignified, secondary walls, and the protoplasts of the first-

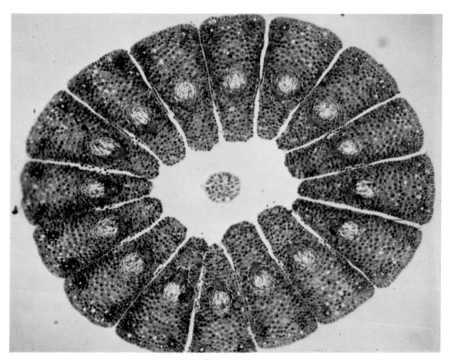

FIG. 32. Cross section of sugar pine shoot apex 4 days after germination. Magnification: × 45. (From Berlyn, 1967.)

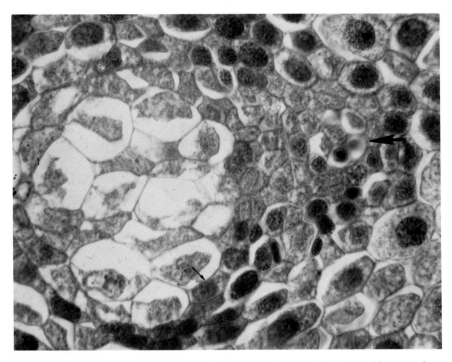

FIG. 33. Transverse section of cotyledonary vascular strand of 4-day-old sugar pine seedling. Large arrow indicates position of first-formed protoxylem tracheids and small arrow indicates wall thickenings of incipient metaphloem. Magnification: × 600. (From Berlyn, 1967.)

formed protoxylem tracheids are degenerating; but the demarcation between xylic and phloic procambium in the central region of the bundle is not visually distinct. Proceeding basipetally into the 4-day hypocotyl (Figs. 34 to 36), the vascular strands are initially separate (Fig. 34) and are similar in anatomy to those depicted for the cotyledon (Fig. 33). These illustrations are constructed by means of partially polarized light which accentuates the pattern of xylem tracheid differentiation and also provides information on the molecular architecture of the cell walls. Two hundred microns further into the hypocotyl the primary vascular tissue appears as a hollow cylinder (Fig. 35) as a result of insertion of the cotyledonary traces. One hundred microns lower the cylindrical symmetry is altered by the organization of xylem arcs. This manifestation of rootlike affinity is expressed in the form of five to six xylem poles with terminal resin ducts (Fig. 36). As one proceeds basipetally down the hypocotyl the state of differentiation decreases until it reaches a minimum in the root initial zone. Thus, starting at the shoot apex, there is first a pattern of

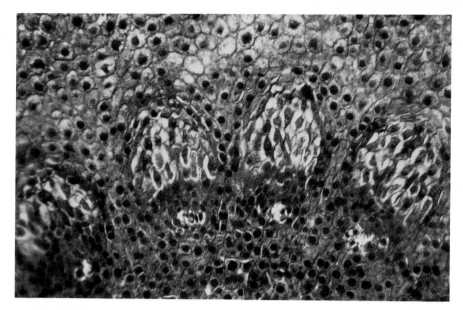

Fig. 34. Transverse section of hypocotyl of 4-day-old sugar pine seedling. Partially polarized light. Magnification: × 150. (From Berlyn, 1967.)

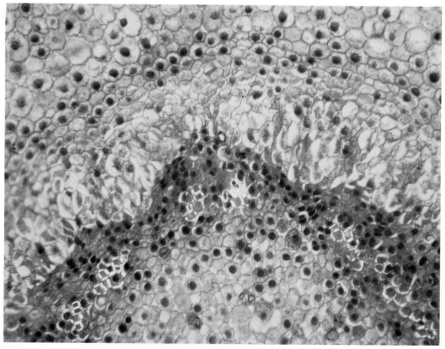

Fig. 35. Transverse section of hypocotyl 200 μ lower than Fig. 24. Magnification: × 150. (From Berlyn, 1967.)

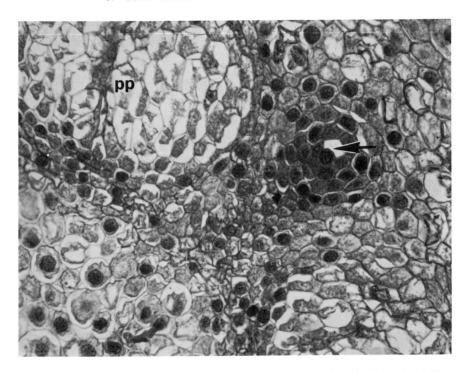

FIG. 36. Transverse section of hypocotyl of same seedling depicted in Figs. 34 and 35. Xylem pole with terminal resin duct (arrow). (pp) Protoxylem. Magnification: × 300. (From Berlyn, 1967.)

increasing differentiation (with a transition from stemlike to rootlike anatomy) reaching a maximum below mid-hypocotyl; subsequently the pattern of differentiation reverses and terminates in the minimally differentiated root meristem region.

According to Greenwood (1969), polar auxin transport is established in sugar pine hypocotyl slices a few hours after explantation from dormant seeds. These hypocotyl slices undergo extensive vascular differentiation and regenerate a root at the basal tip of the tissue slice. The site of regeneration is always opposite a xylem pole. These data suggest that polar auxin transport may also be a key component in the intact germinating seedling.

In addition to the intense mitotic activity in the apical meristems at the 4-day stage, there is also considerable intercalary cell division and elongation in the hypocotyl (Fig. 37). Note the cytoplasmic degeneration in the intracortical secretory cell.

With increasing plastochrons the cytohistological zonation is accen-

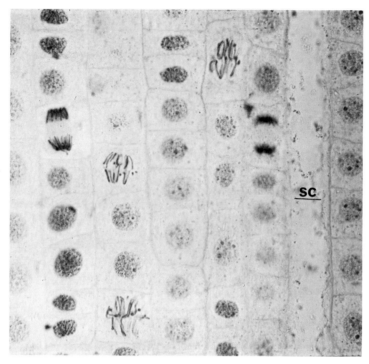

FIG. 37. Longitudinal section of hypocotyl of a 4-day-old sugar pine seedling, showing intercalary meristematic activity. Magnification: × 750. (From Berlyn, 1967.)

tuated (Fig. 30). Cell divisions in the peripheral zone are randomly oriented (unpolarized), but in the rib meristem the preponderance of transverse divisions is obvious. However, Tepper (1962) notes that some longitudinal divisions do occur in the rib meristem, particularly along the outer periphery of the region. These longitudinal divisions result in the contribution of cells to the peripheral zone, but they also add columns of cells to the rib meristem which widens that zone and the meristem. Tepper also notes that there are differences in the relative growth rates of the different zones of the 4–10-day stem apex; there is little vertical growth and thus little or no internodal expansion. This is quite evident in Fig. 38 which shows sixteen needle primordia in various stages of development but nearly in the same plane although some expansion has begun. From 8 to 15 days the seedlings of sugar pine undergo a phase of extremely rapid needle formation coupled with a low rate of internodal expansion. This results in a more rounded and less regularly shaped apical dome (Fig. 39) with mitosis evident in all zones of the apex. Data for the rate of leaf production are not available for this stage, but leaf

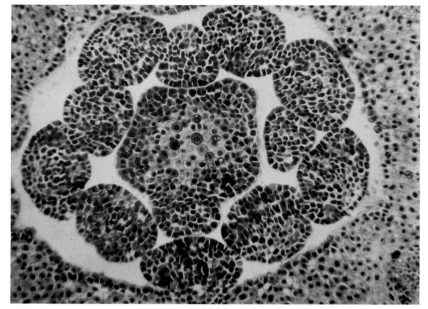

FIG. 38. Transverse section of shoot apex of a 7-day-old sugar pine seedling. Magnification: × 150. (From Berlyn, 1967.)

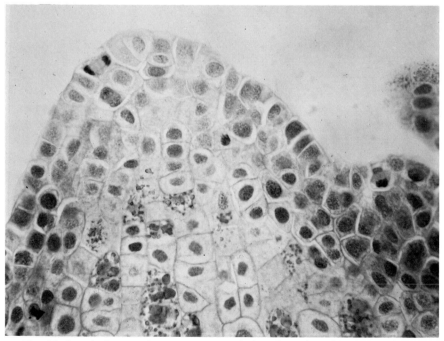

FIG. 39. Longitudinal section of sugar pine apical meristem in rapid organogenic phase. Magnification: × 300. (From Berlyn, 1967.)

production rates for 18 to 87-day apices are given in Table II for *Pinus ponderosa* (Tepper, 1962).

During the 8–15-day period there is considerable primary vascularization in the sugar pine hypocotyl. A maturing vascular bundle from the upper portion of the hypocotyl (just above insertion) is shown in Fig. 40.

TABLE II

LEAF PRODUCTION BY YOUNG *Pinus ponderosa* SEEDLINGS[a]

Age of seedling (days)	Range in number of leaves per seedling	Average number of leaves per seedling	Average number of leaves produced per day
18	12–20	16.4	—
25	22–33	27.0	1.5
32	29–41	34.6	1.1
39	45–56	48.4	2.0
87	82–130	100.0	1.1

[a] From Tepper, 1962.

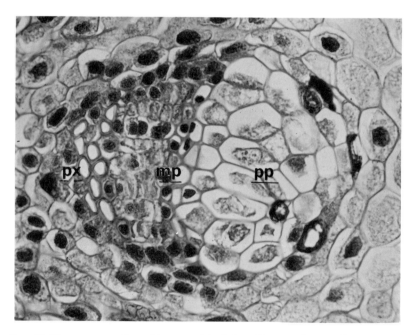

FIG. 40. Transverse section of hypocotyl in 7-day-old sugar pine seedling. (px) Protoxylem; (mp) metaphloem; (pp) protophloem. Magnification: × 600. (From Berlyn, 1967.)

The metaphloem is well-differentiated and the cambial region is becoming evident through boundary differentiation; it is not functional at this stage and, therefore, there is no secondary xylem or phloem. However, the protophloem cells are already beginning to show evidence of senescence, e.g., the protoplasts are highly vacuolate and lightly staining, and the nucleoli are often choked with masses of birefringent material (Fig. 41).

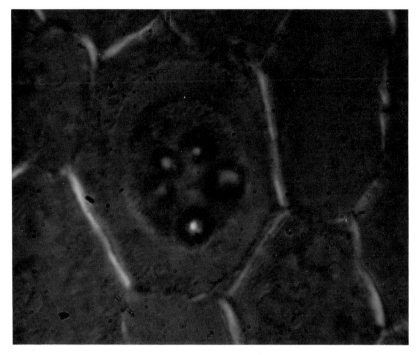

FIG. 41. Transverse section of protophloem nucleus of sugar pine seedling showing birefringent bodies in nucleoli. Polarized light. Magnification: × 2500. (From Berlyn, 1967.)

Note that the cross section of the thickened primary wall exhibits nearly maximum birefringence which indicates a transverse or statistically transverse orientation of the cellulose micelles [see Berlyn (1964, 1969, 1971) and Mark (1967) for a discussion of cell wall fine structure and the analytical procedures used in deducing these phenomena].

Like the protophloic nucleoli, secretory cell nucleoli (Fig. 42) are often packed with birefringent material (starch and unidentified materials, part of which may be calcium oxalate); some nuclei are completely degenerated at this stage. Birefringent material is not limited to the nucleoli but has been observed as small particles throughout the nucleus; however, no large birefringent bodies have been observed outside of the nucleoli.

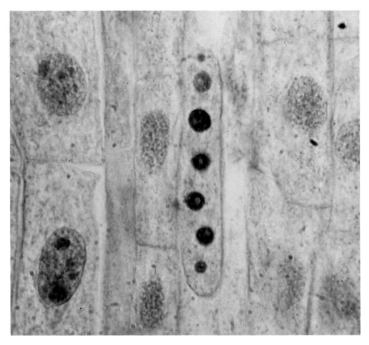

FIG. 42. Longitudinal section of multinucleolar secretory cell nucleus from 4-day sugar pine seedling. Magnification: × 1000. (From Berlyn, 1967.)

In several instances small birefringent bodies have been observed on both sides of the nuclear envelope. According to Kordan and Preston (1967) this type of birefringence may be present in the living state (see discussion of aleurone grains and protein bodies below).

The cytology of cell division in fusiform initials of *Pinus* was investigated by Bailey (1919). He noted that the mitotic figure is placed diagonally across the cell at a 20°–40° angle. Bailey observed that the cell plate was necessarily greatly extended in these cells, which in *Pinus strobus* may attain lengths of 4000 μ. The term kinoplasm was used for a form of fibrous cytoplasm of special function in cell wall formation. The peripheral fibers added to the cell plate were called kinoplasmic fibers and their addition formed a structure referred to as kinoplasma-somes — an alternative term for the phragmoplast. The concept was once thought to be obsolete, but some recent work on microtubules would support Bailey's view that a special fibrous cytoplasm was associated with wall synthesis [Ledbetter and Porter (1963, 1964) and, more recently, Esau *et al.* (1966a,b)]. In fusiform initials, mitosis requires about

5 hours, whereas subsequent cytokinesis (largely phragmoplast movement) requires about 19 hours (Wilson, 1964; Kennedy and Farrar, 1965). Thus not only are karyokinesis and cytokinesis distinctly separated, but also cell plate formation is greatly extended with respect to time and space. The remarkable cell plate formation has been separated into disc, halo, and frame stages by Bailey (1920). This type of division occurs not only in the fusiform initials, but also in the xylem and phloem mother cells. There is also a second fundamental type of cell division in the cambial zone of pine. This type is associated with increase in cambial girth required by the increase in the circumference of the wood body. The problem was first correctly investigated by Robert Hartig (1895, 1901), but the phenomenon was only fully clarified by the work of Bailey (1923). In pine (as in gymnosperms and vesselless dicots in general) the increase in girth is due to pseudotransverse divisions of the overlapping fusiform initials followed by sliding (intrusive) growth as the daughter initials elongate (see Bailey, 1923; Figs. 2 and 3; or Bailey, 1954, p. 26). Approximately 1 to 2% of the cell divisions in a radial file of the cambial zone are pseudotransverse divisions (Wilson, 1964). By the term "radial file" Wilson implies that the pseudotransverse divisions are not limited to the fusiform initial tier. In any case his original model did not recognize the existence of a unicellular initiating layer (see comments by I. W. Bailey following Wilson's paper, 1964, p. 34) but for his specific purpose this was deemed unnecessary.

The xylic procambial cell is an elongated cell that is in fact quite similar in structure to a fusiform initial. Bailey (1919) hypothesized that the dynamics of cell division as observed in the fusiform initials are of a general type and would occur in any cell of dimension large enough to display the phenomenon. This argument may be perhaps more logically stated in reverse in that increased cellular magnitude may require additional complexity in cytokinesis. At any rate the xylary procambial cell exhibits both fundamental types of division described for fusiform initials (viz., tangential-longitudinal and pseudotransverse; Figs. 43 to 45). In Fig. 43 there is the typically canted, telophase division figure. The massed spindle fibers have sufficient birefringence to be resolved by Ortholux-Pol optics used in this study. On one side of this mitotic cell is a protoxylem tracheid with the typical elongated and highly structured nucleus. The secondary wall helices of this cell are already lignified, as evidenced by their autofluorescence (Frey-Wyssling, 1964) at an exciting wavelength of 405 μ (Fig. 44; Frey-Wyssling, 1964). Autofluorescence is also observed in the cytoplasm of cortical parenchyma and in the degenerated secretory cell. The extended phragmoplast (or kinoplasmasomes) (Esau, 1965a, pp. 56–63) is illustrated in Fig. 45. It is conceivable that the

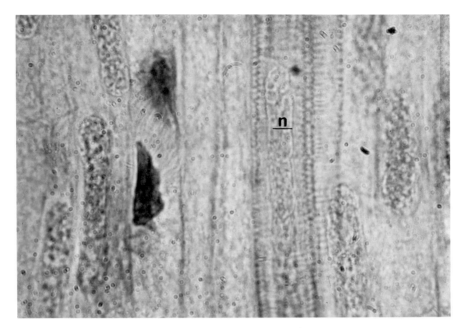

FIG. 43. Longitudinal section of canted division figure in a procambial cell of a 4-day sugar pine seedling. (n) Elongated tracheid nucleus. Magnification: × 1500. (From Berlyn, 1967.)

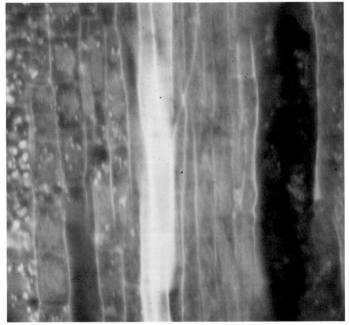

FIG. 44. Fluorescence microscopic view of cells shown in Fig. 43. Magnification: × 500. (From Berlyn, 1967.)

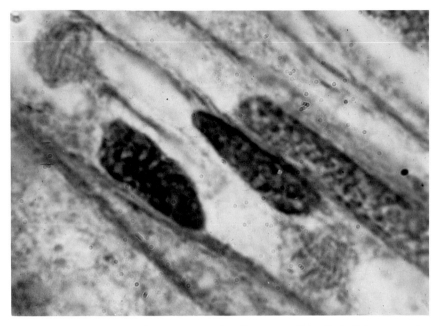

FIG. 45. Longitudinal section of phragmoplast in a procambial cell of a 7-day sugar pine seedling. Magnification: × 2500. (From Berlyn, 1967.)

phragmoplast fibers may give rise to the microtubules. Cytoplasmic microtubules and Golgi vesicles are now thought to be very much concerned with cell wall formation (Ledbetter and Porter, 1963, 1964; Hepler and Newcomb, 1963, 1964; Cronshaw, 1965; Mühlethaler, 1965; Wardrop, 1965; Esau *et al.,* 1966a,b; Mollenhauer and Morré, 1966). Microtubules might possibly function in micellar orientation by directing the movement of sites of synthesis, perhaps the Golgi vesicles, along the wall–cytoplasmic interfaces; the vesicles themselves are postulated to perform the actual synthetic functions.

The cellular morphology, as well as the time of differentiation, of protoxylem tracheids is to some extent a function of their position in the organism; for example, tracheids of the hypocotyl node (Figs. 46 to 48) are often short and irregular in form, whereas those lower in the hypocotyl are generally long and narrow (Fig. 49). In Figs. 46 to 48, a single tracheid (6 days after initiation of germination) is shown under various forms of microscopy. This tracheid was macerated enzymically with pectinase (4% for 10 hours at 37°C). Figure 46 is a phase-contrast view and shows the mixed bordered pits and helical thickenings. In Fig. 47 the use of polarized light shows that the walls are heavily cellulosic and, further-

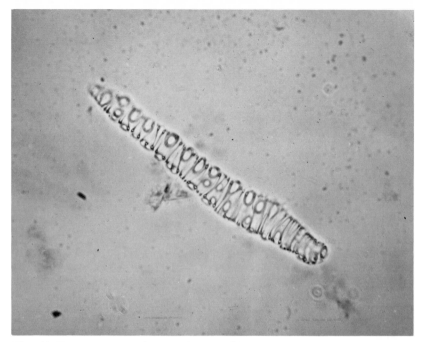

FIG. 46. Enzyme-macerated protoxylem tracheid from 6-day-old sugar pine seedling. Phase contrast microscopy. Magnification: × 600. (From Berlyn, 1967.)

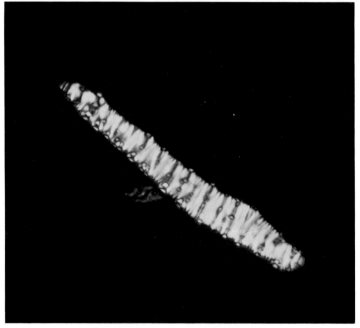

FIG. 47. Same tracheid as in Fig. 46. Polarized light microscopy. Magnification: × 600. (From Berlyn, 1967.)

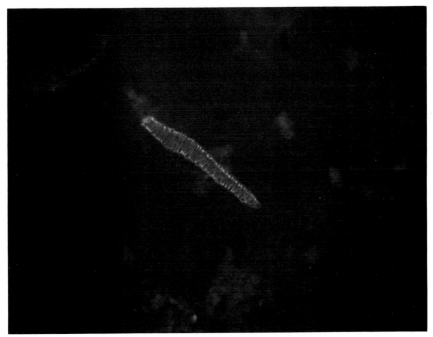

Fig. 48. Same tracheid as in Fig. 46. Fluorescence microscopy. Magnification: × 300. (From Berlyn, 1967.)

more, the near maximum birefringence $(N_\gamma - N_\alpha)$ at this orientation indicates that the direction of the cellulose crystallites is predominantly transverse. In Figure 48 the wall is shown to be autofluorescent which indicates that development has proceeded through the terminal phase of lignification. The protoplast is no longer present and, thus, this tracheid has attained functional maturity.

In contrast the tracheid shown in Fig. 49 is in a much earlier stage of development. The protoplast is still visible and, although helical secondary thickenings have formed, the wall filaments are not as yet lignified. The nucleus of this tracheid is beginning to elongate and is developing a characteristic reticulate appearance. The cytoplasm is highly vacuolate as are the prominent nucleoli. The nucleoli and the cytoplasm still contain RNA at this stage as evidenced by acridine orange staining in combination with fluorescence microscopy. Often, in addition, they contain birefringent materials of unknown composition. Prior to protoplasmic senescence the nuclei of protoxylem tracheids often become enormously elongated (Figs. 50 and 51), attaining lengths of 75 μ or more.

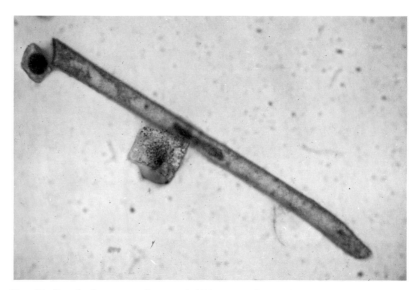

FIG. 49. Developing protoxylem tracheid in sugar pine seedling. Magnification: × 400. (From Berlyn, 1967.)

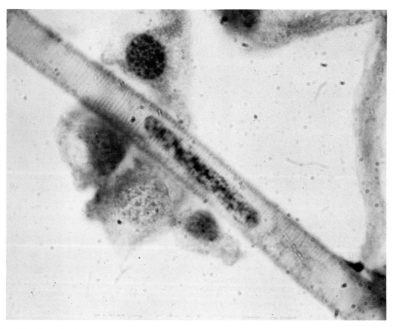

FIG. 50. Developing protoxylem tracheid with partially elongated nucleus. *Pinus lambertiana.* Magnification: × 1000. (From Berlyn, 1967.)

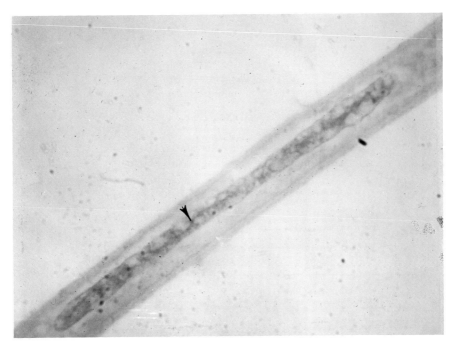

FIG. 51. Developing protoxylem tracheid with fully elongated nucleus. *Pinus lamberti-ana*. Magnification: × 1500. (From Berlyn, 1967.)

3. CELL WALLS AND GROWTH

Cell wall formation and its regulation are critical components of ger-mination and morphogenesis. The higher plant cell is encased in a cel-lulosic cell wall and if the plant cell is to grow and divide so must the wall. In fact, much current research has linked the cell wall to the immediate and basic growth responses of plant cells (Baker and Ray, 1965a,b; Ray and Baker, 1965; Ray, 1967).

Only in rare instances and under special circumstances do cell walls get thinner during growth (Roelofsen, 1959; Baker and Ray, 1965a,b), and in general it can be concluded that both the walls and the cell proper increase in size during growth. Thus it is clear that regulation of growth in higher plants has special properties associated with cell walls, for even if the wall plays no role in induction of growth, provision for wall syn-thesis would have to be concomitant with any growth activity.

The structure and chemistry of the cell wall has been recently reviewed by many authors (Roelofsen, 1959; Berlyn, 1964, 1971; Wardrop, 1964; Frey-Wyssling and Mühlethaler, 1965; Mark, 1967; Jutte, 1969). All higher plant cells are enveloped by at least primary walls (P), and certain types with secondary walls (S). The primary wall can be defined as that part of the cell wall laid down at a point in space and time where cell elongation or surface expansion is occurring. Wall layers laid down in space-time where expansion is not occurring are termed secondary; the latter wall is usually much thicker and composed of several subunits. The walls enclosing plant cells are themselves embedded in an isotropic, amorphous matrix termed the middle lamella (M) The secondary wall of most normal fibers including tracheids, fiber tracheids, and libriform fibers, is further subdivided into three concentric layers: a thin outer shell termed S_1; a thicker middle layer S_2; and a thin inner shell bordering the cytoplasm designated S_3. Vessels may have additional secondary layers termed S_4, S_5, and so forth. Each S layer may itself consist of several lamellae, i.e., 4–6 for S_1 and S_3 and 30–150 for S_2 (Jutte, 1969). In xylary cells that are devoid of protoplasts at functional maturity, there may be a warty layer or terminal lamella adpressed to the inner most layer of the wall (Liese, 1951). The warty layer is considered by some workers to consist of cytoplasmic debris adhering to the inner layer of the secondary wall (Frey-Wyssling, 1959; Liese, 1960; Liese and Ledbetter, 1963). However, Wardrop (1963, 1964) believes that, at least in part, the warts are localized thickenings of the S_3 layer. The warts are quite characteristic for a given species and this fact supports Wardrop's view. The walls are also provided with pits which permit intercellular communication and transport. This communication system has special relevance for germination and morphogenesis. The germination process is heavily dependent on imbibition, hydration, and water transport and the wall–pit system is of crucial importance in these water relations (Berlyn, 1964, 1969).

The cellulose chain molecules are aggregated into elementary fibrils (28–37 Å in diameter; Heyn, 1969) which microscopically appear to form larger threads or microfibrils (100–250 Å) which, in turn, form macrofibrils (ca. 2–5×10^3 Å). Mühlethaler (1967), however, questions this breakdown. He feels that the 35-Å elementary fibrils are the ultimate structural units and *in vivo* are not aggregated into microfibrils or macrofibrils. The cellulose molecules are predominantly crystalline (less than 10% amorphous) and can be conceived of as a one-phase crystalline with order defects system (Berlyn, 1964; Mark, 1967; Heyn, 1969). This cellulosic portion of the wall is termed the framework, and, in addition to the framework, there is the matrix (noncellulosic polysaccharides, hemicelluloses) and the encrusting substances (mainly lignin). The framework functions like the steel bands of reinforced concrete, and the fibrils are wound around the cell in helices. The primary wall has a statically transverse orientation of the helices. However, the outer part of the primary wall shows a good deal of axial dispersion due to stretching of the wall during cell expansion. Layer S_1 has a crossed helical structure (S and Z helices), whereas S_2 consists of steep helices which are unidirectional. They may be either S or Z helices but the S_2 layer commonly has a Z helix. Layer S_3 typically has an S helix with a rather flat winding angle, but Z helices have also been demonstrated (Mark, 1967) and even a crossed helical structure has been reported (Berlyn, 1964).

The cell wall is quite permeable to water, and large amounts of water and perhaps other substances are transported through the wall (Berlyn, 1964, 1969). Berlyn (1969) found that there was about 25% free space in the S_2 layers of *Pinus resinosa* cell walls and the magnitude of free space is probably even greater in the other layers of the wall (Mark, 1967).

A plant cell undergoes a series of phases during its developmental cycle which may be summarized as follows: (*1*) cell division; (*2*) primary wall formation; (*3*) surface area growth; (*4*) cell wall thickening; (*5*) lignification; and (*6*) cell death. Phases *4* and *5* occur in those plant cells forming secondary walls, and phase *5* is further restricted in that even some secondary walls may not lignify (e.g., in sieve cells). The first five phases are reversible in plant cells, and the phases last various lengths of time. In some cases a xylem ray parenchyma cell may not divide for a hundred years or so and then be stimulated to divide and/or even differentiate as a lignified tracheid. This can occur due to heat or frost cracks which create pockets in the xylem into which the parenchyma can proliferate (see, e.g., Berlyn, 1963).

There is good evidence that the carbohydrate fraction of the cell wall is synthesized through sugar nucleotide pathways (Hassid, 1969; Ordin and Hall, 1968). Also, a detailed explanation of lignin biosynthesis is emerging through the elucidation of the shikimic acid and chorismic acid pathways leading to ammonia lyase function and also from the coniferyl alcohol hypothesis of Freudenberg (Freudenberg and Neish, 1968; Gamborg, 1967; S. A. Brown, 1969).

As a seed germinates the processes of cell elongation and wall formation become of central importance. The currently most logical physical concept of primary wall formation is the multinet hypothesis of Roelofsen (1959). As mentioned previously the primary walls have a statistically transverse orientation on their inner side but a more axial distribution on their outer surface. When a layer of the primary wall is laid down the new cellulose microfibrils are oriented transversely (or largely so). As the next layer of wall material is laid down the previous layer will be influenced in three different ways: (*1*) it will be shifted in an outward direction; (*2*) it will be stretched in the axial direction due to continued elongation of the cell and this stretching will cause reorientation of the microfibrils—first into a more-or-less obliquely crossed configuration and, then, gradually into a more axial disposition; and (3) the layer will decrease in thickness as it is stretched and the density of the texture will decrease, resulting in a tenuous, wide-meshed fibrillar network. Roelofsen (1959) likens this situation to a set of superposed fishing nets which are successively stretched in the same direction. In reality the layers may not exist as distinct entities and there may be gradual transitions of microfibril orientation and width of meshes across the primary wall. The multinet theory requires the microfibrils to slide along one another as the nets are stretched and the reality of this concept is open to question.

During elongation, wall material is laid down throughout the thickness of the primary wall, but the cellulose itself is added by apposition. The intercalated material is mostly hemicellulose (Ray, 1967). Unlike animal cells where enlargement is largely protein synthesis, the primary material added to plant cells is water, and water uptake by developing plant cells is accompanied by an irreversible enlargement of the plant cell wall. Peter Ray (1961, 1962, 1967) and co-workers (Baker and Ray, 1965a,b; Ray and Baker, 1965) have done much to elucidate the *modus operandi* of wall formation during cell expansion. By working with oat coleoptiles, these workers have shown a direct effect of auxin on cell wall formation. This effect is a promotion of matrix polysaccharides. Cellulose production (crystalline α-cellulose) is not directly promoted by auxin but occurs as a consequence of the cell elongation induced by auxin. The elongation promotes cellulose synthesis by increasing sugar uptake. In older, nonelongating coleoptiles, secondary wall synthesis may be high but it is not promoted by auxin. Ray states that there are two types of wall formation—extensial synthesis (wall expansion) and intensial synthesis (wall strengthening)—and he concludes that auxin only affects extensial synthesis. However, in woody plants this conclusion is clearly not correct. Reaction wood in conifers can be induced by auxin, and in woody dicots it can be induced by auxin antagonists (Berlyn, 1961; Kennedy and Farrar, 1965; Morey and Cronshaw, 1968; Westing, 1968). A great deal of intensial wall formation is involved in reaction tissue formation. Nevertheless, the biosynthesis and intussusception into the wall of matrix polysaccharides is the primary basis for cell expansion.

The next question that must be answered is where in space-time does cell wall synthesis of germinating plants take place and what is the extent of protoplasm involvement in wall formation? Various locations for wall synthesis have been proposed including: (1) the outer cortex of the protoplast; (2) the surface of the plasmalemma, i.e., the cell wall–cytoplasm interface; (3) the wall itself; and (4) a combination of locations of 2 and 3. The previously discussed physiological researches would indicate that location *4* is the most probable and/or the most common condition. It is well known that lignification begins in the middle lamella and moves centripetally through the primary and secondary wall (Wardrop, 1964, 1965; Berlyn and Mark, 1965; Hepler and Fosket, 1970). Therefore at least part of wall formation occurs within the wall. Hemicelluloses are deposited throughout the wall, but the possibility exists that they are synthesized in the cytoplasm and transported to sites of incorporation within the wall (Mühlethaler, 1967; Marchant and Robards, 1968; Preston, 1964; Robards and Humpherson, 1967; Robards, 1968). Except under unusual circumstances, cellulose is not found within the cytoplasm of plant cells, and it is likely that cellulose synthesis in higher plants occurs at the cell wall–cytoplasm interface. Some recent cytological observations are suggestive of this possibility. In the past few years there have been an increasing number of reports of membrane-bound bodies being incorporated into the cell wall (Arrigoni and Rossi, 1963; Buvat, 1964, 1966; Czaninski, 1966; Esau *et al.,* 1966b; Halperin and Jensen, 1967; Manocha and Shaw, 1964; Marchant and Robards, 1968; O'Brien, 1967; Opik, 1966; Robards, 1968; Skvarla and Larson, 1966; Srivastava and O'Brien, 1966; Walker and Bisalputra, 1967). Of course the role of the Golgi vesicles in the formation and orientation of the cell plate is clearly established (Mollenhauer and Morré, 1966; Mühlethaler, 1967; R. M. Brown, 1969), but whether such vesicles continue to play a role in primary and secondary cell wall formation is open to question. The membranous and vesicular bodies associated with the cytoplasm — cell wall boundary — may, in some instances have their ultimate origin in the Golgi apparatus. These plasmalemma-associated vesicular and membranous structures have been designated paramural bodies by Marchant and Robards (1968). This name does not imply a specific origin and, according to Marchant and Robards, paramural bodies are divided into two classes based on hypothesized derivations: (*1*) lomasomes that are derived from cytoplasmic membranes and (*2*) plasmalemmasomes that are formed entirely from the plasmalemma. Paramural bodies may form discrete membrane-bound structures that contain vesicles, granules, tubules, and/or fibrils. In my observations of germinating pine and lentil seedlings, I have found three types of paramural bodies. The first type is much larger than the other two and contains a fine fibrillar material (Figs. 52 to 55). These large paramural bodies are probably of cytoplasmic origin and, hence, can be designated as lomasomes (see Figs. 52 and 53). They are usually present during the very early stages of primary wall formation and, therefore, could be primarily concerned with pectin and hemicellulose incorporation. Note the long, linear, double membranes of the endoplasmic reticulum with associated polysomes in Fig. 52; these

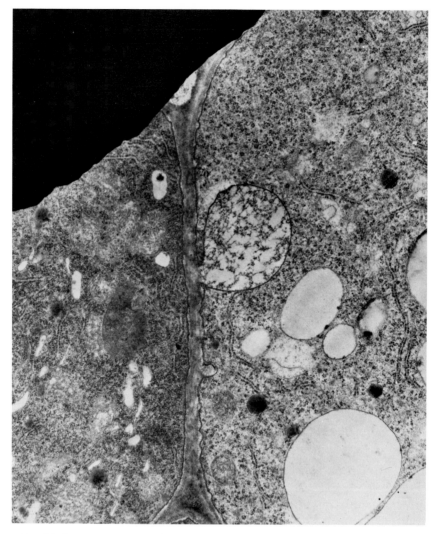

FIG. 52. Large-type paramural body associated with the cell wall in a germinating lentil root. Magnification: × 66,000.

structures are typically present in this form during wall formation. These lomasomes (type 1) are much larger than the Golgi apparatus (Fig. 54), but the fine fibrillar network contained by these lomasomes could be of Golgi origin (Fig. 54). A second type of paramural body is illustrated in Figs. 56 to 58. This type is smaller than type 1 and usually contains tubules or granules. It too is usually associated with

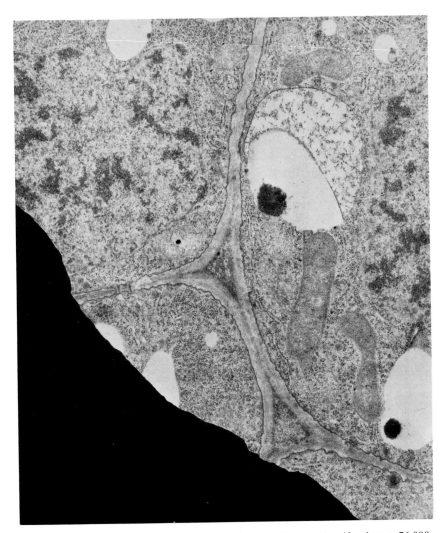

FIG. 53. Large-type paramural body in cytoplasm lentil root. Magnification: × 76,000.

linear polysomes (Fig. 56) and Golgi are often in the vicinity. Occasionally these structures, which could also be of cytoplasmic origin, are seen separated from the wall (Fig. 57). Rarely does this second type of lomasome contain myelinlike membranes (Fig. 58). These membranes have the loose configuration shown by Tamulevich and Evert (1966), but the condensed structures shown by Robards (1968) are also occasionally observed (Fig. 57). The third type of paramural body is the plasmalemmasome which is formed by invagination of the plasmalemma itself (Figs.

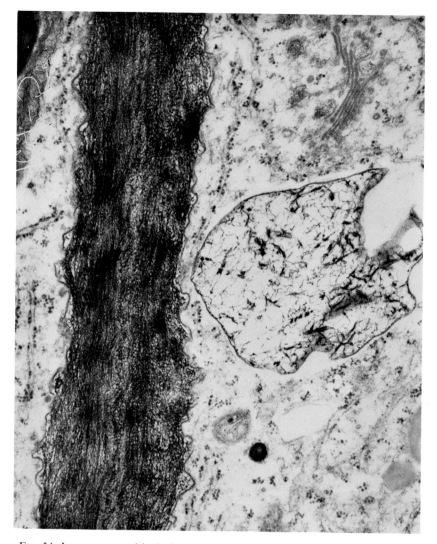

FIG. 54. Large paramural body free in cytoplasm and Golgi apparatus of a germinating sugar pine embryo. Magnification: × 50,000.

59 to 61). The possibility exists, of course, that these structures are simply more advanced stages of the two previously described paramural bodies; however, this is doubtful. Note that the corners of the cell walls have a totally different texture than the rest of the cell wall. The material appears protoplasmic, but no ribosomes have ever been observed in this region and this casts doubt on the possibility that the space is occupied by cyto-plasmic material.

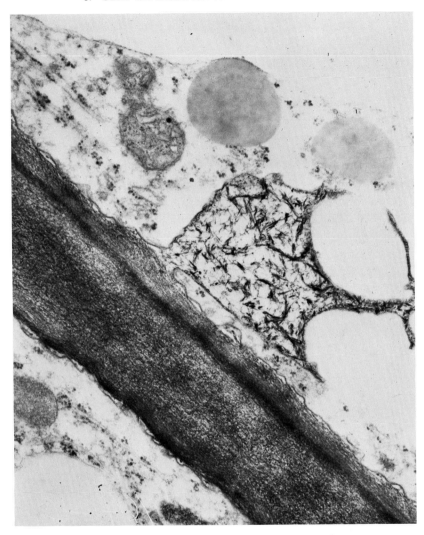

Fig. 55. Large paramural body in contact with cell wall. *Pinus lambertiana*. Magnification: × 60,000.

4. Histochemistry of Stem Apex during Germination

Tepper (1962) has made a detailed histochemical study of the stem apex of the germinating *Pinus ponderosa* seedling. The results were similar to those reported for other species and, consequently, will be discussed in some detail here. Tepper used the periodic acid–Schiff reactions to localize starch and found a rather low starch content in the dormant apex (Fig. 62). Upon germination the starch content first increased (Fig. 63) and

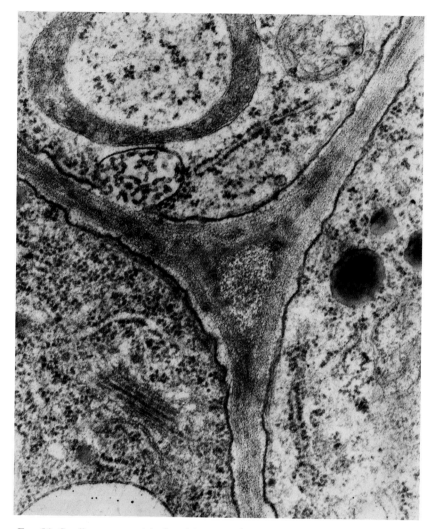

FIG. 56. Smaller paramural body with thicker filaments. Lentil root. Magnification: × 75,000.

then rapidly diminished. By 12 days after initiation of germination, there was only a small amount of starch in the apical and peripheral zones, but starch masses could still be observed in the rib meristem (Fig. 64). Lipoidal material was present in almost all cells of the dormant pine embryo. The use of Oil Red O stain imparted a brillant red coloration to the cells. The summital cells were especially well stained and only the nuclei failed to give a positive test. Tepper postulated that the fat is contained in small

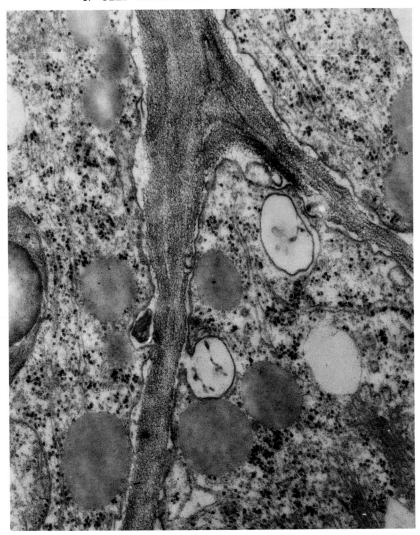

Fig. 57. Smaller paramural bodies in cytoplasm and in communication with cell wall. *Pinus lambertiana*. Magnification: × 54,000.

discrete vacuoles. The rib meristem was significantly lower in fat content than the rest of the meristem and the fat was largely localized in the aleurone grains as was also the case for cells of the hypocotyl–root axis. When the embryos began active germination, the fat content receded, especially at the loci of leaf buttresses and primordia. However, after the onset of germination, the fat content of the summital cells was somewhat higher than the rest of the embryo; however, Tepper could not

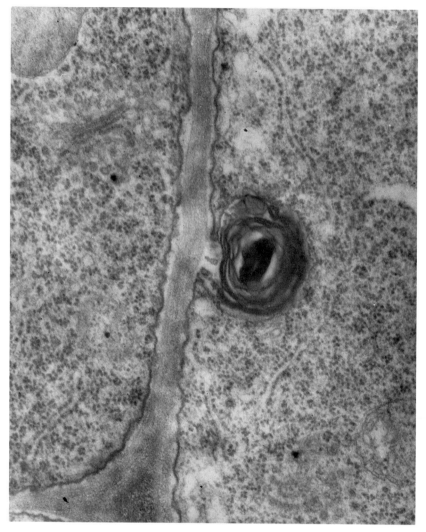

Fɪɢ. 58. Myelinlike figure contained in a paramural body. Lentil root. Magnification: × 57,000.

support Chouinard's (1959) observations on the apical axial cells of 10-day-old *Pinus banksiana* seedlings indicating that the summital cells were rich in lipoidal materials at this time.

Tepper used mercuric bromphenol blue (Mazia *et al.*, 1953) to localize protein. The cytoplasm of the apical initial and peripheral zone cells stained uniformly blue but was not as intensely stained as the aleurone grains of the rib meristem. These results are similar to what Rickson

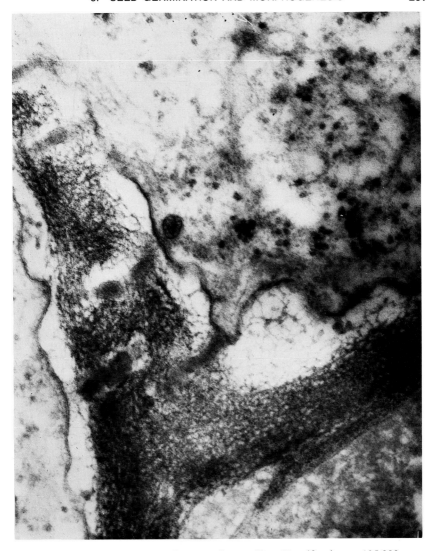

FIG. 59. Plasmalemmasome in sugar pine seedling. Magnification: × 135,000.

(1968) reported for *Paulownia tomentosa* with the exception that Rickson used the term protein bodies instead of aleurone grains; however, it is probable that these are homologous structures. According to Tepper (1969), they may also be identical to the so-called glyoxysomes of Beevers, which are the sites of the glyoxylate pathway in seeds (Breiden-bach *et al.,* 1968). If this is the case, these single membrane-bound struc-tures have multiple functions since they have recently been implicated

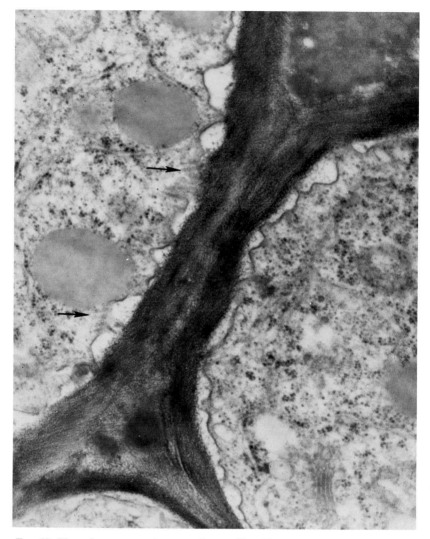

FIG. 60. Plasmalemmasomes in sugar pine seedling. Note microtubules (arrows). Magnification: × 70,000.

in β-oxidation of long-chain fatty acids in germinating castor bean seeds (Hutton and Stumpf, 1969). They have also been termed peroxisomes, and, in barley seed, protein bodies have been shown to contain two acid hydrolases—acid phosphatase and acid protease (Ory and Henningsen, 1969). During germination the protein reserves were mobilized, but as the cells reactivated, they also began to synthesize new protein. This

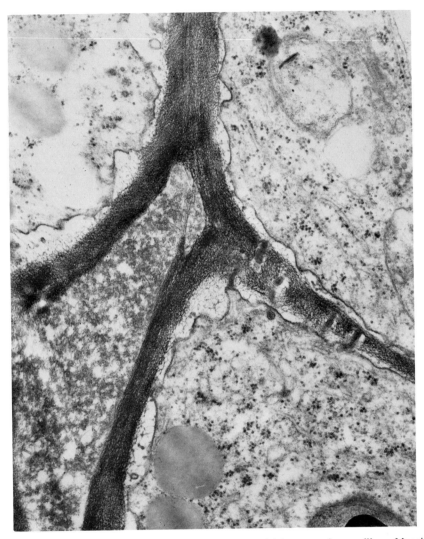

FIG. 61. Plasmalemmasomes and intercellular material in sugar pine seedlings. Magnification: × 63,000.

phase of vigorous protein synthesis culminated in the 12-day seedling and synthesis diminished somewhat thereafter. Tepper used Diazo Blue B to localize protein-bound sulfhydryl groups and, as expected, these localizations followed the pattern of cell divisions. Feulgen staining revealed that nuclei in the apical initial and peripheral zones increased in staining intensity prior to mitosis, whereas stain intensity was relatively constant

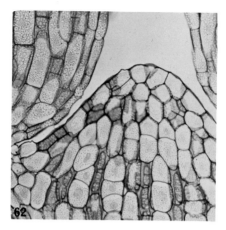

FIG. 62. Distribution of starch in apex of dormant embryo of *Pinus ponderosa*. Magnification: × 270. (From Tepper, 1962.)

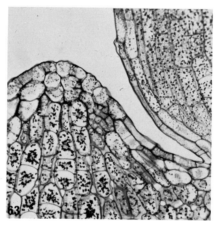

FIG. 63. Distribution of starch in apex of stratified embryo of *Pinus ponderosa*. Magnification: × 270. (From Tepper, 1962.)

in the rib meristem zone. Tepper hypothesized that the summital and peripheral nuclei are in the 2C (i.e., G_1) state in the dormant embryonic apex, whereas the nuclei in the rib meristem are in the 4C (G_2) state.

Histone content was studied using the basic Fast Green procedure of Alfert and Geschwind (1953). Again the histone content paralleled that of DNA (Figs. 65 to 67). Nuclei of the dormant summital cells were extremely low in histone (Fig. 65), but the rest of the embryonic nuclei contained significant amounts of basic protein. There was also some

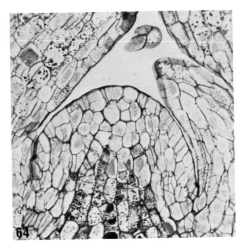

FIG. 64. Distribution of starch in apex of 12-day-old seedling of *Pinus ponderosa*. Magnification: × 270. (From Tepper, 1962.)

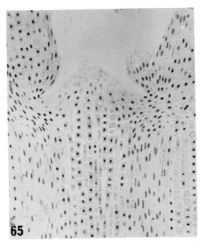

FIG. 65. Localization of histone in apex of dormant *Pinus ponderosa* embryo. Magnification: × 135. (From Tepper, 1962.)

cytoplasmic staining in the dormant apex which was missing at 5 days (Fig. 66) but which reappeared to a lesser degree in older seedlings (Fig. 67). In the older seedlings, the histone concentration of the summital cells appeared lower than that of the rest of the apex, but no quantitative data were recorded. Volume as well as stain intensity determines the total

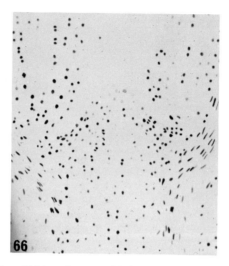

FIG. 66. Localization of histone in apex of 5-day-old seedling of *Pinus ponderosa*. Magnification: × 135. (From Tepper, 1962.)

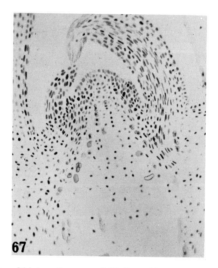

FIG. 67. Localization of histone in apex of 12-day-old *Pinus ponderosa* seedling. Magnification: × 135. (From Tepper, 1962.)

mass of basic protein present in each cell and visual estimates of concentration differences have to be taken with caution.

Pyronine Y was used to determine the distribution and developmental fluctuation of RNA in these germinating pine seedlings. Tepper checked the verisimilitude of the stain with RNase and found that the red stain was removed or greatly reduced by this treatment in many cellular structures, e.g., chromocenters and nucleoli of interphase nuclei, chromosomes, mitotic spindles, and phragmoplasts. The cytoplasm was also found to be rich in RNA, and RNA granules (bodies) were observed in the cytoplasm. These RNA bodies were most notably present in the cotyledons, cortex, and pith; they were, however, absent from the shoot apex. Tepper states that the RNA granules were commonly localized in aleurone grains and likens these RNA bodies to those found in the proteid vacuoles of pine archegonia by Takao (1959). Nevertheless, as noted by Tepper, Takao (1960) did not report these structures as present in the mature embryo of *Pinus thunbergii*. As germination began, there was, as expected, a rapid increase in RNA synthesis (Fig. 68) and the apical zone stained

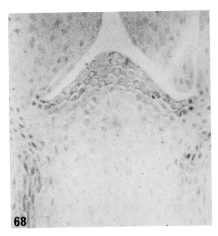

FIG. 68. Distribution of RNA in apex of stratified *Pinus ponderosa* embryo. Magnification: × 135. (From Tepper, 1962.)

somewhat less intensely than the peripheral and rib meristem zones. The RNA bodies were mobilized during the germination process and could not be detected in the 12-day seedling; however, total cytoplasmic RNA staining culminated at this point. Cell walls in the shoot apices, especially in the dormant condition, stained heavily with Pyronine Y, and such staining was removable with RNase leading Tepper to conclude that RNA is in some manner connected with wall development at least in these cells (Figs. 68 and 69).

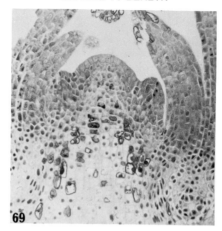

F<small>IG</small>. 69. Distribution of RNA in apex of 12-day-old *Pinus ponderosa* seedling. Magnification: × 135. (From Tepper, 1962.)

REFERENCES

Abbe, E. C., and Stein, O. L. (1954). The growth of the shoot apex in maize: Embryogeny. *Amer. J. Bot.* **41**, 285–293.

Abbe, E. C., Phinney, B. O., and Baer, D. F. (1951). The growth of the shoot apex in maize: Internal features. *Amer. J. Bot.* **38**, 744–751.

Alfert, M., and Geschwind, I. I. (1953). A selective staining method for the basic proteins of cell nuclei. *Proc. Nat. Acad. Sci. U.S.* **39**, 991–999.

Arrigoni, O., and Rossi, G. (1963). I lomasomi & loro probabili rapporti con la crescita per distensione della parcte cellulare. *G. Bot. Ital.* **70**, 476–481.

Avebury, J. L. (1892). "A Contribution to Our Knowledge of Seedlings." Appleton, New York.

Avery, G. S., Jr. (1930). Comparative anatomy and morphology of embryos and seedlings of maize, oats, and wheat. *Bot. Gaz.* **89**, 1–39.

Bailey, I. W. (1919). Phenomena of cell division in the cambium of arborescent gymnosperms and their cytological significance. *Proc. Nat. Acad. Sci. U.S.* **5**, 283–285.

Bailey, I. W. (1920). The formation of the cell plate in the cambium of the higher plants. *Proc. Nat. Acad. Sci U.S.* **6**, 197–200.

Bailey, I. W. (1923). The cambium and its derivative tissues. IV. The increase in girth of the cambium. *Amer. J. Bot.* **10**, 499–509.

Bailey, I. W. (1954). "Contributions to Plant Anatomy." Chronica Botanica, Waltham, Massachusetts.

Baker, D. B., and Ray, P. M. (1965a). Direct and indirect effects of auxin on cell wall synthesis and cell elongation. *Plant Physiol.* **40**, 345–352.

Baker, D. B., and Ray, P. M. (1965b). Relation between the effects of auxin on cell wall synthesis and cell elongation. *Plant Physiol.* **40**, 360–368.

Baldovinos De La Pena, G. (1953). Growth of the root tip. *In* "Growth and Differentiation in Plants." (W. E. Loomis, ed.), pp. 27–54. Iowa State Coll. Press, Ames.

Baron, F. J. (1966). Embryo growth and seed germination of sugar pine (*Pinus lambertiana* Dougl.). *Advan. Front. Plant Sci.* **17**, 1–13.

Bergonié, J., and Tribondeau, L. (1959). Interpretation of some results of radiotherapy and an attempt at determining a logical technique of treatment. *Radiat. Res.* **11,** 587–588.

Berlyn, G. P. (1961). Factors affecting the incidence of reaction tissue in *Populus deltoides* Bartr. *Iowa State J. Sci.* **35,** 367–424.

Berlyn, G. P. (1962). Developmental patterns in pine polyembryony. *Amer. J. Bot.* **49,** 327–333.

Berlyn, G. P. (1963). Methacrylate as an embedding medium for woody tissues. *Stain Technol.* **38,** 23–28.

Berlyn, G. P. (1964). Recent advances in wood anatomy: The cell walls in secondary xylem. *Forest Prod. J.* **14,** 467–476.

Berlyn, G. P. (1967). The structure of germination in *Pinus lambertiana* Dougl. *Yale Univ. Sch. Forest., Bull.* **71.**

Berlyn, G. P. (1969). Microspectrophotometric investigations of free space in plant cell walls. *Amer. J. Bot.* **56,** 498–506.

Berlyn, G. P. (1963). Unpublished data.

Berlyn, G. P. (1966). Unpublished data.

Berlyn, G. P. (1971). Ultrastructural and molecular concepts of cell-wall formation. Wood & Fiber **2,** 196–227.

Berlyn, G. P., and Mark, R. E. (1965). Lignin distribution in wood cell walls. *Forest Prod. J.* **15,** 140–141.

Berlyn, G. P., and Miksche, J. P. (1965). Growth of excised pine embryos and the role of the cotyledons during germination. *Amer. J. Bot.* **52,** 730–736.

Berlyn, G. P., and Passof, P. C. (1965). Cytoplasmic fibrils in proembryo formation in *Pinus. Can. J. Bot.* **43,** 175–176.

Black, T. M. (1960). Abnormal seedlings of *Pinus contorta* Loudon. *Scot. Forest.* **14,** 81–86.

Breidenbach, R. W., Kahn, A., and Beevers, H. (1968). Characterization of glyoxysomes from castor bean endosperm. *Plant Physiol.* **43,** 705–713.

Brown, C. L., and Gifford, E. M. (1958). The relation of the cotyledons to root development of pine embryos grown *in vitro. Plant Physiol.* **33,** 57–64.

Brown, C. L., and Lawrence, R. H. (1968). Culture of pine callus on a defined medium. *Forest Sci.* **14,** 62–64.

Brown, R. M., Jr. (1969). The role of the Golgi apparatus in wall formation in *Pleurochrysis schefferlii. Abstr. Int. Bot. Congr., 11th, 1969,* p. 23.

Brown, S. A. (1969). Biochemistry of lignin formation. *Biol. Sci.* **19,** 115–121.

Brumfield, R. T. (1942). Cell growth and division in living root meristems. *Amer. J. Bot.* **29,** 533–543.

Buchholz, J. T. (1946). Volumetric studies of seeds, endosperms, and embryos in *Pinus ponderosa* during embryonic differentiation. *Bot. Gaz.* **108,** 232–244.

Buvat, R. (1964). Comportement des membranes plasmiques lors della differenciation des parois latérales des vaisseaux (métaxylème de *Cucurbita pepo). C. R. Acad. Sci.* **258,** 5511–5514.

Buvat, R. (1966). Quelques modalités de la differenciation des vaisseaux secondaires du xylème de *Cucurbita pepo. Electron Microsc., Proc. Int. Congr., 6th, 1966* Vol. 2, pp. 311–312.

Byrne, J. M., and Heimsch, C. (1968). The root apex of *Linum. Amer. J. Bot.* **55,** 1011–1019.

Caldecott, R. S., and Smith, L. (1952). A study of x-ray-induced chromosomal aberrations in barley. *Cytologia* **17,** 224–242.

Carpenter, A. M. (1941). Seedling anatomy of certain Pinaceae. *Bull. Pittsburgh Univ.* **37,** 62–69.

Cayford, J. H., and Waldron, R. M. (1965). Multiple jack pine seedlings. *Can. J. Bot.* **43,** 481–482.

Celakovsky, L. J. (1897). Über die Homologien des Grasembryos. *Bot. Z.* **55**, 141–174.

Chamberlain, C. J. (1935). "Gymnosperms Structure and Evolution." Univ. of Chicago Press, Chicago.

Chouinard, L. (1959). I. Structure et fonctionnement de l'apex caulinaire de *Pinus banksiana* Lamb. au cours de la germination. II. Sur l'existence d'un centre quiescent au niveau de l'apex radiculaire juvenile de *Pinus banksiana* Lamb. *Laval Univ. Forest Res. Contrib.* No. 4.

Clare, T. S., and Johnstone, G. R. (1931). Polyembryony and germination of polyembryonic coniferous seed. *Amer. J. Bot.* **18**, 674–683.

Clowes, F. A. L. (1954). The promeristem and the minimal construction centre in grass root apices. *New Phytol.* **53**, 108–116.

Clowes, F. A. L. (1956). Nucleic acids in root apical meristems of *Zea*. *New Phytol.* **55**, 29–34.

Clowes, F. A. L. (1958). Protein synthesis in root meristems. *J. Expt'l. Bot.* **9**, 229–238.

Clowes, F. A. L. (1959). Apical meristems of roots. *Biol. Rev.* **34**, 501–529.

Clowes, F. A. L. (1961). "Apical Meristems." Davis, Philadelphia, Pennsylvania.

Clowes, F. A. L. (1963). The quiescent center in meristems and its behavior after irradiation. *Brookhaven Symp. Biol.* **16**, 46–58.

Clowes, F. A. L. (1969). Anatomical aspects of structure and development. *In* "Root Growth" (W. J. Whittington, ed.), pp. 3–19. Butterworth, London.

Cook, F. S. (1959). The generative cycles of protein and transaminase in the growing corn radicle. *Can. J. Bot.* **37**, 621–639.

Cronshaw, J. (1965). Cytoplasmic fine structure and cell wall development in differentiating xylem elements. *In* "Cellular Ultrastructure of Woody Plants" (W. A. Côté, ed.), pp. 99–124. Syracuse Univ. Press, Syracuse, New York.

Czaninski, Y. (1966). Aspects infrastructuraux de cellules contiguës aux vaisseaux dans le xylème de *Robinia pseudo-acacia*. *C. R. Acad. Sci.* **262**, 2336–2339.

D'Amato, F. (1952). Polyploidy in the differentiation and function of tissues and cells in plants. A critical examination of the literature. *Caryologia* **4**, 311–358.

de Ferré, Y. (1965). Structure des plantules et systematique du genre Pinus. *Trav. Lab. Forest. Toulouse* **3**, 18–50.

de Janczewski, E. (1874). Recherches sur l'accroissement terminal des racines dans les phanérogames. *Ann. Sci. Nat.: Bot. Biol. Veg.* [5] **20**, 162–201.

Eames, A. J. (1961). "Morphology of the Angiosperms." McGraw-Hill, New York.

Erickson, R. O., and Sax, K. B. (1956). Elemental growth rate of the primary root of *Zea mays*. *Proc. Amer. Phil. Soc.* **100**, 487–498.

Esau, K. (1943). Ontogeny of the vascular bundle in *Zea mays*. *Hilgardia* **15**, 327–356.

Esau, K. (1965a). "Plant Anatomy," 2nd ed. Wiley, New York.

Esau, K. (1965b). "Vascular Differentiation in Plants." Holt, New York.

Esau, K., Cheadle, V. I., and Gill, R. H. (1966a). Cytology of differentiating tracheary elements. I. Organelles and membrane systems. *Amer. J. Bot.* **53**, 756–764.

Esau, K., Cheadle, V. I., and Gill, R. H. (1966b). Cytology of differentiating tracheary elements. II. Structures associated with cell surfaces. *Amer. J. Bot.* **53**, 765–771.

Evenari, M., Klein, S., Anchori, H., and Feinbrun, N. (1957). The beginning of cell division and cell elongation in germinating lettuce seeds. *Bull. Res. Counc. Isr., Sect. D* **6**, 33–37.

Evert, R. F., and Alfieri, F. J. (1965). Ontogeny and structure of coniferous sieve cells. *Amer. J. Bot.* **52**, 1058–1066.

Fahn, A. (1967). "Plant Anatomy" (translated from the Hebrew by Sybil Broido-Altman). Pergamon, Oxford.

Flahault, C. (1878). Recherches sur l'accroissement terminal de la racine chez les phanér-ogames. *Ann. Sci. Nat.: Bot Biol. Veg.* [6] **6**, 1–168.

Foard, D. E., and Haber, A. H. (1961). Anatomic studies of gamma-irradiated wheat growing without cell division. *Amer. J. Bot.* **48**, 438–446.

Fosket, D. E., and Miksche, J. P. (1966). A histochemical study of the seedling shoot apical meristem of *Pinus lambertiana*. *Amer. J. Bot.* **53**, 694–701.

Freudenberg, K., and Neish, A. C. (1968). "Constitution and Biosynthesis of Lignin." Springer Publ., New York.

Frey-Wyssling, A. (1959). "Die pflanzliche Zellwand." Springer-Verlag, Berlin and New York.

Frey-Wyssling, A. (1964). Ultraviolet and fluorescence optics of lignified cell walls, p. 153–167. In M. Zimmerman (ed.) The formation of wood in forest trees. Academic Press, New York.

Frey-Wyssling, A., and Mühlethaler, K. (1965). "Ultrastructural Plant Cytology." American Elsevier, New York.

Gamborg, O. L. (1967). Aromatic metabolism in plants. V. The biosynthesis of chlorogenic acid and lignin in potato cell cultures. *Can. J. Biochem.* **45**, 1451–1457.

Gelinas, D., Postlethwait, S. N., and Nelson, O. E. (1969). Characterization of development in maize through the use of mutants. II. The abnormal growth conditioned by the knotted mutant. *Amer. J. Bot.* **56**, 671–678.

Goddard, D. R., and Meeuse, B. J. D. (1950). Respiration of higher plants. *Annu. Rev. Plant Physiol.* **1**, 207–232.

Goo. M. (1951). Water absorption by tree seeds. *Bull. Tokyo Univ. Forests* **39**, 55–60.

Goo, M. (1952). When cell division begins in germinating seeds of *Pinus thunbergii*. *J. Jap. Forest. Soc.* **34**, 3.

Goo, M. (1956). A physiological study of germination of coniferous seed by the application of water absorption curve. *Bull. Tokyo Univ. Forests* **51**, 159–236.

Gravatt, A. R., Latham, D. H., Jackson, L. W. R., Young, G. Y., and Davis, W. C. (1940). Multiple seedlings of pines and Douglas fir. *J. Forest.* **38**, 818.

Greenwood, M. S. (1969). Studies on the regeneration of roots by hypocotyl segments from the dormant, mature embryo of *Pinus lambertiana* Dougl. Unpublished Ph.D. Thesis, Yale University Library, New Haven, Connecticut.

Greenwood, M. S., and Berlyn, G. P. (1965). Regeneration of active root meristems *in vitro* by hypocotyl sections from dormant *Pinus lambertiana* embryos. *Can. J. Bot.* **43**, 173–174.

Greenwood, M. S., and Berlyn, G. P. (1968). Feulgen cytophotometry of pine nuclei; effects of fixation, role of formalin. *Stain Technol.* **43**, 111–117.

Guignard, J. L. (1961). Recherches sur l'embryogênie des Graminées; rapports des Graminées avec l'autres Monocotylédones. *Ann. Sci. Nat.: Bot. Biol. Veg.* [12] **2**, 491–610.

Haber, A. H. (1968). Ionizing radiations as research tools. *Annu. Rev. Plant Physiol.* **19**, 463–489.

Haber, A. H., and Foard, D. E. (1964). Further studies of gamma-irradiated wheat and their relevance to use of mitotic inhibition for developmental studies. *Amer. J. Bot.* **51**, 151–159.

Haber, A. H., and Luippold, H. J. (1960a). Effects of gibberellin on gamma-irradiated wheat. *Amer. J. Bot.* **47**, 140–144.

Haber, A. H., and Luippold, H. J. (1960b). Separation of mechanisms initiating cell division and cell expansion in lettuce seed germination. *Plant Physiol.* **35**, 168–173.

Haber, A. H., Carrier, W. L., and Foard, D. E. (1961). Metabolic studies of gamma-irradiated wheat growing without cell division. *Amer. J. Bot.* **48**, 431–438.

Haddock, P. G. (1954). Sapling sugar pines grown from excised mature embryos. *J. Forest.* **52**, 434–437.

Halperin, W., and Jensen, W. A. (1967). Ultrastructural changes during growth and embryogenesis in carrot cell cultures. *J. Ultrastruct. Res.* **18**, 428–443.

Hanstein, J. (1868). Die Scheitelzellgruppe im Vegetationspunkt der Phanerogomen. Festschr. Niederrhein. Ges. Natur- und Heilkunde 1868: 109–134.

Hartig, R. (1895). Ueber den Drewuchs der Kiefer. *Sitzungsber. Bayer. Akad. Wiss., Math-Phys. Kl.* **25**, 199–217.

Hartig, R. (1901). "Holzuntersuchungen Altes und Neues." Springer-Verlag, Berlin and New York.

Hassid, W. Z. (1969). Biosynthesis of oligosaccharides and polysaccharides in plants. *Science* **165**, 137–144.

Hayward, H. E. (1938). "The Structure of Economic Plants." Macmillan, New York.

Hepler, P. K., and Fosket, D. E. (1970). Lignification during secondary wall formation in *Coleus:* An electron microscopic study. *Amer. J. Bot.* **57**, 85–96.

Hepler, P. K., and Newcomb, E. H. (1963). The fine structure of young tracheary xylem elements arising by redifferentiation of parenchyma in wounded *Coleus* stem. *J. Exp. Bot.* **14**, 496–503.

Hepler, P. K., and Newcomb, E. H. (1964). Microtubules and fibrils in cytoplasm of *Coleus* cells undergoing secondary wall deposition. *J. Cell Biol.* **20**, 529–533.

Heyn, A. N. J. (1969). The elementary fibril and supermolecular structure of cellulose in softwood fiber. *J. Ultrastruct. Res.* **26**, 52–68.

Huskins, C. L. (1947). The subdivisions of the chromosomes and their multiplication in non-dividing tissues: Possible interpretations in terms of gene structure and gene action. *Amer. Natur.* **81**, 401–434.

Hutton, D., and Stumpf, P. K. (1969). Fat metabolism in higher plants. XXXVII. Characterization of the β-oxidation systems from maturing and germinating castor bean seeds. *Plant Physiol.* **44**, 508–516,

Jacobs, A. W. (1924). Polyembryonism in sugar pine. *J. Forest.* **22**, 573–574.

Jacques-Felix, H. (1958). Sur une interprétation nouvelle de l'embryon des Graminées. Conséquences terminologiques et rapports avec les autres types d'embryons. *Acad. des Sci. Compt. Rend.* **246**, 150–153.

Jensen, W. A., and Kavaljian, L. G. (1958). An analysis of cell morphology and the periodicity of division in the root tip of *Allium cepa. Amer. J. Bot.* **45**, 365–372.

Johansen, D. A. (1950). "Plant Embryology." Chronica Botanica, Waltham, Massachusetts.

Johansen, D. A. (1966). Personal communication.

Johnstone, G. R. (1940). Further studies on polyembryony and germination of polyembryonic pine seeds. *Amer. J. Bot.* **27**, 808–811.

Jutte, S. (1969). A comparative study on normal and tension wood fibres in beech (*Fagus sylvatica* L.) and ash (*Fraxinus excelsior* L.). Doctoral Dissertation, Rijks-Universiteit, Leiden (Nederland).

Katsuta, M. (1961). The break-down of reserve protein of pine seeds during germination. *J. Jap. Forest. Soc.* **43**, 241–244.

Kennedy, R. W., and Farrar, J. L. (1965). Tracheid development in tilted seedlings. *In* "Cellular Ultrastructure of Woody Plants" (W. A. Côté, Jr., ed.), pp. 419–453. Syracuse Univ. Press, Syracuse, New York.

Kiesselbach, T. A. (1949). The structure and reproduction of corn. *Univ. Nebr., Agr. Exp. Sta., Bull.* **161.**

Kiesselbach, T. A., and Walker, E. R. (1954). Structure of certain specialized tissues in the kernel of corn. *Amer. J. Bot.* **39**, 561–569.

Kordan, H. A. (1964). Nucleolar birefringence in interphase nuclei of *Zea mays. Phyton* **21**, 191–196.

Kordan, H. A., and Preston, R. D. (1967). Birefringence in unfixed lemon fruit nucleoli. *Nature (London)* **216,** 1105–1106.

Krugman, S. L. (1965). Personal communication.

Lam, S. (1968). Interaction of temperature and gibberellin on potato seed germination. *Amer. J. Bot.* **55,** 193–198.

Ledbetter, M. C., and Porter, K. R. (1963). A microtubule in plant cell fine structure. *J. Cell Biol.* **19,** 239–250.

Ledbetter, M. C., and Porter, K. R. (1964). Morphology of microtubules of plant cells. *Science* **144,** 872–874.

Ledin, R. B. (1954). The vegetative shoot apex in *Zea mays. Amer. J. Bot.* **41,** 11–17.

Liese, W. (1951). Demonstration electronenmikroskopiseher Aufnahmen von Nadelholztüpfeln. *Ber. Deut. Bot. Ges.* **64,** 31–32.

Liese, W. (1960). Die Struktur der Tetiärwand in Tracheiden und Holzfasern. *Holz Roh-Werst.* **18,** 296–303.

Liese, W., and Ledbetter, M. C. (1963). Occurrence of a warty layer in vascular cells of plants. *Nature (London)* **197,** 201–202.

List, A. L. (1963). Some observations on DNA content and cell and nuclear volume growth in the developing xylem cells of certain higher plants. *Amer. J. Bot.* **50,** 320–329.

McArthur, I. C. S. (1967). Unpublished M.S. Thesis, University of Saskatchewan, Saskatoon, Saskatchewan (not seen, cited by Steeves and Sussex, 1970).

Manocha, M. S., and Shaw, M. (1964). Occurrence of lomasomes in mesophyll cells of "Khapli" wheat. *Nature (London)* **203,** 1402–1403.

Marchant, R., and Robards, A. W. (1968). Membrane systems associated with the plasmalemma of plant cells. *Ann. Bot. (London)* [N.S.] **32,** 457–471.

Mark, R. E. (1967). "Cell Wall Mechanics of Tracheids." Yale Univ. Press, New Haven, Connecticut.

Mayer, A. M., and Poljakoff-Mayber, A. (1963). "The Germination of Seeds." Pergamon, Oxford.

Mazia, D., Brewer, P. A., and Alfert, M. (1953). The cytochemical staining and measurement of protein with mercuric bromphenol blue. *Biol. Bull.* **104,** 57–67.

Mericle, L. W. (1950). The developmental genetics of the Rg mutant in maize. *Amer. J. Bot.* **37,** 100–116.

Mertz, D. (1961). Distribution and cellular localization of ascorbic acid oxidase in the maize root tip. *Amer. J. Bot.* **48,** 405–413.

Mirov, N. T. (1967). "The genus *Pinus.*" Ronald Press, New York.

Mollenhauer, H. H., and Morré, D. J. (1966). Golgi apparatus and plant secretion. *Annu. Rev. Plant Physiol.* **17,** 27–46.

Morey, P. R., and Cronshaw, J. (1968). Developmental changes in the secondary xylem of *Acer rubrum* induced by various auxins and 2, 3, 5-triiodobenzoic acid. *Protoplasma* **65,** 287–313.

Morton, R. K., and Raison, J. K. (1963). A complete intracellular unit for incorporation of amino acid into storage utilizing adenosine triphosphate generated from phytate. *Nature (London)* **200,** 429–433.

Moutschen, J., Bacq, Z. M., and Herve, A. (1956). Action du rayonnement X sur la croissance de la plantule d'orge. *Experientia* **12,** 314–315.

Mühlethaler, K. (1965). The fine structure of the cellulose microfibril. *In* "Cellular Ultrastructure of Woody Plants" (W. A. Côté, Jr., ed.), pp. 191–198. Syracuse Univ. Press, Syracuse, New York.

Mühlethaler, K. (1967). Ultrastructure and formation of plant cell walls. *Annu. Rev. Plant Physiol.* **18,** 1–24.

Naylor, J. M., and Simpson, G. M. (1961). Dormancy studies on seed of *Avena fatua.* II. A gibberellin sensitive inhibitory mechanism in the embryo. *Can. J. Bot.* **39,** 281–295.

Nelson, M. L. (1941). Polyembryony in seeds of southern pines. *J. Forest.* **39,** 959–960.

Nelson, O. E. (1967). Biochemical genetics of higher plants. *Annu. Rev. Genet.* **1,** 245–268.

Newcomb, E. H. (1967). Fine structure of protein-storing plastids in bean root tips. *J. Cell Biol.* **33,** 143–163.

O'Brien, T. P. (1967). Observations on the fine structure of the oat coleoptile I. The epidermal cells of the extreme apex. *Protoplasma* **63,** 385–416.

Opik, H. (1966). Changes in cell fine structure in the cotyledons of *Phaseolus vulgaris* during germination. *J. Exp. Bot.* **17,** 427–439.

Ordin, L., and Hall, M. A. (1968). Cellulose synthesis in higher plants from UDP glucose. *Plant Physiol.* **43,** 473–476.

Ory, R. L., and Henningsen, K. W. (1969). Enzymes associated with protein bodies isolated from ungerminated barley seed. *Plant Physiol.* **44,** 1488–1498.

Paleg, L. (1961). Physiological effects of gibberellic acid. III. Observations on its mode of action on barley endosperm. *Plant Physiol.* **36,** 829–837.

Paleg, L., Sparrow, D. H. B., and Jennings, A. (1962). Physiological effects of gibberellic acid. IV. On barley grain with normal, x-irradiated and excised embryos. *Plant Physiol.* **37,** 579–583.

Partanen, C. R. (1959). Quantitative chromosomal changes and differentiation in plants. *In* "Developmental Cytology" (D. Rudnick, ed.), pp. 21–45. Ronald Press, New York.

Picklum, W. E. (1953). Histological and cytological changes in the maize embryo during germination. Ph.D. Thesis, Iowa State University Library, Ames, Iowa.

Pollock, B. M., and Olney, H. O. (1959). Studies of the rest period. I. Growth, translocation, and respiratory changes in the embryonic organs of the after-ripening cherry seed. *Plant Physiol.* **34,** 131–142.

Preston, R. D. (1964). Structural and mechanical aspects of plant cell walls with particular reference to synthesis and growth. *In* "The Formation of Wood in Forest Trees" (M. Zimmerman, ed.), pp. 169–188. Academic Press, New York.

Priestly, J. H. (1964). "An Introduction to Botany," 4th ed. Longmans, Green, New York.

Randolph, L. F. (1936). Developmental morphology of the caryopsis in maize. *J. Agr. Res.* **53,** 881–916.

Ray, P. M. (1961). Hormonal regulation of plant cell growth. *In* "Control Mechanisms in Cellular Processes" (D. M. Bonner, ed.), pp. 185–212. Ronald Press, New York.

Ray, P. M. (1962). Cell wall synthesis and cell elongation in oat coleoptile tissue. *Amer. J. Bot.* **49,** 928–939.

Ray, P. M. (1967). Radioautographic study of cell wall deposition in growing plant cells. *J. Cell Biol.* **35,** 659–674.

Ray, P. M., and Baker, D. B. (1965). The effect of auxin on synthesis of oat coleoptile cell wall constituents. *Plant Physiol.* **40,** 353–360.

Rhoades, M. M. (1955). The cytogenetics of maize. *In* "Corn and Corn Development" (G. F. Sprague, ed.), pp. 123–219. Academic Press, New York.

Rickson, F. R. (1968). Anatomy, histochemistry and fine structure of the germinating embryo of *Paulownia tomentosa. Amer. J. Bot.* **55,** 280–290.

Robards, A. W. (1968). On the ultrastructure of differentiating secondary xylem in willow. *Protoplasma* **65,** 449–464.

Robards, A. W., and Humpherson, P. G. (1967). Microtubules and angiosperm bordered pit formation. *Planta* **77,** 233–238.

Roelofsen, P. A. (1959). "The Plant Cell Wall." Borntraeger, Berlin.

Sacher, J. A. (1954). Structure and seasonal activity of the shoot apices of *Pinus lambertiana* and *Pinus ponderosa. Amer. J. Bot.* **41,** 749–759.

Sacher, J. A. (1956). Observations on pine embryos grown *in vitro. Bot. Gaz.* **117,** 206–214.

Sass, J. E. (1955). Vegetative morphology. *In* "Corn and Corn Improvement" (G. F. Sprague, ed.), pp. 63–87. Academic Press, New York.

Schlising, R. (1969). Seedling morphology in *Marah* (Cucurbitaceae) related to the Californian mediterranean climate. *Amer. J. Bot.* **56**, 552–561.

Schmidt, A. (1924). Histologische Studien an phanerogamen Vegetationspunkten. *Bot. Arch.* **8**, 345–404.

Schuepp, O. (1917). Untersuchungen über Wachstum und Formwechsel von Vegetationspunkten. *Jahrb. Wiss. Bot.* **57**, 17–79.

Schuepp, O. (1926). "Meristeme." Borntraeger, Berlin.

Schwartz, D., and Bay, C. (1956). Further studies on the reversal in the seedling height dose curve at very high levels of ionizing radiations. *Amer. Natur.* **90**, 323–327.

Sharman, B. C. (1942). Developmental anatomy of the shoot of *Zea mays. Ann. Bot. (London)* [N.S.] **6**, 245–282.

Sharman, B. C. (1945). Leaf and bud initiation in the Gramineae. *Bot. Gaz.* **106**, 269–280.

Sharman, B. C. (1947). The biology and developmental morphology of the shoot apex in the Gramineae. *New Phytol.* **46**, 20–34.

Sicard, M. A., and Schwartz, D. (1959). The effect of high doses of radiation on seedling growth. *Radiat. Res.* **10**, 1–5.

Skene, M. (1947). "The Biology of Flowering Plants. Sidgwick & Jackson, London.

Skvarla, J. J., and Larson, D. A. (1966). Fine structural studies of *Zea mays* pollen. I. Cell membranes and exine ontogeny. *Amer. J. Bot.* **53**, 1112–1125.

Smith, F. H. (1958). Anatomical development of the hypocotyl of Douglas-fir. *Forest Sci.* **4**, 61–70.

Smith, F. H., and Silen, R. R. (1963). Anatomy of heat damaged Douglas-fir seedlings. *Forest Sci.* **9**, 15–32.

Sprague, G. F. (1955). Corn breeding. *In* "Corn and Corn Development" (G. F. Sprague, ed.), pp. 221–292. Academic Press, New York.

Sparrow, A. H., and Miksche, J. P. (1961). Correlation of nuclear volume and DNA content with higher plant tolerance to chronic radiation. *Science* **134**, 282–283.

Spurr, A. R. (1949). Histogenesis and organization of the embryo in *Pinus stubus. Amer. J. Bot.* **36**, 629–641.

Spurr, A. R. (1950). Organization of the procambium and development of the secretory cells in the embryo of *Pinus strobus* L. *Amer. J. Bot.* **37**, 185–197.

Srivastava, L. M., and O'Brien, T. P. (1966). On the ultrastructure of cambium and its vascular derivatives. I. Cambium of *Pinus strobus* L. *Protoplasma* **61**, 257–276.

Stanley, R. G. (1958). Gross respiratory and water uptake patterns in germinating sugar pine seed. *Physiol. Plant.* **11**, 503–515.

Stanley, R. G., and Conn, E. E. (1957). Enzyme activity of mitochondria from germinating seedlings of sugar pine (*Pinus lambertiana* Dougl.). *Plant Physiol.* **32**, 412–418.

Steeves, T. A., and Sussex, I. M. (1970). "Patterns in Plant Development." Prentice-Hall, Englewood Cliffs, New Jersey.

Steffensen, D. (1968). A reconstruction of cell development in the shoot apex of maize. *Amer. J. Bot.* **55**, 354–369.

Stein, O. L., and Quastler, H. (1963). The use of tritiated thymidine in the study of tissue activation during germination in *Zea mays. Amer. J. Bot.* **50**, 1006–1011.

Stone, E. C. (1957). Embryo dormancy and embryo vigor of sugar pine as affected by length of storage and storage temperature. *Forest Sci.* **3**, 357–371.

Swift, H. (1950). The constancy of desoxyribose nucleic acid in plant nuclei. *Proc. Nat. Acad. Sci. U.S.* **36**, 643–654.

Takao, A. (1959). A cytochemical study on the proteid vacuoles in the egg of *Pinus thunbergii.* Parl. *Bot. Mag.* **72**, 853–854.

Takao, A. (1960). Histochemical studies on embryogenesis of *Pinus thunbergii* Parl. *Bot. Mag.* **73**, 379–388.

Tamulevich, S. R., and Evert, R. F. (1966). Aspects of sieve element ultrastructure in *Primula obconica. Planta* **69**, 319–339.

Tepper, H. B. (1962). Ontogeny and histochemistry of the shoot apex of seedlings of *Pinus ponderosa.* Ph.D. Thesis, University of California, Davis.

Tepper, H. B. (1964). Ontogeny of the shoot apex of seedlings of *Pinus ponderosa. Amer. J. Bot.* **51**, 859–865.

Tepper, H. B. (1969). Personal communication.

Toole, E. H. (1924). The transformations and course of development of germinating maize. *Amer. J. Bot.* **11**, 325–350.

Torrey, J. (1967). "Development in Flowering Plants." Macmillan, New York.

Toumey, J. W. (1923). Multiple pine embryos. *Bot. Gaz.* **76**, 426.

Unger, J. W. (1954). A study of the Jack Pine seedling and the origin and morphology of callus grown *in vitro* from seedling segments. Ph.D. Thesis, University of Wisconsin, Madison.

Walker, W. S., and Bisalputra, T. (1967). The fine structure of vesicles associated with the cell surface in *Helianthus* shoot tissue. *Can. J. Bot.* **45**, 2103–2108.

Wardlaw, C. W. (1946). Experimental and analytical studies of pteridophytes. VII. Stelar morphology: The effect of defoliation on the stele of *Osmunda* and *Todea. Ann. Bot. (London)* [N.S.] **10**, 97–107.

Wardlaw, C. W. (1950). The comparative investigation of apices of vascular plants by experimental methods. *Phil. Trans. Roy. Soc. London, Ser. B* **234**, 583–604.

Wardrop, A. B. (1963). Morphological factors involved in the pulping and beating of wood fibres. *Sv. Papperstidn.* **66**, 1–17.

Wardrop, A. B. (1964). The structure and formation of the cell wall in xylem. *In* "The Formation of Wood in Forest Trees" (M. H. Zimmermann, ed.), pp. 87–134. Academic Press, New York.

Wardrop, A. B. (1965). Cellular differentiation in xylem. *In* "Cellular Ultrastructure of Woody Plants" (W. A. Côté, Jr., ed.), pp. 61–97. Syracuse Univ. Press, Syracuse, New York.

Wareing, P. F. (1969). Germination and dormancy. *In* "The Physiology of Plant Growth and Development" (M. B. Wilkins, ed.), pp. 605–644. McGraw-Hill, New York.

Weatherwax, P. (1920). Position of scutellum and homology in coleoptile in maize. *Bot. Gaz.* **69**, 179–182.

Weatherwax, P. (1923). "The Story of the Maize Plant." Univ. of Chicago Press, Chicago.

Westing, A. (1968). Formation and function of compression wood in gymnosperms. II. *Bot. Rev.* **34**, 51–78.

Wilcox, H. (1962). Growth studies of the root of incense cedar, *Libocedrus decurrens.* I. The origin and development of primary tissues. *Amer. J. Bot.* **49**, 221–236.

Wilson, B. F. (1964). A model for cell production by the cambium of conifers. *In* "The Formation of Wood in Forest Trees" (M. H. Zimmermann, ed.), pp. 19–36. Academic Press, New York.

Winton, A. L., and Winton, K. B. (1932). "The Structure and Composition of Foods" Vol. I. Wiley, New York.

Wolff, S. (1954). Some aspects of the chemical protection against radiation damage to *Vicia faba* chromosomes. *Genetics* **39**, 356–364.

Worsdell, W. C. (1916). The morphology of the monocotyledonous embryo and that of the grass in particular. *Ann. Bot. (London)* **30**, 509–524.

6

SEED AND SEEDLING VIGOR

Bruce M. Pollock and Eric E. Roos

I. Introduction

The biological role of the seed is to protect and nourish the living cells of the embryo until the seedling is established. The value of this process is measured by the effectiveness with which a new plant becomes established and completes its life cycle. Even when plants of a single genetic strain are grown under identical conditions, great differences in yield are found. These differences are related to the history of the seed.

Over 50 years ago, Kidd and West (1918) recognized physiological predetermination. They described it as follows: "Presuming always a given equal number of plants, what are the limits of variation in growth and yield which may be pre-determined by the action of environmental conditions during the seed stage both previous to and during the course of germination; and how far may later environmental conditions during the course of growth affect the latitude of such variation." They further stated, "In a survey of the literature of the last 50 years dealing with the seed it is interesting to note that attention has been almost entirely concentrated upon the question of germination. The effect upon germination of every manner of treatment and of every sort of condition has been investigated in the case of the seeds of a great range of cultivated and wild plants."

Kidd and West's description of the status of research in 1918 also describes accurately the status of research accomplishments in 1972. We are, therefore, forced to ask whether this rather extended and intensive research effort has been fully utilized in agricultural practice. The answer to this question is largely negative. We must then ask whether this situation can be expected to change in the future. Here the answer is positive. A recent popular article described advances in agricultural technology which demand the exploitation of physiological predetermination (Billard and Blair, 1970, especially, pp. 153 and 165).

In this chapter, we shall attempt to review the scientific basis for the concept of physiological predetermination or seedling vigor as it is now more commonly designated. We shall also attempt to point out the potential for application of "seed and seedling vigor" in agriculture, forestry, and management of natural resources.

A.   *Terminology*

The literature on seeds is characterized by a multiplicity of terms for the same biological phenomenon and by uncritical usage of this terminology. One of the major problems concerns the use of the word *germination* itself (Lang, 1965). The botanist defines and measures germination as emergence of the radicle from the seed coat; insofar as possible, this is the definition which we shall use in this chapter. The seed technologist is required by law to measure germination (hereafter *technologist's germination*) after the embryo has grown to an appreciable size; he adds to the botanist's definition an attempt to evaluate whether or not the seedling will produce a normal plant. The difference between the seed technologist's and the botanist's definition of germination is thus part of what we are considering to be seedling vigor.

There are two aspects of vigor–genetic and physiological. Genetic vigor can be seen in heterosis (hybrid vigor) or the difference in vigor between two genetic lines. Physiological vigor can be seen in the difference in vigor between two seed lots from the same genetic line. It is primarily physiological vigor with which we shall be dealing in this chapter. However, it is necessary to recognize that physiological vigor has its basis in genetic vigor (or lack thereof). It is also necessary to recognize the difficulty of identifying the cause of vigor, especially when differences in vigor are seen in genetically heterogeneous populations (Fig. 1).

B.   *Importance of Seedling Vigor*

The importance of seedling vigor to modern agricultural technology is outlined in the National Agricultural Program (U.S. Department of Agriculture, 1969) and can be seen by examining divergent situations such as a commercial field of lettuce or a reseeded forest. In the lettuce field, economics demand that the grower plant a single lettuce seed every 12 in. in the row, i.e., "planting to stand." These single seeds must each germinate and establish a plant; all plants in the field must grow at a uniform rate so that all lettuce heads simultaneously reach maturity for once-over machine harvesting. These demands are as yet unsatisfied.

In the lettuce field, the grower has a degree of control over environmental factors such as physical condition of the soil, depth the seed is planted, soil fertility, and soil moisture, but he is still unable to obtain the desired result. By contrast, the forester concerned with replanting in an arid region has no control over seedbed environment, especially water supply. Under these conditions, plant establishment can succeed only if the seedlings have sufficient vigor to maintain rapid root growth and contact with the water supply which rapidly recedes during the growing season. In this natural ecological situation, vigor is also necessary if the seedling is to compete with neighboring plants for light, water, and mineral nutrients.

Examination of a field of any crop will show that a number of the plants are nonproductive (Figs 1 and 2). These nonproductive plants may be the result of abnormally low vigor seeds. Increasing seedling vigor to eliminate these nonproductive plants could permit higher yields at a lower cost per acre.

C. Vigor and Environment

In theory, vigor can be influenced by any one of a large number of environmental variables during seed development, and responds to any

FIG. 1. *Pinus ponderosa* seedlings in a forest tree nursery. Vigor differences may result from differences in genetic background, physiological history, or soil microclimate. Thus, although these seedlings obviously represent different vigor levels, it is very difficult to identify the reason for the differences in vigor. (Author photo.)

FIG. 2. A low vigor plant in a corn field. Such plants are frequently called barren and can be seen in varying numbers in all fields of corn. (Author photo.)

one of a large number of environmental variables during subsequent plant establishment. The number of potential environment–vigor interactions is enormous. The question should be asked: If growth of a seed-bearing plant is retarded at any stage during seed development or germination, can the resulting plant ever resume a "normal" growth rate?

In the literature on vigor, we find many cases where workers failed to find evidence of vigor–environment interactions, or where one worker was unable to repeat another's results. In each case there exists the possibility that the scientist failed to identify or control some of the environmental conditions associated with vigor development or expression. We have attempted to identify some of these conditions in our search of the literature.

II. Expression of Vigor

A. *Plant Growth and Crop Yield*

Although we discuss vigor as a seed or seedling characteristic, the significance of the phenomenon is that vigor can persist throughout the life cycle of the plant and be expressed in crop yield. For example, Brenchley (1923) found that size differences in seeds of peas and barley were expressed in yield.

In agricultural practice, yield differences are noted between plants of the same genotype but of different seed lots. Comparing hybrid corn seed lots produced by different growers, Fleming (1966) found economically significant differences in yield between lots with the same laboratory germination. Similar results have been found in oats (Grabe and Frey, 1966), peas (Perry, 1969a), and tomatoes (Clark and Kline, 1965), to cite only a few of many cases.

Evaluation of vigor by measuring crop yield is complicated by the fact that seed or seedling vigor can produce major differences in stand density. These differences by themselves can produce yield differences as shown by Austin (1963) with onions. The effect of vigor on stand density can be separated from the effect of vigor on productivity of individual plants by greenhouse planting and vigor classification, followed by replanting to standard spacing in the field. This was the technique used by Scheer and Ellison (1960) who found that low vigor in asparagus seedlings was correlated with survival and yield in the field. However, thinning to stand or planting to stand are techniques which are more commonly used [for example, Perry (1969a) with peas and Clark and Kline (1965) with tomatoes].

The competitive interaction between plants is one of the major aspects of vigor. Whalley *et al.* (1966) showed for a number of species that seedling length was correlated with seed size (Fig. 3). Black (1958) working with subterranean clover, *Trifolium subterraneum* L., separated seeds by size, assuming that large seeds would produce seedlings of higher vigor than would small seeds. When seeds of uniform size were planted at equal stand densities, the number of plants surviving decreased throughout the growing season as competition between plants eliminated a percentage of the population. In these uniform plantings, there was no difference between large and small seeds in the percentage of seedlings which died. However, when large and small seeds were planted together, the number of plants from the large seeds remained approximately constant, whereas two-thirds of the plants from small seeds were eliminated by competition. He found that the plants from small seeds received only 2% of the sunlight at the time a foliage canopy was developed and concluded that the elimination of plants was caused by shading.

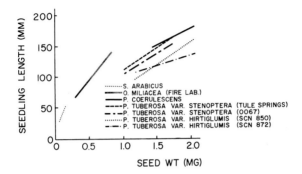

FIG. 3. Regression lines of total seedling length with seed weight for several species (*Schismus, Oryzopis, Phalaris*). The length was measured after 14 days germination in petri dishes at alternating temperatures of 20° and 30°C. [Courtesy of Whalley *et al.* (1966) and *Crop Science.*]

By contrast, W. A. Williams *et al.* (1968) studied a mixed community of *Trifolium incarnatum* L. and *Trifolium subterraneum* L. planted with seeds of different sizes. These were compared with monocultures of plants from the same seed size; plots were adjusted to the same initial stand density. In this case, competition did not eliminate any plants. However, *T. subterraneum* repressed the development of *T. incarnatum* except when the smallest seeds of the former species were mixed with the largest seeds of the latter species. Measurements of plant height showed that these two species developed their leaf canopies at different levels and appeared to form a buffered community.

Competition between species has been suggested as a means of weed control. Pavlychenko and Harrington (1934) found that cereal seeds competed well with weed seeds under adverse moisture conditions; under adequate moisture the weed seeds competed much more effectively. Guneyli *et al.* (1969) found that the competitive advantage of sorghum over weeds was related to rapid germination and early seedling vigor. Although such studies have been concerned with competition related to genetic characteristics of seed germination, they can be extrapolated to include the potential role of physiological seedling vigor in weed control. In recent years, relatively little consideration has been given to this type of question because of the extensive use of chemical weed control methods. However, this approach offers a potential method of non-chemical weed control.

B.   Seed Germination

Although vigor can be expressed throughout the life cycle of the plant, evaluation of vigor at plant maturity is too complex and expensive to be a common experimental parameter. For this reason, and because of the importance of vigor in plant establishment, most research has been concerned with the expression of vigor during seed germination and early seedling development.

Seeds that are capable of extending a root (botanical germination) will not necessarily have the vigor to establish a plant under field conditions. This fact led to the development of regulatory test procedures to evaluate the potential of a seed for field establishment. However, regulatory tests are also the basis on which the marketability of seed is determined (Wellington, 1965), and this fact has inevitably resulted in an attempt to maximize germination by utilizing optimal conditions for germination (de Tempe, 1963). It is, therefore, not surprising that regulatory germination values frequently fail to provide an estimate of field emergence.

The problem of predicting field emergence is much more complex than its relationship to the economics of seed marketing. Although it is obvious that environmental conditions for field germination will seldom be optimal, it is not possible to predict which environmental conditions (temperature, water, physical structure of the soil, etc.) will be unfavorable (Heydecker, 1962). Indeed, it is not yet clear that two lots of different vigor will retain the same relative vigor rating when exposed to different kinds of environmental stresses (Heydecker, 1960).

Developments in agricultural technology are directed toward increasing efficiency of production. Efficient production requires plant populations to be as large as possible without being so crowded that total yield is reduced. We can expect intensification of research efforts to develop methods to predict the planting rates necessary to optimize plant populations. Although there is no published material on the subject, our experience suggests that many vegetable processors now obtain estimates of vigor on seed lots by measuring germination rates. The use of cold testing of corn to predict emergence of early plantings is a common procedure.

1.   SPEED OF GERMINATION

There is fairly good agreement among agriculturalists and research workers that speed of germination (sometimes called germination energy) is an extremely important aspect of vigor. As an example, Wanjura et al. (1969) showed that early emergence of cotton is correlated with both greatest survival of seedlings and highest yield of lint cotton per plant. Whalley et al. (1966) showed that, as a genetic character, quick emergence of *Schismus arabicus* and *Phalaris* spp. was essential to plant establishment and competitive effectiveness. Rohmeder (1962) found that early germination of forest tree seedlings resulted in the lowest mortality and produced rapidly growing seedlings the superiority in size of which could be measured even after 3 or 4 years. Under the harsh growing conditions of Arizona, Larson (1961) found that only the rapidly germinating *Pinus ponderosa* seedlings could survive until the second growing season.

Among vegetable producers, we have noted that "first counts" in technologist's germination tests are frequently utilized to evaluate the relative vigor among seed lots. Unfortunately, the term rate or speed of germination is often used imprecisely to mean the time required for a lot of seeds to reach maximum technologist's germination. This is at least partially the result of the fact that there is no point in seedling development beyond radicle emergence which can be precisely timed. Radicle emergence is a single event which can be timed with some precision. When germination is measured by radicle emergence, and counts are made at short time intervals, typical S-shaped germination curves are obtained (Fig. 4). The "best straight line" slope of these curves is fre-

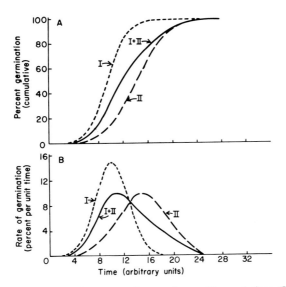

FIG. 4. Theoretical germination curves for two "normal" populations (I and II) and a mixed population (50% I and 50% II). The germination rate of population II is two-thirds that of population I. (A) Percentage germination vs. time. (B) Germination rate vs. time.

quently taken as the rate of germination, and 50% germination or other calculated values are obtained from this slope (Tucker and Wright, 1965). Mathematically, rate is defined as germination per unit time (dg/dt).

During germination, this value increases from zero to some finite value and then decreases again to zero. If the seed population is normal, or homogeneous, the rate curve follows a symmetrical Gaussian distribution (Fig. 4B, I and II). This rate curve is distinct from the usual germination percentage curve, which is actually a plot of cumulative germination (Fig. 4A). Investigators sometimes assume that germination curves are typical of those for homogeneous populations (Fig. 4, I and II). However, it is difficult to evaluate a cumulative germination curve. Examination of rate curves may be more useful and frequently shows that the curves are positively skewed (Nichols and Heydecker, 1968).

Figure 4B shows theoretical rate curves for seed germination of two homogeneous populations. The rate of germination of population II is two-thirds that of population I. This difference has the dual effect of increasing the time required to reach the maximum germination rate and of decreasing that maximum rate; it does not change the symmetry of the curve. However, if populations I and II are mixed in equal proportions, the resulting rate curve is positively skewed.

The problem of evaluating slowly germinating seeds in a lot was recognized by Kotowski (1926) who defined a coefficient of velocity in which the late germinating seeds contributed less than did early germinating seeds, i.e.,

$$\text{coefficient of velocity} = 100 \cdot \frac{A_1 + A_2 + \cdots + A_x}{A_1 T_1 + A_2 T_2 + \cdots + A_x T_x}$$

where A = the number of seeds germinating and T = time corresponding to A.

More recently, similar types of formulas have been proposed by Czabator (1962) working with forest tree seeds, Tucker and Wright (1965) for range grass seeds, and Heydecker (1966) and Nichols and Heydecker (1968) for vegetable seeds.

The effect of slowly germinating seeds on cumulative germination curves can be seen clearly in Abbott's (1956) data on apple seeds (Fig. 5) and in summaries of germination data such as those used by Czabator (1962) in developing his germination value formula for pine seeds. The same type of cumulative germination curve is shown in Fig. 4A; this theoretical curve was generated from an equal mixture of populations I and II which differed only in the rate of germination.

The fact that a theoretical curve developed by mixing different populations can mimic actual germination data indicates a seed lot should be considered to be a mixture of different populations. In later parts of this chapter we shall examine this question further.

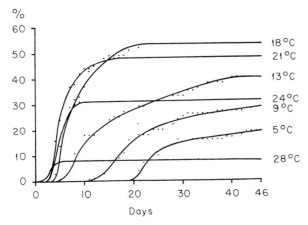

Fɪɢ. 5. Apple seeds stratified at 5°C for 6 days. The embryos were then excised and grown at the temperatures indicated. Germination below 18°C resulted in additional stratification of partially after-ripened seeds, thus resulting in a slowly germinating portion of the population. Exposure of partially after-ripened seeds to temperatures above 18°C reversed after-ripening and prevented germination of the portion of the population in which after-ripening was not complete. [Data courtesy of Abbott (1956) and *Report of the 14th International Horticultural Congress.*]

These observations raise further questions about the rationale behind much current research on seedling vigor. Because of the formidable statistical questions involved in studying mixed populations, most investigators of seedling vigor are attempting to obtain populations of uniformly high or low vigor by techniques such as selection (Pollock and Toole, 1966), radiation treatment (Woodstock and Combs, 1965), and artificial aging (Delouche *et al.,* 1967). We must eventually consider the extent to which the results of such experiments with homogeneous populations can be extrapolated to the natural mixed populations in normal seed lots. Furthermore, quick tests such as those for respiration (Woodstock and Grabe, 1967), glutamic decarboxylase (Linko and Milner, 1959; Grabe, 1964), and glucose metabolism (Abdul-Baki, 1969) are being developed to measure the vigor of bulked, relatively uniform, seed lots. We need to examine carefully the extent to which such tests can be expected to provide practical information about potential expression of vigor in normal, nonhomogeneous seed lots under field conditions.

2. GROWTH RATE

Rate of seedling growth is frequently used to evaluate seedling vigor. The most common method is to sort seedlings into size classes at the time technologist's germination counts are made and count the number of seedlings judged to be vigorous or nonvigorous (Perry, 1969b) (Fig. 6). The sensitivity of the method may be increased by assigning numbers, such as 1 through 10, to a graded (but arbitrary) series of vigor ratings (Allen and Donnelly, 1965). Although this technique is simple to use, it is arbitrary and does not provide precise data which can be reproduced.

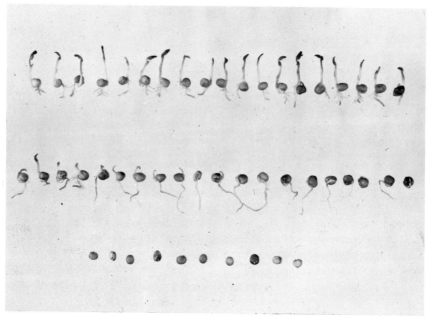

FIG. 6. An arbitrary classification of pea seedlings into vigor classes. Top row, vigorous germination; middle row, low vigor germination; bottom row, ungerminated. [Courtesy of Perry (1969a) and *Proceedings of the International Seed Testing Association.*]

Seedling growth can be objectively evaluated by a continuous variable such as plant height as used by Kerr (1961) to study the relation between seedling vigor and size of sweet cherry trees. More commonly, growth is evaluated by weight. For example, Nanda *et al.* (1959) used seedling dry weight as a measure of growth rate in wheat seedlings, Christiansen (1962) used decrease in cotyledon dry weight as a measure of seedling growth in cotton, Pollock and Toole (1966) used fresh weight of excised lima bean axes as a measure of seedling vigor, and Pollock *et al.* (1969) and Roos and Manalo (1971) used epicotyl and total seedling weight to measure vigor as a function of stress conditions (Fig. 7).

It should be noted that length measurements and, in certain cases, weight measurements (Pollock and Toole, 1966) can be used to obtain

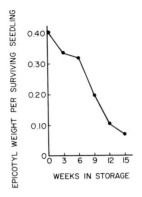

FIG. 7. Effect of adverse storage conditions (32°C, 90% relative humidity) on the epicotyl size of garden beans after 7 days germination [Courtesy of E. E. Roos and J. R. Manalo (1971) and *HortScience*.]

repeated measurements on the same seedlings. When this is done, growth rates can be calculated for the same seedlings. However, if the measurement is destructive, such as is frequently the case in weight measurements, then rate curves can be constructed only by the use of statistical sampling techniques. Again, as in the case of germination rate, much of the literature on rate refers only to single point measurements and thus cannot be interpreted as true rates.

C. Uniformity of Performance

Although the major emphasis in research on vigor has been on stand establishment and gross yield, some data are available on the effect of seed and seedling vigor on the variability of the plant population produced. Hoffman (1925) noted that large and small seeds of sweet corn germinated at about the same rate, but the seedlings from large seeds were larger and reached canning maturity five days earlier than plants from small seeds. Therefore, he recommended the use of size-graded seeds to obtain the greatest uniformity of ears for canning. Cameron *et al.* (1962) confirmed these results for winter-planted sweet corn in California. Tseng and Lin (1962) noted that yield of rice could be greatly affected if nonuniform seedlings were planted in the same hill. Tompkins (1966) showed that seed sizing in broccoli produced more uniform yielding plants. Christiansen and Thomas (1969) suggested that variability in yield between cotton plants is the result of environmental conditions during seed germination.

Current developments in agriculture are directed toward the reduction of costs by replacing manual steps in production and harvesting with machines. However, effective utilization of machines requires a uniformity of plant development which is in opposition to the diversity we expect to find in a natural plant population. Therefore, we can expect the emphasis of research on vigor to be shifted to its effect on uniformity of plant development as a prerequisite to attaining a maximum yield of usable product. This is an aspect of vigor which has not been studied extensively. To illustrate, Fig. 8 shows pods picked from three randomly selected plants in a row of garden beans. Note that the yield of harvestable pods on plant A was 10 times that on plant C. Plant B produced only slightly fewer pods than plant A, but those pods were about 4 days later in maturity.

Prior to mechanical harvesting, yield from late producing plants was included in late pickings. Now, with once-over machine harvesting, late producing plants are not only nonproductive but serve as functional weeds. Furthermore, nonuniformity in development lowers the quality of the processed beans, since harvest at optimal yield means that the early producing plants will yield overmature pods. Similar problems are now critical in production of lettuce and other vegetables.

D. *Presence of Morphological Abnormalities*

In practice, the major difference between botanical germination and technologist's germination is the classification of structurally abnormal seedlings as nongerminating in regulatory tests. This classification causes three serious problems. (*1*) Living but nongerminating seeds produce late or unproductive plants which contribute to variability and serve as functional weeds in competition with normal plants. (*2*) Inclusion of abnormal with nonviable seeds in the report of a germination test makes it impossible to identify the cause of low germination or vigor in a seed lot; the causes for these two categories are frequently different (see Sections V, VI). The results of seed tests would be much more valuable if the report included at least three classifications of seedlings: normal, structurally abnormal, and nonviable. (*3*) Different germination laboratories may provide different values on the germination of the same seed lots because of different interpretations of abnormal seedlings.

The rules for seed testing, and various manuals on seed testing provide detailed descriptions and figures of the various major structural abnormalities, and these are discussed in detail in Chapter 5, Volume III, of this treatise, so that we need to outline only the types here. However, it should also be emphasized that there are many structural abnormalities known which do not adversely affect vigor, and may even produce desirable plant types. Tricotyledonous seedlings, for example, are common in small percentages in most dicotyledonous seed lots. There appears to be

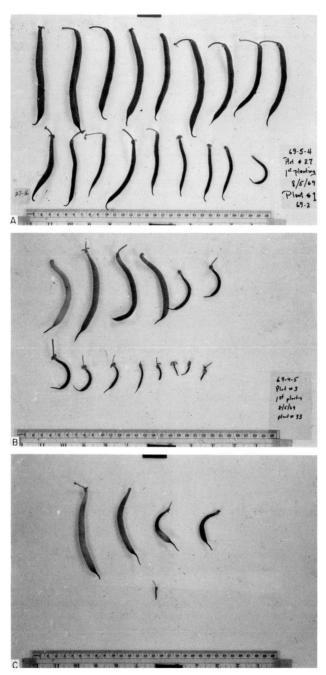

FIG. 8. Differences in yield among three plants (A,B, and C) in a single planting of bean seeds. This ten-fold difference in productivity, as well as time of maturity among plants in a row, can be considered normal variability. (Author photo.)

no evidence that these seedlings are low in vigor, and Rodionenko (1955) has pointed out their value in ornamental plants such as *Antirrhinum majus*.

1. LOSS OF STORAGE TISSUE

Among the most common structural abnormalities are those involving loss of storage tissue. These are obviously serious because they reduce the amount of reserve nutrients available to the developing seedling. Seedlings affected never develop into a full-sized plant. From available data the reason for this is not clear. It is possible that the affected seedling is later and thus grows under less favorable environmental conditions or grows in an unfavorable competitive position relative to other plants. It is also possible that the relationship between the cotyledon nutrient supply and the seedling includes a mechanism, as yet unknown, which controls the subsequent growth rate of the plant.

2. MERISTEMATIC DAMAGE

Injury to meristematic tissue of either the plumule (Fig. 9, plant A) or

A   B   C   D

FIG. 9. Damage to garden bean seedlings which results in lowered vigor. (A) Loss of apical meristem of shoot; (B) transverse cracking of cotyledons; (C) vascular damage caused by broken hypocotyl; and (D) undamaged seedling. [Photograph courtesy of Pollock and Manalo (1970) and *Journal of the American Society for Horticultural Science*.]

radicle is a common result of mechanical damage or attack by micro-organisms. In the case of mechanical damage, the seedling may have the ability to regenerate the lost meristem so that the only effect is to delay plant development. However, in the case of attack by microorganisms, the effect is frequently to weaken the seedling so that it cannot survive.

3. VASCULAR DAMAGE

Breaking of vascular connections is also a common abnormality result-ing from mechanical injury. In cotyledons, transverse breaks (Fig. 9, plant B) frequently cannot be healed so that the nutrient reserves distal to the point of breakage are lost to the plant. In the stem (Fig. 9, plant C) or root, vascular tissue can frequently regenerate, but the resulting delay in development condemns the seedling to a life of low vigor.

4. OTHER ABNORMALITIES

In gymnosperms, many seeds are empty or contain abnormal embryos. In such cases, germination is slow and seedlings are low in vigor. This problem has been studied extensively in Scandinavia and India by the use of the X-ray contrast method (Simak, 1957). Weak solutions of radiopaque salts are vacuum injected into seeds which are then photo-graphed with soft X-rays. Abnormal seeds contain cavities which show on the X-ray film; size and shape of the cavities can be used to approxi-mate roughly seed quality (Simak and Kamra, 1963). Dogra (1967) has studied embryo development in relation to the types of abnormalities which can be detected by X-rays and has reviewed the literature on the subject. In the gymnosperms, multiple archegonia are normally formed and may be fertilized to produce embryos by a process of cleavage. In normal seed development, one embryo becomes dominant and the others degenerate. However, in northern latitudes, physiological disturbances sometimes destroy dominance and polyembryony results. Since all embryos in a seed must depend on the single endosperm (female gameto-phyte in a gymnosperm), competition for nutrients results either in the failure of the seed to germinate or in the production of nonvigorous seedings.

III. Evaluation of Vigor

A. Stress Conditions

For spring-planted crops, cold, wet, compact soils are a common stress condition. However, for many trees, forage crops, range grasses, and fall-planted grains, high temperature coupled with lack of water is the environmental hazard which must be overcome. One method of evaluating vigor is to subject the seed to one or more of the environmental stresses it might encounter under soil conditions and then measure its response in terms of survival, rate of germination, rate of growth, and presence of structural abnormalities.

1. TEMPERATURE

For seeds of many crops, the temperatures used for regulatory tests, as well as most experimental purposes, are higher than those under which the seeds would germinate in the soil. For example, germination rules (U.S. Department of Agriculture, 1968) for garden beans specify either a constant temperature of 25°C (76°F) or an alternating 20°–30°C (68°–86°F) temperature. We have measured the soil temperature at seed depth (1–2 in.) for early spring planting in Colorado and found an average daily temperature of 10° to 20°C, with a maximum temperature seldom reaching the temperature used in laboratory germination tests (Fig. 10).

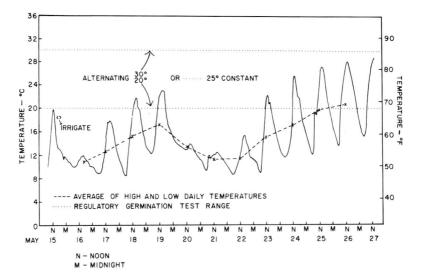

FIG. 10. Thermocouple recording of soil temperatures in a bean field at 1 to 2 in. depth at Fort Collins, Colorado, during May, 1969, as compared with temperatures used in technologist's germination tests for beans. Approximately 1 in. of water was applied by sprinkler irrigation immediately after inserting the thermocouples.

Low temperature has at least two adverse effects on germinating seeds: (*1*) it may cause direct low-temperature injury and (*2*) it may slow germination so that the seeds remain for a longer period in a soil environment favoring competition and attack by microorganisms. Because low temperature is such a common stress under field conditions, it has been the factor most frequently studied.

Perhaps the best-documented and most widely used stress test is the cold test for corn (Isely, 1950; Clark, 1954; Rice, 1960). Although germination of corn in cold soils had been a problem, Tatum and Zuber (1943) noted that the problem seemed to become more serious with the development of artificial drying and centralized processing by the hybrid seed corn industry. Subsequently, a laboratory test was developed which involved exposure of seeds to a temperature of approximately 10°C for 5 to 7 days, followed by germination at 30°C. Critical factors in the test are the presence of fungi, particularly *Pythium* (which is introduced through nonsterile soil or sometimes by adding ground seeds which previously failed to germinate) and moisture content of the medium (which must be relatively high). The extreme to which this test may influence germination was shown by Rice (1960) for one lot of corn seeds which germinated 99.5% in autoclaved sand at 30°C and only 11% using the cold test procedure.

The effect of cold test conditions on corn germination can be modified by seed treatment with chemical fungicides. Thus, in practice, the test evaluates the seed–chemical system rather than the seed alone. Cold test results are valuable and widely utilized by the corn industry. However, the results have been very difficult to standardize. This probably arises from the difficulty in obtaining and maintaining a soil containing a uniformly acting fungus population (Hooks and Zuber, 1963).

The value of the cold test for crops such as soybeans, spinach, and peas has also been investigated (Clark and Baldauf, 1958; Rice, 1960; Perry, 1969a). In addition, high temperature during germination is also a stress condition, and its use has been investigated for peas (Caldwell, 1956) and many other crops (de Tempe, 1962, 1963).

2. SOIL MOISTURE AND OXYGEN

Soil moisture, in deficiency or excess, is probably the most common stress encountered by germinating seeds. Soil moisture and soil atmosphere compete for the same physical space in the germination medium. The solubility of oxygen in water is extremely low, with the consequence that the rate of oxygen diffusion in air is 300,000 times that in water (Goddard and Bonner, 1960). Thus, an excess of water also includes a stress from restriction of oxygen supply to the germinating seedling.

As an added complication, many seeds absorb water (and possibly oxygen and nutrients) at restricted localities on the surface, such as the micropyle, hilum, or raphe. This means that the physical position of the seed relative to the air and water spaces in the surrounding medium is extremely important. Oxygen and water supply can be influenced not only by gross quantities of air and water in the medium, but also by geometry and the intimacy of the seed–medium contact.

In germination research, soil moisture is frequently measured as the amount of water removed by oven drying; no consideration is given to the physical structure of the soil. However, there are considerable data on seed–soil–moisture relationships which could be of great value in the study of vigor. Collis-George and Sands (1959) studied the relationship between soil moisture and germination in an attempt to explain the colonization of pasture lands in New South Wales, Australia, by *Juncus vaginatus* R.Br. For this purpose, they germinated seeds on tension plates consisting of sintered glass discs in contact with a column of water, which, in turn, was attached to a mercury leveling bulb used to place the water under tension. They found that germination of seeds of both *Juncus* and *Medicago* spp. decreased in rate until, at 10 atm suction, germination practically ceased. Seeds of *Medicago* were much less sensitive to tension than were those of *Juncus,* and they concluded that invasion of pastures by *Juncus* was related to abnormally wet soils. Since the relationships between soil moisture and suction, and moisture and hydraulic conduction of water to seeds are both physical properties of soil, they concluded that these as well as soil moisture content must be defined in germination experiments.

In further work, Collis-George and Sands (1962) noted that the total potential of soil water, ϕ, can be considered to have three components: (*1*) a solute (osmotic) potential, (*2*) gravitational potential resulting from the vertical position of the water, and (*3*) a suction component (capillary or matric potential). They found that increases in matric potential decreased the rate of germination. Further studies by Collis-George and Williams (1968) showed that the influence of matric potential in the range 0–400 cm of water can be attributed to its effect on the isotropic stress in the solid framework of the soil rather than to its effect on the free energy of the soil water.

Sedgley (1963) in studying the effect of matric potential on germination, found that the liquid–seed contact was extremely critical – if contact was established before a matric potential was applied, then the effect of matric potential was greatly reduced. Collis-George and Hector (1966) confirmed for both *Medicago tribluoides* and *Lactuca sativa* that the wetted area of the seed surface is a factor controlling germination, but indicated that this is of greatest significance at matric potentials near that of free water. Manohar and Heydecker (1964) extended these data to the larger seeds of *Pisum sativum* and also showed that both the area of the seed in contact with water and the anatomical localization of that area of contact

were important. Thus, not only soil moisture, but the intimacy of seed–
soil contact, is critical in controlling germination.

In an attempt to provide a means of applying a reproducible level of
stress in the germination medium, Pollock and Manalo (1969) developed
a method in which seeds are placed in sand in a plastic box (Fig. 11). The

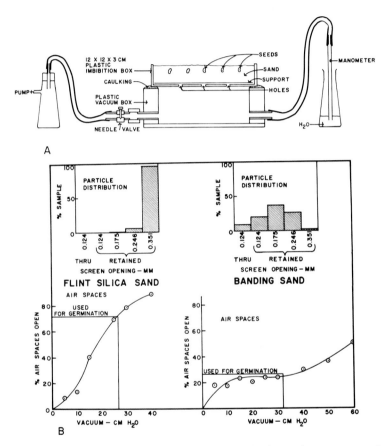

FIG. 11. (A) Diagrammatic view of apparatus for adjusting moisture content of sand.
(B) Effect of particle size and vacuum on air-holding capacity of sand. [Courtesy of Pollock
and Manalo (1969) and *Journal of the American Society for Horticultural Science.*]

box is quickly submerged in water, and the excess water is removed by
applying vacuum from below. The air–water relationships of the medium
are controlled by varying the particle size of the sand used and the amount
of vacuum applied. The initial submergence creates a uniform seed–sand
contact and temperature may be precisely controlled. This method has
been used to show that the effect of stress conditions on garden beans

varies with vigor of the seed lot (Pollock *et al.,* 1969). However, this method has the basic problem that the water in pure sand is at essentially zero potential, the potential of free water. The same is true of all laboratory germination methods involving blotters, rolled towels, or sand. By contrast, water potential in the soil is normally less than zero. In addition, it is not possible to vary water and oxygen supply independently, but neither is this possible in the tension plate method (Sedgley, 1963).

Recently, two methods have been developed to evaluate the influence of moisture stress on germination. Kaufmann (1969) and Kaufmann and Ross (1970) have used an osmotic medium (polyethylene glycol-6000) separated from soil by a dialysis membrane to apply a controllable moisture stress to the soil. Use of the dialysis membrane effectively converts the osmotic potential to a matric potential. Pollock and Manalo (1971) have used a modification of the soil physicist's tension table to develop reproducible moisture stresses for evaluating the response of lettuce seed lots to moisture stress (see Section VI,B). Both of these methods offer the possibility of extending our knowledge of vigor to conditions which simulate field environment.

3. OSMOTIC STRESS

Many workers have utilized osmotic stress in an attempt to select genetic strains that will be tolerant to drought. Similar techniques have related seed vigor to osmotic stress. For example, Parmar and Moore (1968) used Carbowax polyethylene glycol 6000, mannitol, and NaCl as osmotic agents and found a relationship between osmotic sensitivity and a vigor rating based on the tetrazolium test. This work was based on the assumption that the osmotic stress simulated matric potential since the osmotic potential of most agricultural soils is negligible. However, Collis-George and Sands (1962) separated osmotic and matric potentials and concluded that they were not biologically equivalent. Later, Collis-George and Williams (1968) showed that the matric potential is important through its effect on the physical stress of the soil structure on the germinating seed rather than through its influence on the free energy of soil water. Manohar and Heydecker (1964) showed that in peas mannitol entered the seed through the micropyle, but not through other areas of the seed coat. Germination was somewhat slow when seeds were germinated in a mannitol solution of 15 atm. However, if the micropyles were blocked with Vaseline, germination was much slower and many seeds failed to germinate. This result showed that in seeds without Vaseline the mannitol entered through the micropyle and penetrated the cells, thus relieving the osmotic stress. Therefore, the assumption that a high molecular weight compound acts as an inert osmotic agent can only be verified by testing, and the interpretation of osmotic data relative to soil moisture stresses must be accepted with some caution. Nevertheless, osmotic agents do offer potential tools for testing one aspect of seed quality. Manohar and Mathur (1967) used polyethylene glycol 4000 as an osmotic agent and found that it inhibited germination of pea seeds. Inhibition could be relieved by interrupting the osmotic treatment by a 4-hour exposure to pure water; relief was greater if the water was applied after 20 hours than after 68 hours. Such a system might be extremely useful if it were applied to a study of vigor in different seed lots.

4. PHYSICAL PROPERTIES OF SOIL

In spring-planted crops, soil compaction in the form of normally poor soil structure, crusting (especially in irrigated soils), and mechanical pressure from machine wheels are common sources of stress in the field (Martens, 1960). One of the early attempts to evaluate seedling vigor was to cover the seeds with a layer of brick dust (more properly brick sand) and count the number of seeds with sufficient vigor to emerge from this covering (Isley, 1957). A more recent modification has been to place seeds on a layer of wet sand and cover them with a layer of filter paper and then another layer of wet sand; good correlation with soil emergence was found (Fritz, 1965). W. A. Williams (1956) developed a method for stress testing individual seedlings of small-seeded legumes. Lengths of glass tubing were placed upright on Vermiculite. Into each was placed a single seed and a piece of glass rod of known weight which slid smoothly within the tubing. Mathematical probit analysis was used to measure the median emergence force, i.e., the force which 50% of the population is too weak to exert.

Many crops are planted in soils in which water supply is limited. Under these dry land conditions, crusting is seldom a problem, but it is advantageous to plant the seed at a maximum depth to take greatest possible advantage of the limited moisture supply. Selection of genetic strains adapted to emerge from deep plantings has been used in crops such as alfalfa (Beveridge and Wilsie, 1959), winter wheat (Kolp et al., 1967), and intermediate wheat grass (Hunt and Miller, 1965). However, this factor does not seem to have been applied to evaluation of physiological vigor.

Because of the increasing importance of precision planting, agricultural engineers have studied the physics of forces encountered by a seedling during emergence. Morton and Buchele (1960) constructed a mechanical seedling consisting of interchangeable probes with tips of different diameters attached to a strain gauge and located so that the tip could be forced upward through soil in a plastic box. They found that the energy required for emergence increased with compaction pressure and initial soil moisture content. When the soil was compacted from the top and allowed to dry slightly, the force required for emergence increased greatly. However, if the soil was compacted at seed level, and uncompacted soil was placed on top, the required increase in emergence force was eliminated. We believe that these and similar observations on the physics of seed–soil–moisture relationships could be effectively utilized in evaluating vigor.

5. Interactions of Soil Stresses

In nature, stresses due to soil temperature, structure, or water–oxygen content do not occur singly but rather as a complex of environmental stresses which tends toward two extremes: (1) low-temperature, wet, compact soil with poor aeration and (2) high-temperature, dry soil with adequate aeration.

Heydecker (1961) developed a seedling test by which soil characteristics can be determined. He showed that different genera of plants are tolerant to different soil stresses, and within one genus two seed lots responded differently to the same soil stresses (1962). Obviously, if this method could be sufficiently standardized, it could be used to evaluate seed lot differences.

In evaluating the complex of soil stress conditions, it is essential to appreciate that all types of stress do not change in the same direction as the change in total stress. For example, although a cold, wet, soil generally constitutes a stress condition, lowering the temperature increases solubility of oxygen in water and, hence, its movement by diffusion. In addition, respiration of seeds decreases at lower temperatures, decreasing the rate of oxygen consumption. The total effect is that, in cold, wet soil the supply of oxygen might be more adequate for the seed than in warm, wet soil.

The seasonal trend in soil stress is from the cold, wet soil to the hot, dry soil. Therefore, if seed lots differ in their response to the soil stress complex, it might be feasible to plant lots according to their potential response to the most likely soil stresses, i.e., plant the most vigorous lots under the greatest stress conditions.

Some seeds can avoid soil stresses. For example, Pollock and Toole (1966) found that hard seeds in lima beans remained impervious to water longer at low temperature and high osmotic stress than at high temperature in pure water. Such hard seeds would avoid the damaging effect of stress conditions in cold, wet soil.

B.   Quick Tests

The ultimate expression of vigor is the productivity of the plant. Because of the time and expense involved in evaluation of vigor in terms of productivity, measurement of seedling vigor is frequently used as a substitute. For many purposes, determination of vigor by germination has been found to be slow and expensive, so chemical, or quick tests, have been developed as a substitute.

However, since the ultimate goal of evaluating seeds for vigor is to select seed lots that will produce maximum yield, the validity of any vigor or quick test is only as good as the correlation which has been established with yield. Although some work has been done to correlate germination tests with yield, quick tests have most commonly been correlated with germination percentages in the laboratory or field. Rarely have these tests been extended to include the effect of stress conditions on yield.

1.   TETRAZOLIUM TEST

Tetrazolium is used as a viability test and, as such, is discussed in Chapter 5, Volume III, of this treatise. It can also be used in some cases to evaluate vigor in terms of death or damage to specific tissues within the seed (Moore, 1966).

2.   LEACHING

Another method of evaluating relative vigor of seed lots is to measure the amount of material leached from seeds soaked in water; the lower the vigor the greater the amount of leaching. This has been shown to be true for corn (Tatum, 1954), cotton (Presley, 1958), lima beans (Pollock and Toole, 1966), peas (Matthews and Bradnock, 1967), and other crops.

3.   ENZYMES

Glutamic decarboxylase was shown by Linko and Milner (1959) to be activated in wheat seeds by the addition of water, and Linko (1961) showed that the activity of this enzyme was a better indicator of the storage condition of wheat than was fat acidity (Baker *et al.,* 1957). Glutamic decarboxylase activity in the seed is also correlated with seedling growth in corn (Grabe, 1965; Woodstock and Grabe, 1967). However, Grabe (1965) showed that, although glutamic decarboxylase activity was a good indication of longevity in storage, germination tests gave a better indication of potential field emergence. This observation suggests that it may not be possible to use a single vigor test to evaluate the potential expression of vigor under different conditions.

The reason for the lack of correlation between quick tests and potential field emergence is probably related to the way in which the kind and quantity of the enzyme systems are linked with the environmental response of the seed. In the past, seed physiologists have believed that the enzymes in seeds were a more-or-less fixed component of these seeds. If this is true, then it should be possible to relate seed quality to the amount of some enzymes present. However, it has more recently been shown that enzyme systems change with environmental stress. For example, Engelsma (1967) found that phenylalanine deaminase in gherkin seedlings increases in response to light. Both the formation and subsequent destruction of the enzyme require synthesis of new protein and can be blocked by treatment with cycloheximide. Kollöffel (1968) has shown that alcohol dehydrogenase increases in the cotyledons of peas germinated under wet conditions, and Sherwin and Simon (1969) made similar observations with lactic dehydrogenase in beans. Thus the conceptual basis for the development of biochemical tests for vigor in seeds probably requires reevaluation.

4. RESPIRATION

Respiration during the first few hours of imbibition has been shown to be correlated with subsequent growth of seedlings of lima beans (Woodstock and Pollock, 1965), corn (Woodstock and Grabe, 1967), and wheat (Kittock and Law, 1968). However, Abdul-Baki (1969) has developed a glucose metabolism test which appears to be a more sensitive indicator of vigor in barley than is the respiration test (see Chapter 4, Volume II, of this treatise).

The topographical tetrazolium test, because it permits evaluation of individual seeds, has potential use in evaluating the variability between seeds within a seed lot. The other tests, which bulk together many seeds, can be used to evaluate only the relative vigor of different seed lots. These tests have generally been developed with, and tested on, lots containing seeds at uniform vigor levels. Thus, their value would seem to be restricted to the situation where a measurement of the variation within a seed lot is not important (see, also, Sections II, B, 1 and II, C).

Because of the rapidly expanding interest in seed quality and observations suggesting that different tests may be needed for different purposes, there is considerable interest in the development of new and more sophisticated types of tests. An alternative to this approach would be to obtain more information from the tests already available. For example, in evaluating the vigor of bean seedlings, Pollock et al. (1969) measured number of seeds (1) decaying, (2) emerging, (3) abnormal, (4) mechanically damaged, (5) with cracked cotyledons, and (6) hard at the end of the experiment, as well as epicotyl fresh weights, total fresh weights, and number of cracks per cotyledon. Many of these parameters were shown to be related to vigor and germination conditions. However, their relative importance varied with the germination conditions used; there was no way to combine them into a single value to compare the vigor of the seed lots.

The common problem of seed physiologists is how to combine dissimilar types of measurements to evaluate one seed lot relative to other seed lots. The problem is analogous to that of a taxonomist who must combine the morphological characteristics of one group of plants and show that the sum total of these characteristics establishes that group as distinct from a second group of plants. Historically, taxonomists have arrived at plant classification schemes by subjective observation. More recently, taxonomists have utilized computers to combine many bits of mathematically unrelated objective data into clusters of related values which can be used to describe objectively the characteristics of genera (Wirth *et al.*, 1966; Hawksworth *et al.*, 1968; Solbrig, 1970). Applications of similar computer techniques to existing laboratory methods of seed evaluation might have more value than the development of new laboratory methods.

IV. Seed Development and Vigor

Yield is an extremely important and common parameter for measuring genetic differences. To obtain yield differences among varieties, several varieties are usually grown together under essentially identical environmental conditions and the yields compared. However, as shown by McFadden (1963), this is not an adequate test of genetic capability. Working with barley, he grew seeds of several varieties at different locations and then planted these seeds together for yield trials. He found differences in yield associated with the location in which the seed was produced. These differences were as high as 16%, large enough to alter genetic yield rankings had not the importance of seed source been appreciated.

Available data show that vigor of germinating seedlings can be influenced at any time during development of the seed. It is relatively simple to demonstrate the effect of conditions during the last stages of seed development on the vigor of the resulting seedlings. However, because of the longer time involved and the greater opportunity for complex interactions, it is much more difficult to show the relationship between environmental conditions during flower formation or fertilization and the germination vigor of the next generation.

A. *Flower Formation and Fertilization*

Hardesty and Elliot (1956) showed that wheat seeds germinated immediately after harvest showed appreciable differences in primary root length after 96 hours. The differences were related to the sequence of anthesis and maturation of florets on the parental spike, with the florets that reached anthesis first producing seeds with the slowest rates of germination. This observation is perhaps typical in demonstrating the complexity of experiments relating flowering events to seed quality. In most species there is a developmental sequence of floral initiation and development on each plant. In nature, the plant develops under a regular progression of seasonal environmental changes overlaid by a highly irregular series of day-to-day weather fluctuations. Therefore, unless the plant is grown under conditions of controlled environment, it is difficult to relate a specific group of seeds to the environmental conditions that prevailed during the initial stages in the development of those seeds. Furthermore, the first seeds in a developmental sequence may have a competitive advantage over later seeds formed in the sequence.

The results of these complexities are well illustrated by Umbelliferae such as the carrot. Borthwick (1931) found that the umbels of carrots are formed terminally on branches and denoted the umbels in the order in which they were formed. Each plant has a single first-order umbel at the apex of the flower stalk. The second-order umbels are on branches from this main stalk, the third-order are on branches from the second order, etc. Anthesis in one order of umbels is completed before it begins on the next order; thus, flowering on the plant as a whole occurs as a series of periodic waves. Data in the first five columns of Table I were obtained by Borthwick. Hawthorn *et al.* (1962) performed a similar experiment, with

TABLE I

INFLUENCE OF DISTRIBUTION OF CARROT PLANT FLOWERING
ON YIELD AND POTENTIAL VIGOR OF RESULTANT SEEDS[a]

Order	Day of first anthesis	No. of umbels per plant	Potential yield (% of total)	Germination (% seed)	Seed age (days) at harvest for maximum vigor (50 days)		
					1st Order	2nd Order	3rd Order
1st	1	1	3.9	73	50	59	70[d]
2nd	10	15	46.7	67	41[c]	50	61[d]
3rd	21	53	47.2	54	30[c]	39	50
4th	34	13	2.2	[b]	—	—	—

[a]Data adapted from Borthwick (1931) and Hawthorn *et al.* (1962).

[b]Although 4th or higher-order umbels are possible, these are so small and late that the seeds seldom mature.

[c]Low vigor.

[d]Some seeds lost due to shattering.

comparable results. In addition, they found that maximum germination was reached 30 days after flowering. However, maximum vigor, as judged by speed of germination, was not reached until 50 days after flowering. Thus, seeds harvested between 30 and 50 days after flowering were viable but low in vigor.

Carrot seeds are easily lost by shattering about 60 days after flowering or 10 days after they have attained maximum vigor. Therefore, as shown in columns 6–8 of Table I, maximum yield can be obtained only by sacrificing some of the seeds from the first-order umbels; even so, seeds of the third order will be of low vigor. This shows one mechanism for the formation of nonhomogeneous seed lots, as discussed in Section II,B,1.

The effect of temperature stress during anthesis was studied by Dotzenko (1967) using Russian wild rye (*Elymus junceus* Fisch.) and white wonder millet (*Setaria italica* Beauv.). He exposed *Elymus* plants to temperatures of $-1°$ to $21°C$ for 24 hours during anthesis and found a delay in subsequent embryo development at temperatures above and below the optimum of $16°C$. Seed weights, indicative of potential vigor, were also maximal at $16°C$. However, although the data suggest a temperature optimum for the reproductive process, the types of injury noted at high- and low-temperature extremes were different. Anther injury and collapsed embryo sacs occurred in plants exposed to $-1°$ and $4°C$, whereas style damage occurred at $21°C$. Similarly, *Setaria* plants were subjected to stress at $2°$ or $35°C$ for periods of 4 to 24 hours during anthesis. Seed weight decreased as exposure increased at either high or low temperature.

The effect of soil moisture stress on ovule development in corn was studied by P. R. Rowe and Andrew (1964) who found that a serious water shortage imposed during the six-to-nine leaf stage could decrease the number of rows of kernels—a relatively stable phenotypic character in corn. Their data showed that ovule development was sensitive during only a relatively short period in ontogeny, whereas increase in plant height and root dry weight were sensitive at all times. Campbell *et al.* (1969) studied the effect of excess moisture on seed set in wheat. They found that excess moisture reduced oxygen diffusion in the soil and decreased seed set and yield. Under conditions of excess soil moisture, low relative humidity in the air increased seed set. Pollen development was especially sensitive to poor soil aeration. Unfortunately, germination and seedling vigor were not examined in these experiments. However, since low vigor frequently accompanies poor seed set, the possible role of soil moisture during anthesis on subsequent seedling vigor should be given careful consideration.

B. *Seed Development*

A major feature of postfertilization seed development is accumulation of nutrient reserves. The greater the supply of stored nutrients in the seed, the greater the vigor of the seedling and its potential for survival. This fact is the basis for the art of seed cleaning in which large and/or high-density seeds are removed from small and/or light seeds of the same crop as well as from contaminating weed seeds and foreign material. In cleaning, two parameters of seed size are used for the basic separation of high and low vigor seeds — volume and density. For small seed lots, high vigor seeds may be separated by their greater density using an upward flow of air to remove the light, low vigor, seeds (Harmond *et al.,* 1961). For large commercial seed lots, the cleaning process normally includes a density separation such as that achieved by a vibratory deck on which seeds are stratified by air passing upward through the porous surface of the deck. In some cases, seeds can be separated by specific gravity in water — good seeds sink to the bottom and the floating seeds, which are most frequently of low vigor, can be discarded. In the case of large coconut seeds, it has been shown (Marar and Shambhu, 1961) that seedlings from seeds that float vertically are of higher vigor than those from seeds that float horizontally.

The general subject of seed size has been discussed by Lang (1965) and is described relative to specific topics throughout this chapter. The following general conclusions can be drawn:

a. Any reserve nutrient that can control the rate of seedling development, under any set of conditions, is a potential factor in seedling vigor.

b. Any environmental condition that influences accumulation of nutrient reserves in seeds has the potential for influencing vigor in the following generation.

We need now to examine the experimental basis for these broad conclusions and later to examine the potential for research and development which is offered by these conclusions.

1. ENDOSPERM DEVELOPMENT

Seed development is characterized by a complex of structural and nutritional interrelationships among the embryo and the maternal and endosperm tissues. Central to these interrelationships is the development of the endosperm and its subsequent digestion, either before or after germination, by the developing embryo. Endosperm development is obligatory to seed development; failure to do so results in death of the embryo (Brink and Cooper, 1957; Rangaswamy, 1963). In plants having the endosperm as the seed storage tissue (particularly gymnosperms and monocotyledonous angiosperms), any abnormality leading to reduced size or loss of endosperm tissue reduces the potential vigor of the seeds.

2. INORGANIC NUTRIENTS

Inorganic nutrients stored in the seed provide the elements needed during the early stages of plant establishment, but the degree to which these nutrients actually influence vigor depends on the postgermination environment. Austin and Longden (1965) and Austin (1966) studied vigor of peas, carrots, and watercress plants obtained from seeds which were produced by plants growing in nutrient solutions with different concentrations of phosphorus. When the seedlings were grown on nutrient containing a high phosphorus content, they showed no evidence of the mother plant's nutrition. However, when seedlings were grown on low concentrations of phosphorus, the size of the seedlings was directly related to the concentration of phosphorus in the nutrient medium of the mother plant; this phosphorus concentration was also reflected in the phosphorus content of the seed. Szukalski (1961) obtained a similar result with phosphorus on flax, and Gurley and Giddens (1969) showed that soybean seeds having a high molybdenum content could overcome a molybdenum deficiency in the soil. Austin (1966) and Austin and Longden (1965) failed to find a similar effect with nitrogen fertilization of the mother plant. They noted, however, that the pea plants were nodulated, suggesting that nitrogen level in the nutrient medium of the parent plant actually might not have been limiting.

By contrast, Schweizer and Ries (1969), working with wheat, found that application of nitrogen to mother plants increased seed nitrogen and protein content resulting in increased vigor of seedlings. However, Terman *et al.* (1969), working with wheat under irrigated conditions found an inverse relationship between yield and grain protein content at all levels of applied nitrogen. This relationship did not hold under dry land conditions, suggesting that yield–quality relationships may vary according to the type of agriculture. Hawthorn and Pollard (1966) found that addition of nitrogen fertilizer to peas decreased seed yield and germinability. This was probably an indirect effect whereby vegetative growth was stimulated by nitrogen which, in turn, prevented normal curing of the seeds and encouraged attack by parasitic microorganisms during curing.

Trace element and other nutrient deficiencies in the soil may result in deficiency symptoms in seeds and seedlings. In peas, an abnormality consisting of necrotic areas in the center of the cotyledons coupled with plumule damage, known as "marsh spot," is believed to be the result of manganese deficiency (Heintze, 1956). In peanuts, boron deficiency of the mother plant auses "hollow heart" of the cotyledons, whereas calcium deficiency of the mother plant causes "dark plumule" with a visible effect on the vascular system at the base of the plumule (Cox and Reid, 1964; Harris and Brolmann, 1966). A somewhat similar effect is seen in certain lots of the Tendercrop cultivar of garden beans which show collapse and decay of an area of the hypocotyl (hypocotyl collar rot). This can be prevented if the seeds are germinated in water containing a calcium salt (F. J. Williams *et al.,* 1966). There has been discussion as to whether this abnormality should be considered a true seed weakness for planting purposes, since it appears to be expressed only in certain laboratories under rather abnormal conditions in which the substratum and water are free of calcium.

Borderline trace element deficiencies may have complex effects on seed quality. Beans are sensitive to zinc deficiency, and it has been repeatedly noted that one symptom is a delay in maturity or irregular maturity within a field (LeBaron, 1966; Boawn *et al.,* 1969). This delay can be as much as 30 days. Since beans are grown in areas where the length of the growing season is marginal, this delay could result in the maturation of the seeds under unfavorable environmental conditions. Thus, zinc deficiency could affect seed vigor indirectly.

3. ENVIRONMENTAL CONDITIONS

Environment during seed development can influence the vigor of the resulting seeds. Probably the most extreme case was documented by Highkin (1958) who grew peas under constant temperature for several generations. The growth rate of the plants decreased in each succeeding generation, and the original vigor could be restored only if the plants were grown under alternating temperatures for two or three generations.

Stearns (1960) grew *Plantago aristata* Michx. to maturity at a 16-hour photoperiod under constant temperatures of 60°, 70°, and 80°F and germinated and grew the resultant seeds at 60° and 80°F. Cotyledon length at 21 days and leaf length at 65 and 120 days were all greatest from seeds from 80°F plants and least from seeds produced by plants at 60°F. Seed size did not appear to be responsible for the effect on vigor. In contrast to most available data, the greatest growth was associated with the smallest seeds.

Although the role of temperature on seed development has been studied frequently, there is serious reason to question the reliability of air temperature data obtained from thermometer readings under field conditions. Hawthorn *et al.* (1966) inserted thermocouples into developing pea seeds and measured seed temperature. They found that seeds in pods exposed to the sun averaged 2.8°–6.6°C higher than pods in the shade, and those in pods resting on the soil, 1.8°–4.9°C higher than those in the air. The average temperature of pods on the soil surface exposed to sun was 6°–7°C higher than that of shaded pods or pods in air; on 1 day they recorded a temperature 15°C higher for pods on dry soil in the sun than for pods on moist soil in the shade. The highest temperature recorded was 68°C in a sun-exposed pod on the surface of dry soil. Two of the seeds in this pod appeared to be damaged, but the others appeared to be normal. We have also examined (Pollock, 1962) the temperature of pea pods curing in windrows in the field under the arid conditions of Utah. The pods on the outside of the windrows experienced a large diurnal temperature change from the high temperature of exposure under midday sun to very low (almost freezing) temperature caused by radiation to the cloudless sky at night. However, seeds toward the center of the windrow cured with only slight diurnal temperature changes. The potential significance of these extreme seed temperature changes in affecting vigor has not been carefully studied.

There have been many papers published relating chemical constituents of seeds to environmental conditions during seed development. For example, Howell and Cartter (1953) showed that oil content of soybeans was correlated with temperature during the periods 20–30 and 30–40 days before maturity. Hoshikawa (1962) noted that temperature during development of wheat seeds affected starch deposition, shape of endosperm cells, and seed size. Nagato and Ebata (1965) found that high temperature during development of rice seed accelerated starch accumulation but increased variability among seeds on a plant and altered the shape and color of the resulting seeds. Paquet (1968), working with wheat, found that a sharp rise in temperature during seed development had a differential effect on yield and protein content, depending on whether the rise occurred during early or late development. Unfortunately, none of these experiments were extended to include the effect on germination potential of the seeds.

The literature also contains papers indicating differences in potential for seed performance which cannot be related readily to specific environmental factors or to stage of seed development. J. S. Rowe (1964) reviewed some of these and pointed out their implications for forestry research. For example, Weibel (1958) showed that immature seeds from an obligate winter wheat variety could be vernalized during development. Bodganov et al. (1967) found major differences in the state of mitotic development in peas from two different harvests. Marshall (1969) studied freezing resistance of young winter oat plants (*Avena byzantina* K. Koch and *Avena sativa* L.) and found differences in frost hardiness of young plants to be related to seed source.

It is unwise to attempt to evaluate the effect of environmental conditions during seed development on seed vigor independent of their effects on seed yield. Adams (1967) pointed out that yield of a crop is the summation of a number of yield components, e.g., seed size, number of seeds per pod, pods per plant, and plants per unit area. Negative correlations among yield components are common for crop plants and presumably also for noncrop plants.

Adams (1967) considered that a field bean (*Phaseolus vulgaris*) plant consists of a number of relatively independent nutritional units, each including two trifoliate leaves and several flower buds which open at different times. Within each nutritional unit there is a "peck order" such that, if the plant is exposed to environmental stress, it will abort, in the following order; (*1*) freshly pollinated embryos and very young pods, (*2*) young fertilized ovules in developing pods, (*3*) open flowers, (*4*) unopened flowers, and (*5*) rapidly developing young pods.

The concept of peck order can be extended to include the viability and potential vigor of the surviving seeds. Although not supported by experimental evidence, it is our impression that, under conditions of stress, a plant will tend to produce a minimum number of seeds of the highest possible vigor rather than a larger number of low vigor seeds. If this is true, the germination rate curves shown in Fig. 4B can be interpreted to mean that the early germinating seeds are those of the maximum germination potential provided by the genetic capability of the plant. The late germinating seeds, those represented by the skewed portion of the curve, represent the low vigor seeds in the population and may, in the case of crop plants, be mostly the result of harvest conditions. This interpretation of vigor relative to seed production would be compatible with maximum potential for survival of a species under unfavorable environmental conditions.

C. MATURITY AND CURING

A zygote must undergo a complex series of developmental processes culminating in the ability to grow independently of the mother plant. The ability of a seed to germinate marks a major transition point in the life cycle of a plant. This transition cannot be considered to be a gradual sequence of progressive quantitative changes, instead it marks a major qualitative discontinuity in development. S. Klein and Pollock (1968) studied development of lima bean seeds and found that embryonic axes attained only a minor capacity for growth before they also attained the ability to withstand desiccation — the physiological capacity which marks maturity of most seeds. It was only after the seeds had attained the capacity to withstand desiccation that rapid growth of the embryonic axis, characteristic of vigor, was possible.

The general conclusion which can be drawn from the vast literature on seed maturity is that the more mature a seed when harvested, the greater its vigor and, therefore, potential for establishment of a new seedling. However, the problem (both conceptual and practical) is that each seed lot consists of seeds removed from the parent plants at different stages of maturity. During curing, seeds of different stages of maturity are exposed to different sets of environmental conditions. The result is a heterogeneous population which may be difficult to interpret.

Variability begins with the sequence of floral initiation and opening (see Section IV,A, above). After fertilization, seeds borne on different parts of a single plant may be exposed to different sets of environmental conditions. For example, Jensen *et al.* (1967) noted large differences in maturity and chemical composition in seeds from different levels and sides of individual *Picea abies* trees. Variation tended to be associated with exposure to the sun. This contrasts somewhat with the case of *Phaseolus vulgaris* plants in which Adams (1967) was able to note a pattern of nutritional units; variations within a nutritional unit tended to be larger than variations between units.

The causes of variation in seedling vigor associated with maturity and curing of seeds differ according to the growth habit of the plant. For convenience in discussing the subject, we can recognize three general types of plants.

Type I: plants with dry fruits that dehisce readily at maturity and also plants that distribute single, dry propagules. This group would include carrots, onions, certain members of the Cruciferae, Compositae, and Leguminosae families and most trees such as pine. These plants generally are noncultivated or, if cultivated, grown for structures other than fruits or seeds. Thus, the flower and seed have not been subjected to extensive selection for agricultural purposes. Because, in nature, these seeds are disseminated as soon as they are mature, for agricultural use they must be harvested before they have become fully mature on the mother plant. Then they are subjected to some sort of curing and drying procedure. In the case of forest tree seeds, this frequently involves kiln-drying. Vegetable seed stalks are cut and allowed to dry on canvas or are placed in hot-air dryers before threshing. Therefore, these seeds usually cure and dry rather quickly after harvest in the presence of a minimum amount of leaf and stem material. Under these conditions, seeds that were immature when harvested do not have time to mature during curing and are usually small, light, and easily cleaned out and discarded. Also, the absence of vegetative material minimizes attack by saprophytic microorganisms during curing—one of the major problems of seed harvest.

Type II: plants with dry fruits that do not dehisce and disseminate seeds immediately upon reaching maturity. This is the common situation in most crop plants such as beans, peas, corn, wheat, and other grains and can be considered to be the result of man's selection of the seed for agricultural purposes over the centuries. In some cases, seeds are harvested directly from the plant. However, the more mature the plant the greater the tendency for its stem to bend or "lodge." In other crops, seed harvest normally involves cutting entire plants and piling them in windrows for drying and curing. Because immature seeds cure rather slowly in contact with a bulk of plant material, they may partially mature and, thus, are not so easily separated from seeds that have matured under more favorable conditions before harvest. Furthermore, the bulk of vegetative material included with the seeds results in relatively slow drying increasing the potential for attack by saprophytic microorganisms. For this reason, wet weather during curing is a potential cause of low vigor.

Type III: seeds produced in fleshy fruits. In nature, many of these seeds are harvested when birds eat the fruits. The seeds are separated from the fruit tissue in the digestive tract of the birds, and are simultaneously distributed to new locations. In agricultural practice, seeds of such fruits as tomato and cucurbits are separated mechanically from the bulk of the fruit tissue. The tissue remaining in intimate contact with the seeds must be removed either by acid treatment or fermentation; following this the seeds are washed and dried. Most of the process of maturation and curing is completed in an aqueous environment, protected from extremes of temperature and from rapid environmental changes but limited in available oxygen and exposed to a multitude of microorganisms.

Specific examples of the relationship between seed maturity, curing conditions, and subsequent vigor can best be considered by grouping the examples according to the maturation types described above.

Type I. Increased use of crops, such as crested wheatgrass (*Agropyron cristatum* L. Gaertn.) led to studies on methods for harvesting seeds. Since these seeds are lost by shattering shortly after they mature, the practical problem is to find the most immature stage at which the seeds may be harvested and still have the vigor required to establish a plant under the harsh environmental conditions where these crops are of agronomic value. Hermann and Hermann (1939) found that, although seeds harvested in the early dough stage have high viability, vigor increases up to the time of full ripeness. Some of these data showing the relationship of several measures of vigor to seed maturity are reproduced in Fig. 12. These results were confirmed and extended by McAlister (1943) to demonstrate the same general type of response with species of *Bromus, Elymus,* and *Stipa* and to show that longevity in storage was

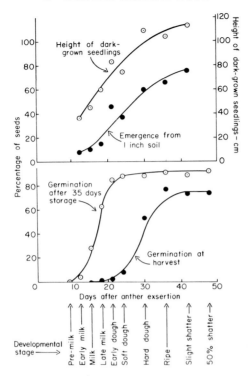

FIG. 12. Effect of seed maturity on germination vigor of crested wheat grass (*Agropyron cristatum* L.). (Data from Hermann and Hermann, 1939.)

also a function of maturity. In practice, harvest must be timed to give a compromise between yield and maximum vigor. Lawrence (1967) found that different harvest methods could affect yield and vigor for *Elymus*. S. R. Anderson (1955) working with bird's-foot trefoil (*Lotus cornicalatus* L.) noted that losses due to dehiscence became very high if relative humidity dropped below 40% before harvesting was completed. Bennett and Marchbanks (1966) found that immature seeds of rough pea (*Lathyrus hirsutus* L.) were more sensitive than mature seeds to drying temperature.

The problem of timing seed harvest in forest trees, such as pines, is extremely critical because of the large amount of labor necessary and the high vigor of the seeds required for direct seeding purposes. This problem was studied by McLemore (1959) for *Pinus palustris* growing in Louisiana. He found that seeds from cones harvested on September 9 averaged 46% germination compared to 66% for cones gathered September 24 and 88% for cones harvested at the height of the commercial cone-harvesting season, October 7. By October 14, cones were beginning to open and disperse seeds naturally, indicating the critical nature of the harvest operation.

Type II. In plants in which the pods do not readily dehisce to release the seeds, a certain amount of force must be used in threshing to break the fruits. This force has the unfortunate effect of introducing mechanical damage which may be very serious. (This problem is considered separately in Section V.)

In discussions of seed maturation, frequently the term "drying" is used in a very uncritical manner, implying a simple, physical loss of water. As shown by Fig. 13, the situation with respect to seed water content is relatively complex. Following fertilization, there is a period during which cell division, expansion, and differentiation proceed to form the visible structure of the seed. During this period, the size of the seed increases but its moisture content remains constant and high. Deposition of reserve materials continues beyond the time of seed enlargement so that seed dry weight increases more rapidly than fresh weight; thus, the percent of water in the seed declines. During later stages of seed development, there is an actual loss of water, but only during the last few days or hours is this a simple drying process, not accompanied by accumulation of nutrient reserves. Fig. 13 shows these relationships for oats. Although other species may vary in detail, it is our impression that most seeds show a similar pattern of changes in weight and water content.

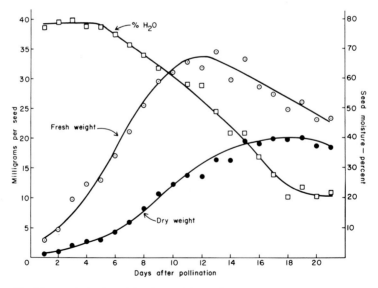

FIG. 13. Changes in fresh weight, dry weight, and percent water of oat seeds during development. (Data from Booth, 1929.)

Many papers report age of seeds in terms of days from flowering, a value which can vary widely with environmental conditions. We would like to suggest that such papers include seed moisture among their data so that it would be possible to evaluate different experiments and species on a common basis.

As in the case of Type I plants, the more mature seeds are at harvest the higher is their potential vigor (Bartel, 1941; Robertson and Curtis, 1967; Oelke et al., 1969). In many crops, the plants are cut and piled in windrows to dry. However, because of variability in maturity of seeds within a crop, physical drying of immature seeds may be preceded by a period of maturation during which water is lost rather slowly (Pollock et al., 1970). During this period of maturation, the immature portion of the seed population must depend on nutrients which can be translocated from only the cut portions of the plant to which they are attached. Furthermore, these seeds may be exposed to moisture and temperature stresses which exceed those to which they would have been subjected during maturation on the intact plant. For example, Wester and Jorgensen (1956) noted fading of chlorophyll from cotyledons of lima beans which is indicative of low vigor (Pollock and Toole, 1966). Seedsmen recognize that the numbers of bleached seeds increase when a crop is allowed to remain too long in the windrow. Exposure to high atmospheric humidity while seeds are drying on the plant (Wester and Magruder, 1938) or rain on the windrows (Fields and King, 1963) results in damage to the seed coat and infection of seed by fungi. Wetting and drying may cause a natural crushing of seed tissues due to uneven absorption of water (Moore, 1965).

Corn and sorghum normally mature on the plant before harvest, although much corn is now being harvested at relatively high seed moisture and then kiln dried. Because these crops are grown in areas in which the length of the growing season is marginal, seeds are frequently exposed to freezing injury while on the plant; the more immature the seed (the higher the moisture content), the greater the freezing damage (Rush and Neal, 1951; Rosenow et al., 1962).

The rate of reduction of seed moisture is partially determined by atmospheric conditions. Dodds and Pelton (1967) have described extensively such conditions as they influence drying in wheat.

Type III. Although seeds in fruits, such as tomato, develop and mature in an aqueous environment, McIlrath et al. (1963) have shown that they undergo the same changes in seed moisture as seeds on other plants which mature in air. This indicates that the loss of water during seed maturation is more an inherent phase of seed development than implied by the passive concept of seed drying.

A relationship between fruit maturity and germinability of seeds has been demonstrated for butternut squash (Holmes, 1953), cucumber (Bowers, 1958), muskmelon (Harrington, 1959), and mango (Giri, 1967). By using mature green fruit as the youngest stage tested, Kerr (1963) failed to find an effect of immaturity on seed quality in tomatoes. He did find a tendency for germination to decrease when overripe fruits were used as a seed source. Toole *et al.* (1941) found that rotting of the fruit caused by *Phomopsis vexans* resulted in decreased germination, slower germination, and small seeds in eggplant. Holmes (1953) showed that both mature and immature fruits of butternut squash could be stored for up to 211 days and still yield highly viable seed. It seems reasonable to believe that the decrease in seed quality with increasing age of fruits is probably a secondary effect of damage by microorganisms rather than a direct effect of age on the seed.

Seeds of fleshy fruits are surrounded by a layer of tissue which is very difficult to remove and which may interfere with germination. Although this material may be removed mechanically, Harrington (1959) showed that in muskmelon fermentation produced seeds with higher germination. Kerr (1963) studied the effect of fermentation on tomato seed quality. The locular jelly plus seeds were squeezed into 4-gal clay pots and allowed to ferment. He found that fermentation was rapid; the pulp and seeds rose to the top within 24 hours, and within 4 days most of the seeds had settled to the bottom. Germination increased for the first day as did vigor (rate of germination). Thereafter both declined rapidly.

In Kerr's experiment, fermentation was bacterial, probably caused by *Erwinia carotovora* which was a naturally occurring organism. Commercial tomato seed is fermented in wooden barrels or large wooden tanks (Hawthorn and Pollard, 1954) which may build up a microflora specific to the fermentation process.

Although the fermentation process is widely used to remove fruit tissue and to control bacterial canker in tomato (Hawthorn and Pollard, 1954), acid treatment may also be used to remove the fruit pulp. In spite of the potentially interesting physiology of seeds in the fermentation tank, apparently no detailed study of these processes has been attempted.

V.   Mechanical Damage and Vigor Reduction

One of the most serious, and probably the best understood of the causes of low seed vigor is mechanical injury during threshing, cleaning, handling, or planting. One of the earliest references to this form of injury is that of Nobbe (1872). Hurd (1921) stated, "Nobbe first recognized the fact that machine-thrashed seed was more injured by copper sulphate than was hand-thrashed seed, because damage done to the seed coats by the machine allowed the copper sulphate to penetrate to the embryo. He pointed out that the drier and more brittle the crop, the greater the thrashing injury." Since that time, increasing attention has been paid to the problem of mechanical damage. Some workers have suggested that the problem may increase in severity as new techniques of seed harvest are introduced (Tatum and Zuber, 1943). Although there is no way to evaluate the accuracy of this suggestion, three facets of the problem should be recognized: (*1*) the increasing use of machinery and the introduction of new types of machines; (*2*) the introduction of new varieties which may be more susceptible to damage; and (*3*) an increasing awareness of the problem as technology and technologists become more sophisticated.

A. Sources of Mechanical Damage

Seeds may be damaged at several stages during harvesting, processing, and planting.

1. THRESHING

The essential operation in seed harvest is the separation of the seed from the parent plant or from the fruit. The importance of this process in causing mechanical damage is indicated by the large amount of published research on many crops: corn (Alberts, 1927), beans (Borthwick, 1932; Bainer and Borthwick, 1934; Harter, 1930; Toole *et al.,* 1951), crested wheatgrass (Conard, 1946), wheat (Davies, 1964), soybean (Green *et al.,* 1966), sorghum (Kantor and Webster, 1967), clover and grasses (L. M. Klein and Harmond, 1966); see, also, U.S. Department of Agriculture (1948) for references on tree seeds.

Although details of the threshing operation vary from crop to crop, the most common type of equipment utilizes one or more cylinders rotating in relation to a fixed bar. The dried plants and fruits are disrupted to release the seeds which fall through a screen and are collected. Threshing is possible only when the fruits are dry enough to be broken between the cylinder and bar; this requires not only mature plants, but ones which have been exposed to favorable drying conditions. To obtain maximum seed yields it is necessary that the threshing process be relatively severe, and economic considerations require that threshing be rapid (Bainer and Borthwick, 1934). However, to obtain optimal seed quality, it is essential that the process be as gentle as possible. Because fruit moisture content changes drastically and quickly throughout the day, threshability changes just as quickly, so that threshing efficiently with minimal damage to seeds requires skill and experience.

Some of the variables in threshing are well illustrated by the data of Bainer and Borthwick (1934) obtained for lima beans grown and threshed under the arid conditions of the central valley of California (Fig. 14 and 15). Figure 14 shows that the faster the speed of the revolving cylinder, the better the threshing, but also the greater the mechanical damage.

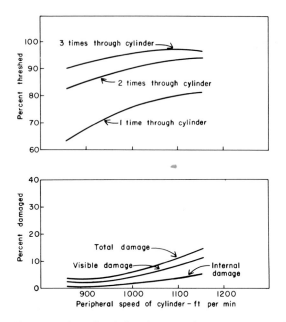

FIG. 14. Threshing tests of Fordhook lima beans (moisture content 15.9%). (Data from Bainer and Borthwick, 1934.)

Figure 15 shows that the higher the seed moisture content, the less the damage; unfortunately, the data do not include values on the maximum moisture content at which threshing is possible. Because of the importance of the threshing process, there has been continual progress in design of equipment (L. M. Klein and Harmond, 1966).

2. CLEANING AND HANDLING

In a modern seed-cleaning plant the number of processing steps a particular lot of seeds may encounter varies from two or three to as many as six or seven. Some of the basic cleaning and handling operations include air screen cleaning, specific gravity separation, size and shape sorting, and color separation (Harmond *et al.,* 1961; U.S. Department of Agriculture, 1948).

Seed-cleaning plants are of two basic types: (*1*) vertical, in which seeds

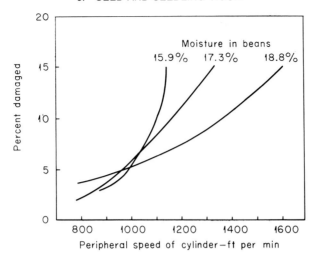

Fig. 15. Cylinder speeds required to produce 15.0% damage in Fordhook lima beans as a function of seed moisture. (Data from Bainer and Borthwick, 1934.)

are elevated to the top of a multistoried building and allowed to flow by gravity from machines on one floor to the cleaning machines for the next operation on the floor below, and (2) horizontal, in which seeds are elevated into a tank above a particular machine, pass through that machine by gravity, and then are reelevated for the next cleaning process. In either case, the seeds often come in contact with hard surfaces or other seeds.

It has been clearly demonstrated that mechanical damage to seeds may result from rather minor falls. The effect of the number of falls is cumulative, as shown by the data in Fig. 16. As a result of these and many similar studies, seed companies have redesigned cleaning and conveying equipment to reduce the distances seeds drop and to cushion any falls which must occur, although no way has yet been developed to eliminate this problem completely.

3. PLANTING

Consideration of vigor as affected by mechanical damage is usually confined to threshing and processing damage, probably because seeds damaged during these processes are conspicuous in laboratory germination tests. Unfortunately, this obscures the fact that serious mechanical damage can occur during planting. This was studied by Hulbert and Whitney (1934) who found up to 10% injury to a large, smooth seeded pea variety during planting with a grain drill; they recommended mixing seeds with graphite to minimize damage. Dexter (1966) found that if the moisture content of beans could be increased from 11 to about 16% before planting, emergence of seedlings could be increased from 39 to 78%.

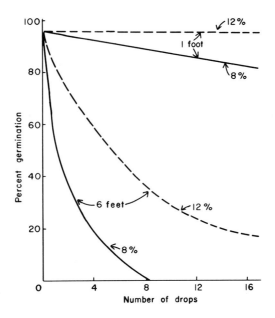

FIG. 16. The effect on germination of repeatedly dropping garden bean seeds from different heights, and the influence of seed moisture on this damage. (Data from Asgrow, 1949.)

4. Nonmechanical "Mechanical" Damage

In the case of beans, mechanical damage is such a common phenomenon that it is assumed that any abnormal seedlings are the result of machine injury during threshing or processing. Nevertheless, in recent years it has become obvious that rapid absorption of water by seeds low in moisture may cause transverse cracking and loss of cotyledons that resembles the effect of machine damage (McCollum, 1953; Pollock and Manalo, 1969). There is reason to believe that this type of damage has caused serious economic losses due to rejection of seed lots by regulatory germination laboratories. The possible importance of this phenomenon in the establishment of field stands has not been determined, nor has the possibility of its importance in other species been tested.

Breaking in milled rice is frequently attributed to mechanical damage. However, Stermer (1968) found that cracking can be caused by rapid changes in seed hydration due to changes in relative humidity and temperature. Rice with a high moisture content was more susceptible to damage than rice with low moisture. Rice subjected to a change in relative humidity of 20% or more showed stress-crack formation in less than 15 minutes.

B. *Expression of Mechanical Damage*

In large seeds, such as those of corn or bean, three levels of mechanical damage are recognized: (*1*) gross damage to the seed and seed coat so that the seed is visibly damaged—seed-cleaning techniques are sufficiently effective to remove most of this type of seed; (*2*) internal damage which becomes visible only after the seed has germinated—these are abnormalities which were previously noted (Section II, D) as being important in classifying seedlings as abnormal in regulatory tests; and (*3*) microscopic breaks, particularly in the seed coat, which make the seed susceptible to attack by microorganisms.

1. LOSS OF STORAGE AND MERISTEMATIC TISSUE

This has been studied extensively for corn (Brown, 1920, 1922) and beans (Ching and Pierpoint, 1957; Clark and Peck, 1968; Crosier, 1942; Harter, 1930; Waters and Atkin, 1959). The general conclusion is that the greater the loss of storage tissue, the lower the amount of growth and vigor of the resulting seedling. This is in agreement with other data relating seed size to vigor.

2. INVASION BY FUNGI

Working with wheat and barley, Hurd (1921) found that intact seed coats prevented attack by saprophytic fungi. Small breaks over the endosperm permitted attack by fungi which resulted in fewer and smaller seedlings. Tatum (1954) found that damaged corn seeds were highly susceptible to microorganisms in the cold test. Meyers (1924) found that an unbroken pericarp of corn was an absolute protection against *Penicillium,* whereas this fungus reduced the stand from seeds with broken seed coats by 53%; only 55% of those emerging survived until harvest. Because of the necessity of protecting slightly damaged seeds against invasion by saprophytic fungi, chemical seed treatments have been routinely applied to most agricultural seeds for many years.

3. SHORTENING OF STORAGE LIFE

Alberts (1927) examined the rates of water vapor exchanged between seeds and the atmosphere and found that the moisture content of damaged corn seeds fluctuated much more rapidly than did that of intact seeds. Thus, the pericarp not only provides a barrier against attack by fungi but also against exchange of water vapor. The combined effect might be expected to facilitate the attack of storage fungi on the seeds. These observations are in agreement with the finding of Toole and Toole (1960) that viability of garden bean seeds in storage decreases most rapidly in lots with the largest amount of mechanical damage.

C. Factors Modifying Susceptibility to Mechanical Damage

1. PRODUCTION CONDITIONS

On the basis of tetrazolium tests, Moore (1963, 1965) has suggested that both mechanical damage and natural crushing of tissue during development of a seed can precondition it to a rapid decline in germinability under storage conditions. The experience of seedsmen shows that viability of some seed lots declines very rapidly in storage and that the reason for this rapid decline is not related to obvious events in the previous history of the seed lot. McMaster et al. (1960, 1965) studied the sensitivity of garden beans to mechanical injury and the possibility that this sensitivity was controlled by irrigation frequency during seed development. Their data showed marginal significance; but, unfortunately, there is no record that the moisture content of the seeds was controlled or maintained at a uniform level at the time of applying mechanical stress.

Farooqui and McCollum (1954) described a developmental abnormality in beans, called "fishface" or "fishmouth" which they suggested results from a more rapid development of cotyledons than of seed coat, resulting in the splitting of the seed and protrusion of the cotyledons. They found that the viability of these seeds declined more rapidly than that of normal seeds. We have noted that these seeds are especially sensitive to mechanical damage because the lack of an intact seed coat permits more frequent breakage at the cotyledonary node.

In 1969, in Michigan, a normally very sensitive variety of navy bean, Sanilac, was found to be resistant to mechanical damage. The reason seemed to be a small seed size which resulted from bacterial blight and consequent premature seed maturation (Adams, 1970).

2. SEED MOISTURE

The most important factor controlling sensitivity to mechanical damage is seed moisture content (Asgrow, 1949; Webster and Dexter, 1961; Barriga, 1961; Green et al., 1966; Dexter, 1966). An example of the effect of seed moisture on sensitivity to threshing is shown in Fig. 15 and on the sensitivity to dropping in Fig. 16. As discussed in Section IV,C, maturity is closely associated with seed moisture, and seeds do not become fully dry until they are mature. Immature seeds which are exposed to the beating action of the threshing process may be bruised instead of broken (Conrad, 1946), probably accounting for the bruising noted by the tetrazolium test (Moore, 1963).

Dexter (1966) has suggested that threshing damage to beans may be minimized by spraying windrows with water shortly before the threshing process. Even without this treatment, it is obvious from the previous discussion that care must be taken during the threshing operation to adjust the machinery according to the moisture content of the seed. The problem of proper seed moisture extends to both milling and planting operations. As Dexter has pointed out, both of these may be improved by handling the seed at a slightly higher moisture level than is normally the case; this is now being done commercially, as discussed in Section VII, C. McCollum (1953) and Pollock and Manalo (1969) have shown that breakage during seed germination can be minimized by raising seed moisture content slightly.

The requirement for relatively high seed moisture during threshing, processing, and planting is in conflict with the general philosophy of seed technologists: "the drier the better." This philosophy seems to have its origin in the recognition that the longevity of seeds in storage is greatest when the seeds are very dry. However, examination of the data on which that philosophy is based suggests that the data may have been extrapolated beyond a reasonable limit. Toole and Toole (1960) found a 6-month storage life in an open warehouse in Mercedes, Texas, for bean seeds which had moisture contents of 12.5 to 15.5%. By contrast, most lots of seeds stored in San Antonio, Texas, at moisture contents of 10.8 to 11.7% had a storage life of 27 months—much longer than would normally be necessary for commercial storage of a bulk crop such as beans. There has been little or no consideration of the possibility of controlling seed moisture at a level which would provide a reasonable compromise between storage life and optimal handling and planting. This may be a fruitful field for future research.

3.   GENETICS

There is a genetic component for susceptibility of seeds to mechanical damage. Davies (1964) found that susceptibility to visible damage was a genetic factor in wheat, whereas susceptibility to microscopic damage was not. Atkin (1958) found that susceptibility to mechanical damage in garden beans was related to genotype and suggested that it might be possible to develop a damage-resistant white-seeded variety, even though most white-seeded varieties then available were extremely sensitive to damage.

Because the problem of mechanical damage is so serious in beans, it has been studied by investigators who have applied a uniform impact either by dropping the seeds (Bainer and Borthwick, 1934) or by striking them with a rotating blade in a device developed by the Asgrow Seed Company (pictured in McMaster *et al.,* 1960). By using the latter method, Dorrell and Adams (1969) studied mechanical damage to beans and found that a seed moisture of 12% (which could be obtained by equilibration at 20°C over a saturated solution of manganous chloride) was critical for use in distinguishing between seeds tolerant or susceptible to impact. Seed coat cracking increased as seed weight increased, as the shape became less spherical, and as seeds became less dense. Examination of the beans revealed that the embryonic axis is weakly attached to the two cotyledons and lies close to the seed coat, thus placing it in an extremely vulnerable position. Seeds that withstood mechanical abuse best were those which had a very tightly adhering seed coat; this is also a common observation with peas. Although some of these factors can be influenced during seed development and curing, some of them are genetically controlled, thus offering plant breeders an opportunity to develop varieties which are less sensitive to mechanical damage and which can be expected to produce seeds with higher potential vigor.

VI.   Other Factors Influencing Vigor

A.   Seed–Microorganism Interactions

The literature contains references to four areas of seed–microorganism interactions: (*1*) storage longevity, (*2*) seed-borne diseases, (*3*) germination physiology, and (*4*) pathology of soil microorganisms. We tend to think of seeds as higher plants which, after germination, compete with other higher plants. However, the competition of seeds and seedlings with microorganisms is of equal importance. Our purpose here is to consider the ways by which the vigor of seeds can influence their interaction with microorganisms.

Fungi play a dominant role in the destruction of seeds in storage. This subject has been reviewed for grain (Christensen and Kauffman, 1969) and is also covered in detail in Chapter 3 of Volume III, as is the subject of seed-borne diseases in Chapter 5 of Volume II. Storage fungi can be a major factor in the low vigor of seeds after storage. Seed-borne diseases can contribute to low vigor by weakening the parent plants. Our major point of concern, however, is the mutual interaction by which low vigor seeds provide nutrients that stimulate attack by microorganisms.

The fact that seeds lose nutrients to the surrounding environment has long been utilized to detect low vigor seed lots. Fick and Hibbard (1925) and Hibbard and Miller (1928) established the use of conductivity measurements on seed extracts and found a correlation with viability. A relationship between leaching and vigor has been established for corn (Hottes and Huelsen, 1927; Tatum, 1954), cotton (Presley, 1958; A. M. Anderson *et al.,* 1964), castor beans (Thomas, 1960), lima beans (Pollock and Toole, 1966), crimson clover and ryegrass (Ching and Schoolcraft, 1968), peas and beans (Matthews and Bradnock, 1968), and rape (Takayanagi and Murakami, 1968). Although conductivity has been the method most frequently utilized, optical and chemical methods have shown that the leachate can contain all of the metabolites that exist in plant cells (Schroth and Hildebrand, 1964). Recently, however, Abdul-Baki and Anderson (1970) found that leaching from barley seeds increased with seed age and mechanical injury, but not with accelerated aging. They suggested that leaching was regulated more by utilization of sugars than by permeability. However, there is an anatomical difference between the monocotyledonous and dicotyledonous seeds which may be important in this result. In monocots (and gymnosperms), the bulk of the storage reserve is in endosperm cells (or megagametophytes in gymnosperms) which are isolated from the embryonic cells. In dicots the storage occurs in living cells of the embryo itself.

The term *leaching* has been applied by physiologists with the implication that it is a passive process resulting from a degree of disorganization of membrane structure. Pathologists have used the term *exudation* to refer to a normal, and apparently similar, phenomenon in roots and seeds but without the implication that it results from a decrease in vigor. The effect of exudates, or leachates, on increasing microbial activity in the vicinity of roots and seeds is well established and has been extensively reviewed by Schroth and Hildebrand (1964). The chemical compounds involved include normal cell metabolites, and the effect on the microorganisms appears to be more as a source of carbon and nitrogen compounds than of specific stimulating compounds. As leaching proceeds, a diffusion gradient is established which serves to direct the microorganisms to the source of nutrient supply. In germinating *Vicia faba* seeds, Pearson and Parkinson (1961) found the first evidence of ninhydrin-positive substances in the micropylar region. With root emergence, these substances were found around the root.

Schroth *et al.* (1966) examined the effect of temperature on ninhydrin-positive substances and soluble carbohydrates from beans, cotton, and peas. For beans and cotton they found a minimum of exudation at 25° to 30°C, with greater loss at both higher and lower temperatures. Peas showed a similar, although not very pronounced, minimum at 15°C for ninhydrin-positive compounds, but no obvious minimum for carbohydrate loss which was very high at all temperatures. Fritz (1966) found that the germination capacity of wheat was reduced at lower temperatures because of the severity of attack by *Fusarium nivale*. Flentje and Saksena (1963) found that preemergence rotting of peas was due to exudation of sugars which stimulated growth of *Pythium*. Damage to the testa inflicted either with a needle or by mechanical harvesting greatly increased loss of soluble materials and the resulting growth of *Pythium*. Decay was greatest in wet soil.

In addition to stimulating decay organisms, organic materials from seeds have been shown to stimulate preemergence damping-off in beans caused by *Rhizoctonia solani* and *Pythium* spp. (Schroth and Cook, 1964), bean root rot fungus, *Fusarium solani* f. *phaseoli* (Schroth and Snyder, 1961), and foot rot of rice caused by *Fusarium moniliforme* (Rajagopalan and Bhuvaneswari, 1964).

Although little of this work has been performed using seeds known from other tests to be of low vigor, there is little doubt about the direct applicability of the results. We can, therefore, conclude that one of the major effects of low vigor in seeds is to shift the microenvironmental situation surrounding the germinating seed in favor of microorganisms. The purpose of chemical seed treatments is to shift the balance again in favor of the seed.

In barley, water sensitivity is a vigor condition in which seeds fail to germinate if the water content of the germination medium is too high. Gordon (1968) found very similar activation energies for breaking of dormancy and loss of water sensitivity. Crabb and Kirsop (1969) found that the oxygen consumption of both water-sensitive and water-insensitive seed lots is reduced by the presence of a husk and a surface film of water. Therefore, they suggested that water-sensitive lots have a higher oxygen requirement for germination. However, Gaber and Roberts (1969) found that water sensitivity can be overcome by including a mixture of antibiotics in the germination medium. Both an antibacterial and an antifungal chemical are required; a single compound is not effective. This result emphasizes an aspect of seed research which appears to have been inadequately studied, i.e., the relationship between microorganisms and seeds in normal germination.

B. Dormancy

In the preceding discussion we pointed out that low vigor reduces the survival potential of seeds under adverse environmental conditions, i.e., we have defined low vigor as a condition under which germination and establishment are possible only within a narrow range of environmental conditions. This is essentially the same definition as used by Vegis (1964) for dormancy. We must, therefore, consider the relationships between the phenomenon discussed in this chapter as vigor and the phenomenon of dormancy or rest. Are these identical or are they distinct phenomena which sometimes are expressed in the same way? We must also consider the possibility that dormant seeds may be slow to germinate and, thus, contribute to the skewing of the germination rate curve previously discussed as being caused by low vigor seeds.

In the first place, we must recognize the ubiquitous existence of some degree of dormancy in all seeds. If this were not the case, vivipary would be a common, rather than a relatively rare phenomenon. Thus, we must recognize that most seeds are prevented by some type of dormancy mechanism from germinating until they have matured and are separated from the mother plant. There are, however, some notable exceptions to this generalization, especially in certain wheat varieties.

The question arises, what is the length of time after normal separation from the mother plant before a seed becomes capable of germination? Many seeds, for example most garden beans, are capable of immediate germination. Such seeds can be considered to possess maximum vigor; vigor changes can proceed only in a downward direction until the seed is dead. On the other hand, many seeds possess a moderate degree of dormancy when mature; after-ripening in dry storage is necessary before germination is possible. As after-ripening proceeds, the tolerance of the seed to environmental conditions increases. This situation is well illustrated in the case of lettuce grown in those areas of the United States where winter-harvested crops are planted in autumn when soil temperatures during germination may be very high. Many seed lots fail to germinate under these conditions, creating a serious economic problem. Thompson (1936) noted that percentage germination of these lots decreased rapidly as the temperature increased above 20°C. This temperature sensitivity, or thermal dormancy, decreased with time from seed harvest and could be overcome by exposure of seeds to light. As shown in Fig. 17, this dormancy was associated with immaturity of the seeds at the time of harvest. McCoy (1962–1963) suggested that older lettuce seeds are desirable for warm weather planting because they are less responsive to high-temperature dormancy. However, he has also shown that the vigor of seedlings from older seeds is reduced; apparently the decrease in thermal dormancy and decrease in vigor proceed together.

In the case of lettuce, it is clear that the distinction between dormancy and vigor is one of semantics; the net effect on plant establishment is the same in either case. Although there has been a great deal of research on the mechanism of stimulation of these seeds by light (Toole *et al.,* 1956) and some on the conditions associated with environmental factors controlling production of dormancy (Harrington and Thompson, 1952; Koller, 1962), the practical problem of lettuce germination under high soil temperature conditions has yet to be solved (Pauli and Harriott, 1968).

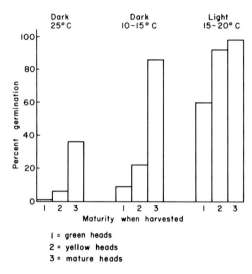

FIG. 17. Germination of *Lactuca sativa* cv. Grand Rapids under different conditions of light and temperature as a function of maturity of seed at time of harvest. (Data from Thompson, 1936.)

Recent findings suggest possible new avenues for approach to the problem. Hsiao and Vidaver (1971a,b) have found that potential germination of lettuce can be controlled by exposure of the seed to red or far-red light at seed moisture levels of 11 to 16%. If seeds are exposed to light at these relatively low moisture levels, subsequent germinability can be fixed by redrying to 7% moisture. Interestingly, this is the same range of seed moisture found by Pollock and others (Section VI,C) to be critical for the elimination of imbibition chilling injury in a number of species. Furthermore, Pollock and Manalo (1971) have found that germination of some lots of lettuce seeds is strongly inhibited by very slight moisture stress at 25°C, while germination of other lots is only slightly affected. Preliminary data suggest that this is a dormancy phenomenon, possibly related to the light-sensitivity of the seeds. If so, the response of seeds to moisture stress, rather than lack of light, may be the critical environmental factor in germination under field conditions.

Dormancy is normally considered to be ended at the time of radicle protrusion, and there is relatively little literature on the possibility of a relationship between partial breaking of dormancy and subsequent seedling growth. Nevertheless, there are several papers which suggest the existence of such a phenomenon. In the case of lettuce, Scheibe and Lang (1967) reported a light-sensitive stage in radicle development following germination of seeds on a medium containing mannitol as an osmotic stress. This light sensitivity of root growth had the same spectral sensitivity to red and far red light as the germination response of intact seeds. In seeds with longer and more profound dormancy, which require a long period of low-temperature after-ripening, some plant growth rates have been found to be related to the degree of after-ripening. In peach (Scott and Waugh, 1941; Zagaja *et al.,* 1960) and cherry (Zagaja, 1961) growth of roots and shoots of embryos from immature seeds is increased by after-ripening. Peach seedlings can be germinated without after-ripening if the seed coat is removed, but the resulting seedlings are dwarfed and may grow in this condition for as long as 10 years (Flemion, 1959). Dwarfing is the result of sensitivity to relatively high temperature during the germination period; this sensitivity decreases with after-ripening (Pollock, 1962). These results, although inadequate to establish the existence of a general phenomenon, suggest that this is an area of research which might be of importance, especially in forest tree establishment.

C. *Imbibition Temperature Sensitivity*

A seed may be considered to be a closed system. It contains all the stored energy and reserve nutrients necessary for initial germination, the metabolic components necessary to utilize these reserves, and the information essential to direct mobilization of these reserves. As seeds require only water and oxygen from the environment, relatively few environmental factors can influence germination: water, oxygen, temperature, soil, microorganisms, and (in some cases) light. We usually think of these factors, for example temperature, in terms of cardinal points — minimum, maximum, and optimum. However, as Lang pointed out (1965), such factors as seed age, genetic variation, and environmental conditions during seed production can influence cardinal temperatures.

Environmental sensitivity changes as seed germination proceeds. Optimal germination temperatures tend to be different from optimal temperatures for later growth (Lang, 1965). Laude (1956) found that the tolerance of several range grasses to freezing decreased as germination progressed. Pollock (1962) found that the expression of physiological dwarfing in peach seedlings from non-after-ripened seeds resulted from brief exposure to relatively high temperature after germination had started. Ibañez (1963) found that cacao seeds were killed by placing them in water at 4°C for 10 minutes. The low-temperature damage could be reversed if the seeds were immediately transferred to a high (37°C) temperature. Reversibility was possible only within the first 15 minutes after exposure to cold. Highkin and Lang (1966) showed that germination temperature could influence the development of the pea plant throughout its entire life cycle.

In comparing the temperature response of green (high vigor) and bleached (low vigor) lima bean seeds, Pollock and Toole (1966) found that lima beans are extremely sensitive to low-temperature injury during the early stage of imbibition. Low vigor seeds are more sensitive than high vigor seeds. For example, almost all of the low vigor seeds imbibed for 18 hours at 5°C were killed. However, if seeds first were allowed to imbibe at 25°C for 6 hours, the subsequent exposure to low temperature was almost without effect (Fig. 18). Later, Pollock (1969) found that temperature sensitivity was controlled by the amount of water in the seed at the time that imbibition of water began. Increasing initial seed moisture to 20% eliminated temperature sensitivity.

Imbibition temperature sensitivity becomes more severe if the seed is imbibed under conditions where available oxygen is limited (Pollock and Manalo, 1969). The combination of stresses—low seed moisture, low oxygen, and low temperature—were utilized by Pollock et al. (1969) to study the differences in vigor between different lots of garden bean seeds. They found very large effects of stresses on some lots and very little effect on others. The different vigor levels appeared to be associated with particular genotypes, but the data were not adequate to establish this with certainty.

More recent work (Roos and Pollock, 1971) has shown that in excised lima bean embryonic axes, the effects of excess moisture and low temperature during imbibition operate on different components of the cell's metabolism. Axis growth can proceed following imbibition conditions of low temperature and excess moisture if the axes have been preconditioned by raising the initial axis moisture to 25%. However, excess moisture is still detrimental to at least one enzyme system (ribonuclease development). In this case initial submergence of preconditioned axes lowers subsequent RNase development by 50% while axis growth is relatively unaffected.

Cotton seedlings are known to be very sensitive to low-temperature injury, and Christiansen (1967) studied the effect of a 96-hour period of germination at 5° or 10°C following different periods of germination at 31°C. He found two periods of temperature sensitivity, one at the beginning of imbibition and the other after 18 to 30 hours of germination. Subsequently (1968), he pregerminated seeds at 31°C, redried them, and found that protection against chilling injury was retained. However,

FIG. 18. Survival and growth of lima bean seeds following 18 hours imbibition at 5° or 15°C preceded by various imbibition periods at 25°C. There were 15 seeds per sample; the number at the right of each sample represents the number of seeds that decayed. The green seeds (top) are high in vigor; the bleached seeds (bottom) are low in vigor. [Photo courtesy of Pollock and Toole (1966) and *Plant Physiology*.]

additional examination (1969) showed that seed moisture was the controlling factor. The seed moisture vs. temperature effect showed that seeds were sensitive to temperature only when imbibition was started at moisture levels below 14%.

The effect of initial seed moisture on emergence and yield of grain sorghum was studied by Phillips and Youngman (1971) who found that, under a mean soil temperature below 20°C, more seeds emerged as moisture content was increased from 8 to 11 or 14%. Under dryland conditions, this effect on emergence was sufficient to increase grain yields up to 30%. Similarly, Obendorf and Hobbs (1970) found that soybeans are sensitive to temperature during imbibition and this sensitivity is controlled by initial seed moisture.

Although not enough data are yet available to evaluate fully the significance of the time–temperature–seed moisture relationship, the wide range of genera already shown to be affected (*Phaseolus, Gossypium, Sorghum,* and *Glycine*) suggest that it may be a phenomenon with appreciable significance when it is utilized to evaluate the vigor of seeds produced under different conditions. Already, the data are adequate to indicate that the relationship between time of exposure to environmental stress and the developmental stage of the seedlings must be given careful consideration in future research.

VII. Modification of Vigor

Recent years have been marked by great changes in agriculture, with a major shift toward mechanization to eliminate hand labor and to increase efficiency of production. Basic to achievement of this goal is development of the ability to "plant to stand," i.e., the development of both machinery and seeds to enable the farmer to plant a single seed for each plant to be harvested, instead of planting heavily and then removing excess and low vigor plants. As a result, there has been a great deal of research on germination and seedling vigor as influenced by planting techniques. Because of the interrelationships between the many factors involved in plant establishment, much of this research has differed from conventional research in seed physiology. For example, seed germination and seedling vigor have been studied as parts of systems which include seed spacing, modification of seed environment during germination, and chemical treatments to influence germination or postgermination plant development. Because of the economic motivation for this research, cost and potential profit have been included as experimental parameters.

A. Planting Systems

Many seeds, such as those of lettuce, beet, and tomato are small or irregularly shaped. They are, therefore, difficult to separate and plant individually; they tend to clump together and be planted and germinate in groups. The resulting competition between plants destroys the economic value of the entire group, so that they must be thinned to one plant per location. In the case of celery, Zink (1967) estimated that 40% of the labor required to produce a crop is expended in the thinning operation. In the case of sugar beet, the plant produced multiple seeds in "balls" from which individual seeds were broken mechanically resulting in serious damage and low seedling vigor. There were two possible solutions to this problem — either to find a method of separating seeds mechanically without damage or to develop plants that produced separated seeds. Research successfully produced monogerm lines.

One of the earliest methods of seed "singulation" was to increase the size of the individual seed by coating it with a layer of bentonite clay to form seed pellets which could be built up to any desired size. These pellets are more regular in shape than the unmodified seeds and, thus, are better adapted to individual planting. Obviously, pelleting offers opportunity for incorporating chemical treatments and for controlling the microenvironment in which the seed germinates. Although pelleted seeds have been widely tested, their performance under soil conditions has not been reliable. Millier and Sooter (1967) found that the rate of emergence in soil of pelleted seeds was much lower than that of unmodified seeds; but the adverse effect of pelleting disappeared when the pellets were cracked, indicating that the problem was the microenvironment of the germinating seed (Fig. 19). Zink (1967) found that coating celery seeds reduced plant population after thinning and also reduced total yield. However, the thinning time in manhours per acre was reduced by approximately half, so the ultimate value of pelleting is one which must be evaluated in economic terms.

A more recent method of seed singulation is the placement of individual seeds on plastic or paper tape (N. A. Smith, 1961; Chancellor, 1969) and the planting of the tape. As in the case of pelleting, the tape modifies the microenvironment of the seed and appears to reduce germination when soil moisture is limited.

At this time, the technical problems of planting to stand have not yet been fully resolved. However, enough progress has been made so that it seems feasible to precision space single seeds and then to remove mechanically the excess plants. For this purpose, a synchronous thinner has been developed (Garrett, 1966) which, sensing the presence of a seedling, mechanically removes all surrounding seedlings, thus eliminating the necessity for hand-thinning. Unfortunately, this device does not distinguish between a crop seedling and a weed seedling and must be utilized as part of a system which also involves an effective weed control component.

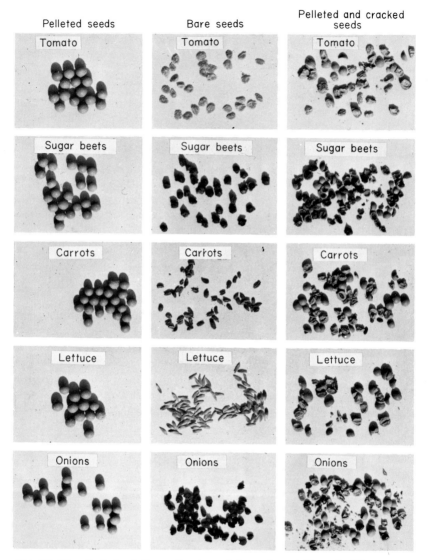

FIG. 19. The effect of pelleting on the size and shape of vegetable seeds. [Photo courtesy of Millier and Sooter (1967) and *Transactions of the American Society of Agricultural Engineers*.]

Compaction of soil, or crusting of irrigated soils, forms a physical barrier to seed emergence which requires either extremely vigorous seedlings or the combined effort of a number of seedlings. Most planting systems include components designed to modify the physical environment in the vicinity of the seed, such as applying Vermiculite with the seeds (Pauli and Harriott, 1968) or planting seeds uncovered, in relatively deep holes (Cary, 1967).

Another component of most planting systems is the selection of seeds by density (Pauli and Harriott, 1968) or size (Alam and Locascio, 1965). We previously discussed (Section IV, B) the importance of seed size in relation to seedling vigor and uniformity of plant development. However, in considering applicability of this concept, it is essential to recognize that an economic question is also important: If only large seeds are to be used or if only a selected portion of seeds from a lot is to be used, then the selected seeds must pay the cost of the entire seed lot. Thus, the question of seed selection requires not only proof of the biological value of the selection but also an estimate of the economic advantage relative to the additional cost. Austin and Longden (1967) studied this question for carrot seeds. They found that at comparable plant densities, large seeds produced bigger seedlings than small seeds and, after 15 to 18 weeks growth, the yield of roots was 15 to 20% higher in plants from large than from small seeds. However, if the crop went to full maturity (24 weeks) no differences in yield could be measured; the advantage of using sized seeds depended entirely on the market for which the crop was intended.

B. Treatments at Time of Planting

Chemical treatment of seeds to control soil microorganisms during germination is a routine technique of particular importance when seeds are planted in cold, wet soils. In addition, many other types of treatments have been applied to modify seedling vigor in a variety of ways. For example, trace elements have been applied in an attempt to minimize or overcome trace element deficiencies in soil. Rasmussen and Boawn (1969) found that zinc applied as a seed treatment could eliminate early season symptoms of zinc deficiency, although it was not adequate to supply the total needs of the crop and influenced seed yield only in terms of an increased uniformity of maturation. Bajpai (1968) suggested that innoculation of seeds with phosphorus-solubilizing bacteria could increase phosphorus uptake by oat seedlings. Ovcharov et al. (1967) treated broad bean seeds with vitamins and found that the resultant plants absorbed more trace elements than did controls.

Chinoy and co-workers (Chinoy, 1957; Abraham et al., 1969) allowed seeds to imbibe a solution containing ascorbic acid and to remain moist for a period of 10 hours or longer. The seeds were then redried and planted. The treatment increased the rate of germination and increased yield of corn, millet (*Pennisetum typhoides*), and raggee (*Eleusine cora-*

cana Gaertn.) as compared to untreated controls. A similar result has been reported for treatment of rice seeds with naphthalene and indole acetic acids (Sinha, 1969). Wheat seeds treated with gibberellin showed stimulated shoot growth; however, vigor of subsequent spring growth was reduced and yield adversely affected (Corns, 1959). Dormant corn seeds treated with 400 kc ultrasonic energy showed increased yield, but these results could not be repeated in a second year, suggesting that the differences were due to factors other than the treatment (Findley and Campbell, 1953). Dry bean seeds subjected to a magnetic field for 240 hours and then planted with the micropyle toward the north pole showed reduced variability and increased yield for a once-over harvest (Pittman and Anstey, 1967). Nelson *et al.* (1964) used infrared, radio-frequency, and gas plasma treatments for reducing hard-seededness in alfalfa.

Several types of treatments have been developed to modify seed responses to temperature extremes. Pauli and Harriott (1968) found that for high-temperature germination, kinetin treatment of lettuce seeds in combination with the use of high-density seeds and modification of soil with Vermiculite, increased yields of marketable heads by almost 30%. O. Smith *et al.* (1968) recommended dipping dry seeds for 3 minutes in 100 mg/liter kinetin solution to overcome thermal dormancy of lettuce. Ells (1963) found that tomato seeds soaked for 6 days in dilute solutions of K_3PO_4 and KNO_3 or NaCl emerged up to 5 days faster at a germination temperature of 50°F (10°C) than did untreated seeds. Oyer and Koehler (1966) found that the effect depended partly on maintaining an osmotic concentration high enough (16.5 atm) to prevent germination during treatment and then drying quickly before germination could occur. They obtained both increased emergence and increased yield under field conditions. Tomato seeds treated by this or a similar process are now commercially marketed in the United States. As pointed out in an earlier section, increased moisture content of beans and some other crops improves performance under cold soil conditions.

Treatments have also been developed to increase drought tolerance of seeds or plants. Bleak and Keller (1970) found that seeds of crested wheat grass which were kept wet but not submerged for 50 hours at 17.2°C and then dried produced better seedling survival and forage yield under arid conditions. Drosdov and Sokolova (1960) found that treatment of flax seeds with boric acid solution before planting increased yield of both straw and seeds. Mart'yanova *et al.* (1961) soaked tomato seeds in water for 30 hours and then dried them before planting. The resulting plants yielded more than twice as much fruit as did plants from untreated seeds.

In the past, a decrease in vigor has been considered to be an inevitable and irreversible part of seed handling and aging. However, Takayanagi and Harrington (1971) have reported that application of exogenous ethylene can accelerate the germination of aged seeds, thus suggesting that at least a partial revival in vigor is possible. Previously, some workers have found germination stimulation by chemical treatments while other workers have failed to find stimulation. Takayanagi and Harrington's observation suggests that such discrepancies may have resulted from differences in vigor of the seed lots used.

C. Treatments during Seed Development, Harvesting, and Cleaning

Gurley and Giddens (1969) found that molybdenum application to soybean leaves could increase the Mo content of seeds. The increased Mo content of the seeds was adequate to provide for plant development and increased yields when those seeds were grown in Mo-deficient soil. Schweizer and Ries (1969) found that subherbicidal (0.07–0.28 kg/ha) concentrations of simazin increased protein content but did not increase size of wheat seeds. This increased protein content was correlated with increased vigor and yield in the next generation.

Chemicals applied for other purposes during the development of a seed crop may have an effect on the vigor of seeds. Austin and Longden (1968) studied the effect of desiccant sprays for defoliating beet seed plants. They found that laboratory germination of the resultant seeds was not affected by the desiccant treatment. However, under the less favorable conditions of field germination, seedling emergence was adversely affected. They believed that this was the result of physiological immaturity of the seeds rather than a direct effect of the chemical carried in small quantity in the seeds. This result emphasizes the possibility that any chemical treatment applied to the mother plant during seed development may influence the vigor of the seed, but this effect becomes visible only when the seeds are germinated under stress conditions.

As previously discussed (Section V), mechanical damage has been a very serious problem in production of garden bean seeds. Machine breakage is most serious when seed hydration is low, and Dexter (1966) suggested moistening seeds with wet sawdust to raise their moisture content to about 17% to minimize damage during handling and planting. On a commercial scale, Northrup, King & Co. of Minneapolis, Minnesota, has developed and patented a technique for adding water to seeds at the time they are delivered to the seed-cleaning plant. In this process, seeds are threshed with a combine directly into trucks and delivered to the seed-cleaning plant. The seeds are elevated and given a preliminary cleaning ("scalping") to remove the large pieces of plant material, dirt, etc., and then stored in boxes 4 × 6 × 4 ft, each holding about 3000 lbs of seed. As the seeds are delivered to the boxes, an inert volcanic glass preparation, Agro-Light, which has been moistened with a known amount of water, is mixed with the seeds. Several weeks normally elapse between the time the seeds are placed in the boxes with the moist material and the seed-cleaning process. This allows time for the moisture to be absorbed and equilibrated within the seeds. During the cleaning process, the Agro-Light is easily removed; the moisture level of the seeds has been raised so that they can be cleaned with minimal damage, thus increasing the average vigor of the final cleaned seeds.

This process can minimize damage to the seeds during cleaning but cannot reduce damage which may have been caused by the threshing process. For this purpose, L. M. Klein (1969) is developing a new type of thresher in which the seeds are separated from the pods between two rubber belts which are driven at different speeds. During planting, increased seed moisture as produced in the Northrup, King & Co. process, will also minimize planter damage. In addition, Pollock *et al.* (1969) have shown that seeds that start to germinate at a relatively high moisture content have a physiological advantage over drier seeds. If we consider these developents in bean harvesting, cleaning, and planting together, we have an example of the type of system of seed production and planting which is likely to become typical of commercial seed utilization in the future. Each part is essential to the final act of placing a seed of maximum vigor into the soil. This is the framework into which future research on seed physiology will ultimately be fitted.

VIII. Summary: Present Status and Future Developments in Seedling Vigor

We believe that the concept of vigor can first be considered as a maximum potential for seedling establishment and, second, as a continuum of potential decrease from that maximum until the seed is dead, i.e., has zero potential for establishment. The maximum is set by the genetic constitution of the plant and is normally attained by part of each seed population (the left side of the rate curve for the mixture of populations I and II in Fig. 4B). From this maximum potential, vigor can only decrease (the positive skewing in Fig. 4B). Many factors can cause a decline in vigor, and large numbers of conditions exist under which this decline can be expressed. However, in that fraction of the seed population in which a decline in vigor has occurred there can be, under certain conditions, recovery toward maximum possible vigor.

This highly abstract picture of vigor is given a mechanistic basis by examining our knowledge of control of biochemical processes in seeds and other biological systems (see Chapter 2, Volume II). For example, we know that, by means of feedback reactions, substrate concentrations control the rate of enzymic reactions and thus permit quick adjustment of the seed to its environment. Substrate concentrations maintained at high or low levels over a period of time will induce or repress synthesis of new enzymes, permitting the seed to adjust to slow changes in its environment. The plasticity of these control mechanisms gives the seed the potential for adapting to a wide range of environmental situations. Conversely, this same plasticity offers opportunity for appreciable damage to one or another component without completely destroying the seed but reducing its potential to respond to the environment.

The environment of a seed must be defined to include competitive pressure exerted by other organisms. After emergence of seedlings, competition with other flowering plants is easily observed. Preemergence competitive pressure of microorganisms is not so obvious. Any decrease in vigor favors development of microorganisms, thus increasing competitive pressure on the developing seedling. Such pressure may include competitition for oxygen, inorganic nutrients, and water in the rhizosphere immediately surrounding the seed. Although these effects per se may be small, they will tend to slow plant development sufficiently to place the plant in a disadvantageous position relative to other plants in its environment. Thus, the total effect of competitive pressure is to amplify the expression of endogenous low vigor.

At the present time, a great deal of research effort is being expended in an attempt to explain mechanisms of development and differentiation in cells and organisms. The processes of seed development and germination offer excellent experimental systems for this type of research and have already been used very effectively for this purpose (Bonner, 1965). The shift from seed development to seed germination is especially interesting because it involves a sharply defined developmental transition which almost certainly includes many changes in kinds of functional genes. Thus, we can expect to find a great increase in seed research by developmental biologists.

At the same time that the seed is a useful tool in developmental biology, it is also the key starting point in much agricultural and forestry production. The technologists in these industries are highly trained and innovative persons who will quickly translate basic seed research results from developmental biology to practical applications. Because of this confluence of interests in the seed, we can reasonably expect rapid and profound advances in the understanding, control, and utilization of vigor in seeds.

REFERENCES

Abbott, D. L. (1956). Temperature and the dormancy of apple seeds. *Int. Hort. Congr., Rep., 14th, 1955* Sect. 2A, p. 746.

Abdul-Baki, A. A. (1969). Relationship of glucose metabolism to germinability and vigor in barley and wheat seeds. *Crop Sci.* **9,** 732.

Abdul-Baki, A. A., and Anderson, J. D. (1970). Viability and leaching of sugars from germinating barley. *Crop Sci.* **10,** 31.

Abraham, P. G., Gurumurti, K., Dave, I. C., Pandya, R. B., and Saxena, O. P. (1969). The yield of some crop plants as affected by ascorbic acid pre-treatment. *Vidya* **12,** 194.

Adams, M. W. (1967). Basis of yield component compensation in crop plants with special reference to the field bean, *Phaseolus vulgaris. Crop Sci.* **7,** 505.

Adams, M. W. (1970). Personal communication.

Alam, Z., and Locascio, S. J. (1965). Effect of seed size and depth of planting effects on broccoli and beans. *Proc. Fla. State Hort. Soc.* **78,** 107.

Alberts, H. W. (1927). Effect of pericarp injury on moisture absorption, fungus attack, and vitality of corn. *J. Amer. Soc. Agron.* **19,** 1021.

Allen, L. R., and Donnelly, E. D. (1965). Effect of seed weight on emergence and seedling vigor in F_4 lines from *Vicia sativa* L. X *V. angustifolia* L. *Crop Sci.* **5,** 167.

Anderson, A. M., Hart, J. R., and French, R. C. (1964). Comparison of germination techniques and conductivity tests of cotton seeds. *Proc. Int. Seed Test Ass.* **29,** 81.

Anderson, S. R. (1955). Development of pods and seeds of birdsfoot trefoil, *Lotus corniculatus* L, as related to maturity and to seed yields. *Iowa, Agr. Exp. Sta., J. Pap.* p. 2771.

Asgrow (Associated Seed Growers, Inc.). (1949). A study of mechanical injury to seed beans. *Asgrow Monogr.* **1.**

Atkin, J. D. (1958). Relative susceptibility of snap bean varieties to mechanical injury of seed. *Proc. Amer. Soc. Hort. Sci.* **72,** 370.

Austin, R. B. (1963). Yield of onions from seed as affected by place and method of seed production. *J. Hort. Sci.* **38,** 277.

Austin, R. B. (1966). The influence of the phosphorus and nitrogen nutrition of pea plants on the growth of their progeny. *Plant Soil* **24,** 359.

Austin, R. B., and Longden, P. C. (1965). Effects of nutritional treatments of seed-bearing plants on the performance of their progeny. *Nature (London)* **205,** 819.

Austin, R. B., and Longden, P. C. (1967). Some effects of seed size and maturity on the yield of carrot crops. *J. Hort. Sci.* **42,** 339.

Austin, R. B., and Longden, P. C. (1968). The yield and quality of red beet seed as affected by desiccant sprays and harvest data. *Weed Res.* **8,** 336.

Bainer, R., and Borthwick, H. A. (1934). Thresher and other mechanical injury to seed beans of the lima type. *Calif., Agr. Exp. Sta., Bull.* **580,** 30.

Bajpai, P. D. (1968). Influence of seed inoculation with phosphorus solubilizing organisms on availability of phosphorus as estimated by Neubauer's technique using P^{32} as tracer. *Indian J. Agr. Sci.* **38,** 696.

Baker, D., Neustadt, M. H., and Zeleny, L. (1957). Application of the fat acidity test as as index of grain deterioration. *Cereal Chem.* **32,** 226.

Barriga, C. (1961). Effects of mechanical abuse of navy bean seed at various moisture levels. *Agron. J.* **53,** 250.

Bartel, A. T. (1941). Green seeds in immature small grains and their relation to germination. *J. Amer. Soc. Agron.* **33,** 732.

Bennett, H. W., and Marchbanks, W. W. (1966). Viability of roughpea seed as affected by time of harvest and drying temperatures. *Agron. J.* **58**, 83.

Beveridge, J. L., and Wilsie, C. P. (1959). Influence of depth of planting, seed size, and variety on emergence and seedling vigor in alfalfa. *Agron. J.* **51**, 731.

Billard, J. B., and Blair, J. P. (1970). The revolution in American agriculture. *Nat. Geogra.* **13**, 147.

Black, J. N. (1958). Competition between plants of different initial seed sizes in swards of subterranean clover (*Trifolium subterraneum* L.) with particular reference to leaf area and the light microclimate. *Aust. J. Agr. Res.* **9**, 299.

Bleak, A. T., and Keller, W. (1970). Field emergence and growth of crested wheatgrass from pretreated vs. non-treated seeds. *Crop Sci.* **10**, 85.

Boawn, L. C., Rasmussen, P. E., and Brown, J. W. (1969). Relationship between tissue zinc levels and maturity period of field beans. *Agron. J.* **61**, 49.

Bodganov, Y. F., Lyapunova, N. A., and Sherudilo, A. I. (1967). Cytophotometric and autoradiographic study of cell population in pea embryos and root tip meristem. *Tsitologiya* **9**, 569.

Bonner, J. (1965). "The Molecular Biology of Development." Oxford Univ. Press, London and New York.

Booth, E. G. (1929). Daily growth of the oat kernel and effect on germination of immaturity and controlled low temperatures. *Minn., Agr. Exp. Sta., Tech. Bull.* **62**, 42 p.

Borthwick, H. A. (1931). Carrot seed germination. *Proc. Amer. Soc. Hort. Sci.* **28**, 310.

Borthwick, H. A. (1932). Thresher injury in baby lima beans. *J. Agr. Res.* **44**, 503.

Bowers, J. L. (1958). Preliminary studies on cucumber seed development as related to viability. *Proc., Ass. S. Agr. Workers* **55**, 163.

Brenchley, W. E. (1923). Effect of weight of seed upon the resulting crop. *Ann. Appl. Biol.* **10**, 223.

Brink, R. A., and Cooper, D. C. (1957). The endosperm in seed development. *Bot. Rev.* **13**, 423 and 479.

Brown, E. B. (1920). Relative yields from broken and entire kernels of seed corn. *J. Amer. Soc. Agron.* **12**, 196.

Brown, E. B. (1922). Effects of mutilating the seeds on the growth and productiveness of corn. *U.S., Dep. Agr., Bull.* **1011**, 13 p.

Caldwell, W. P. (1956). "Laboratory Prediction of Field Emergence of Garden Peas." M.S. Thesis, Iowa State College.

Cameron, J. W., van Maren, A., and Cole, D. A., Jr. (1962). Seed size in relation to plant growth and time of ear maturity of hybrid sweet corn in a winter planting area. *Proc. Amer. Soc. Hort. Sci.* **80**, 481.

Campbell, C. A., McBean, D. S., and Green, D. G. (1969). Influence of moisture stress, relative humidity and oxygen diffusion rate on seed set and yield of wheat. *Can. J. Plant Sci.* **49**, 29.

Cary, J. W. (1967). Punch planting to establish lettuce and carrots under adverse conditions. *Agron. J.* **59**, 406.

Chancellor, W. J. (1969). Seed tape system for precision selection and planting of small vegetable seeds. *Trans. ASAE (Amer. Soc. Agr. Eng.)* **12**, 876.

Ching, T. M., and Pierpoint, M. (1957). Evaluation of germinated seedlings of king green beans by greenhouse planting. *Proc. Ass. Off. Seed Anal.* **47**, 122.

Ching, T. M., and Schoolcraft, I. (1968). Physiological and chemical differences in aged seeds. *Crop Sci.* **8**, 407.

Chinoy, J. J. (1957). Role of ascorbic acid in crop production. *Poona Agr. Coll. Mag.* **57**, 1.

Christensen, C. M., and Kauffman, H. K. (1969). "Grain Storage." Univ. of Minnesota Press, Minneapolis.

Christiansen, M. N. (1962). A method of measuring and expressing epigeous seedling growth rate. *Crop Sci.* **2**, 487.

Christiansen, M. N. (1967). Periods of sensitivity to chilling in germinating cotton. *Plant Physiol.* **42**, 431.

Christiansen, M. N. (1968). Induction and prevention of chilling injury to radicle tips of imbibing cottonseed. *Plant Physiol.* **43**, 743.

Christiansen, M. N. (1969). Seed moisture content and chilling injury to imbibing cottonseed. National Cotton Council. *Beltwide Cotton Prod. Res. Conf.* p. 50.

Christiansen, M. N., and Thomas, R. O. (1969). Season-long effects of chilling treatments applied to germinating cottonseed. *Crop Sci.* **9**, 672.

Clark, B. E. (1954). Factors affecting the germination of sweet corn in low-temperature laboratory tests. *N.Y. Agr. Exp. Sta., Geneva, Bull.* **769**.

Clark, B. E., and Baldauf, D. (1958). A cold test for pea seeds. *Proc. Ass. Off. Seed Anal.* **48**, 133.

Clark, B. E., and Kline, D. B. (1965). Relationship between the vitality and the performance of fireball tomato seed. *N.Y., Agr. Exp. Sta., Geneva, Seed Res. Circ.* **1**.

Clark, B. E., and Peck, N. H. (1968). Relationship between the size and performance of snap bean seeds. *N.Y., Agr. Exp. Sta., Geneva, Bull.* **819**.

Collis-George, N., and Hector, J. B. (1966). Germination of seeds as influenced by matric potential and by area of contact between seed and soil water. *Aust. J. Soil. Res.* **4**, 145.

Collis-George, N., and Sands, J. E. (1959). The control of seed germination by moisture as a soil physical property. *Aust. J. Agr. Res.* **10**, 628.

Collis-George, N., and Sands, J. E. (1962). Comparison of the effects of the physical and chemical components of soil water energy on seed germination. *Aust. J. Agr. Res.* **13**, 575.

Collis-George, N., and Williams, J. (1968). Comparison of the effects of soil matric potential and isotropic effective stress on the germination of *Lactuca sativa*. *Aust. J. Soil Res.* **6**, 179.

Conard, E. C. (1946). The effect of harvesting method on germination of the seed of Russian wild rye, *Elymus junceus*, Fisch. *J. Amer. Soc. Agron.* **38**, 842.

Corns, W. G. (1959). Effects of seed treatments with gibberellin and dates of seeding on winter survival and vegetative yield of Kharkov wheat. *Can. J. Plant Sci.* **39**, 293.

Cox, F. R., and Reid, P. H. (1964). Calcium-boron nutrition as related to concealed damage in peanuts. *Agron. J.* **56**, 173.

Crabb, D., and Kirsop, B. H. (1969). Water-sensitivity in barley. I. Respiration studies and the influence of oxygen availability. *J. Inst. Brew., London* **75**, 254.

Crosier, W. (1942). Baldheads in beans, occurrence and influence on yields. *Proc. Ass. Off. Seed Anal.* **34**, 118.

Czabator, F. J. (1962). Germination value: An index combining speed and completeness of pine seed germination. *Forest Sci.* **8**, 386.

Davies, A. C. W. (1964). The relative susceptibility to threshing damage of six varieties of wheat. *J. Nat. Inst. Agr. Bot.* **10**, 122.

Delouche, J. C., Rushing, T. T., and Baskin, C. C. (1967). Predicting the relative storability of crop seed lots. *Rep. Amer. Seed Res. Found., Mississippi,* 26 p.

de Tempe, J. (1962). Hot-water treatment as a means for determining seed weakness. *Proc. Int. Seed Test. Ass.* **27**, 773.

de Tempe, J. (1963). The use of correlation coefficients in comparing methods for seed vigour testing. *Proc. Int. Seed. Test. Ass.* **28**, 167.

Dexter, S. T. (1966). Conditioning dry bean seed (*Phaseolus vulgaris* L.) for better processing quality and seed germination. *Agron. J.* **58**, 629.

Dodds, M. E., and Pelton, W. L. (1967). Effect of weather factors on the kernel moisture of a standing crop of wheat. *Agron. J.* **59**, 181.

Dogra, P. D. (1967). Seed sterility and disturbances in embryogeny in conifers with particular reference to seed testing and tree breeding in Pinaceae. *Stud. Forest. Suec.* **45**, 97.

Dorrell, D. G., and Adams, M. W. (1969). Effect of some seed characteristics on mechanically induced seedcoat damage in navy beans (*Phaseolus vulgaris* L.). *Agron. J.* **61**, 672.

Dotzenko, A. D., Cooper, C. S., Dobrenz, A. K., Laude, H. M., Massengale, M. A., and Feltner, K. C. (1967). Temperature stress on growth and seed characteristics of grasses and legumes. *Colo., Agr. Exp. Sta., Tech. Bull.* **97**, 27 p.

Drosdov, N. A., and Solokova, G. P. (1960). Influence of boron pre-sowing treatment of seeds on the fiber flax yields and its physiological peculiarities depending on soil moisture content. *Vseso. Nauch.-Issled. Inst. L'na.Tr.* **6**, 96.

Ells, J. E. (1963). The influence of treating tomato seed with nutrient solutions on emergence rate and seedling growth. *Proc. Amer. Soc. Hort. Sci.* **83**, 684.

Engelsma, G. (1967). Effect of cyclohexamide on the inactivation of phenylalanine deaminase by gherkin seedlings. *Naturwissenschaften* **54**, 319.

Farooqui, H. M., and McCollum, J. P. (1954). Relation of morphological structure and development to seed coat rupture in beans (*Phaseolus vulgaris* L.). *Proc. Amer. Soc. Hort. Sci.* **63**, 333.

Fick, G. L., and Hibbard, R. P. (1925). A method for determining seed viability by electrical conductivity measurements. *Mich. Acad. Sci., Arts & Letters* **5**, 95.

Fields, R. W., and King, T. H. (1963). The effects of early harvest and artificial drying on mold deterioration and quality of canning pea seed. *Proc. Minn. Acad. Sci.* **30**, 128.

Findley, W. R., Jr., and Campbell, L. E. (1953). Ultrasonic treatments of dormant hybrid corn seed. *Agron. J.* **45**, 357.

Fleming, A. A. (1966). Effects of seed age, producer, and storage on corn (*Zea mays* L.) production. *Agron. J.* **58**, 227.

Flemion, F. (1959). Effect of temperature, light, and gibberellic acid on stem elongation and leaf development in physiologically dwarfed seedlings of peach and Rhodotypos. *Contrib. Boyce Thompson Inst.* **20**, 57.

Flentje, N. T., and Saksena, H. K. (1963). Pre-emergence rotting of peas in South Australia. *Aust. J. Biol. Sci.* **17**, 665.

Fritz, T. (1965). Germination and vigour tests of cereal seed. *Proc. Int. Seed Test. Ass.* **30**, 923.

Fritz, T. (1966). Influence of temperature and parasitic fungi on germinating capacity of cereal seed. *Proc. Int. Seed Test. Ass.* **31**, 711.

Gaber, S. D., and Roberts, E. H. (1969). Water-sensitivity in barley seeds. II. Association with micro-organism activity. *J. Inst. Brew., London* **75**, 303.

Garrett, R. E. (1966). Development of a synchronous thinner. *Proc. Amer. Soc. Sugar Beet Tech.* **14**, 206.

Giri, A. (1967). Germination percentage, average height and girth of seedling raised from seedstones extracted from syrupy and firm mango fruits. *Pak. J. Sci.* **18**, 79.

Goddard, D. R., and Bonner, W. D. (1960). Cellular respiration. *In* "Plant Physiology" (F. C. Steward, ed.), Vol. 1A, p. 209. Academic Press, New York.

Gordon, A. G. (1968). The interaction of dormancy and water-sensitivity of barley with temperature. *J. Inst. Brew., London* **74**, 355.

Grabe, D. F. (1964). Glutamic acid decarboxylase activity as a measure of seedling vigor. *Proc. Assoc. Off. Seed Anal.* **54,** 100.

Grabe, D. F. (1965). Prediction of relative storability of corn seed lots. *Proc. Ass. Off. Seed Anal.* **55,** 92.

Grabe, D. F., and Frey, K. J. (1966). Seed quality and planting rates affect oat yields. *Iowa Farm Sci.* **20,** 8.

Green, D. E., Cavanah, L. E., and Pinnèll, E. H. (1966). Effect of seed moisture content, field weathering, and combine cylinder speed on soybean seed quality. *Crop. Sci.* **6,** 7.

Guneyli, E., Burnside, O. C., and Nordquist, P. T. (1969). Influence of seedling characteristics on weed competitive ability of sorghum hybrids and inbred lines. *Crop. Sci.* **9,** 713.

Gurley, W. H., and Giddens, J. (1969). Factors affecting uptake, yield response, and carryover of molybdenum in soybean seed. *Agron. J.* **61,** 7.

Hardesty, B., and Elliott, F. C. (1956). Differential post-ripening effects among seeds from the same parental wheat spike. *Agron. J.* **48,** 406.

Harmond, J. E., Klein, L. M., and Brandenburg, N. R. (1961). Seed cleaning and handling. *U.S., Dep. Agr. Agr. Handb.* **179,** 38 p.

Harrington, J. F. (1959). Effect of fruit maturity and harvesting methods on germination of muskmelon seed. *Proc. Amer. Soc. Hort. Sci.* **73,** 422.

Harrington, J. F., and Thompson, R. C. (1952). Effect of variety and area of production on subsequent germination of lettuce seed at high temperatures. *Proc. Amer. Soc. Hort. Sci.* **59,** 445.

Harris, H. C., and Brolmann, J. B. (1966). Comparison of calcium and boron deficiencies of the peanut. II. Seed quality in relation to histology and viability. *Agron. J.* **58,** 578.

Harter, L. L. (1930). Thresher injury a cause of baldhead in beans. *J. Agr. Res.* **40,** 371.

Hawksworth, F. G., Estabrook, G. F., and Rogers, D. J. (1968). Application of an information theory model for character analysis in the genus *Arceuthobius (Viscaceae).* *Taxon* **17,** 605.

Hawthorn, L. R., and Pollard, L. H. (1954). "Vegetable and Flower Seed Production." McGraw-Hill (Blakiston), New York.

Hawthorn, L. R., and Pollard, L. H. (1966). Effect of soil moisture, soil fertility, and rate of seeding on the yield, viability, and quality of seed peas. *Utah, Agr. Exp. Sta., Bull.* **458.**

Hawthorn, L. R., Toole, E. H., and Toole, V. K. (1962). Yield and viability of carrot seeds as affected by position of umbel and time of harvest. *Proc. Amer. Soc. Hort. Sci.* **80,** 401.

Hawthorn, L. R., Kerr, L. B., and Campbell, W. F. (1966). Relation between temperature of developing pods and seeds and scalded seeds in garden peas. *Proc. Amer. Soc. Hort. Sci.* **88,** 437.

Heintze, S. G. (1956). The effects of various soil treatments on the occurrence of marsh spot in peas and on manganese uptake of oats and timothy. *Plant Soil* **7,** 218.

Hermann, E. M., and Hermann, W. (1939). The effect of maturity at time of harvest on certain responses of seed of crested wheatgrass, *Agropyron cristatum* L. Gaertn. *J. Amer. Soc. Agron.* **31,** 876.

Heydecker, W. (1960). Can we measure seedling vigour? *Proc. Int. Seed Test. Ass.* **25,** 498.

Heydecker, W. (1961). The emergence of vegetable seedlings as a standard test of soil quality. *Proc. Int. Hort. Congr., 15th, 1958* Vol. 1, p. 381.

Heydecker, W. (1962). From seed to seedling: Factors affecting the establishment of vegetable crops. *Ann. Appl. Biol.* **50,** 622.

Heydecker, W. (1966). Clarity in recording germination data. *Nature (London)* **210**, 753.

Hibbard, R. P., and Miller, E. V. (1928). Biochemical studies on seed viability. I. Measurements of conductance and reduction. *Plant Physiol.* **3**, 335.

Highkin, H. R. (1958). Temperature-induced variability in peas. *Amer. J. Bot.* **45**, 626.

Highkin, H. R., and Lang, A. (1966). Residual effect of germination temperature on the growth of peas. *Planta* **68**, 94.

Hoffman, I. C. (1925). The relation of size of kernels in sweet corn to evenness of maturity. *J. Agr. Res.* **31**, 1043.

Holmes, A. D. (1953). Germination of seeds removed from mature and immature butternut squashes after seven months of storage. *Proc. Amer. Soc. Hort. Sci.* **62**, 433.

Hooks, J. A., and Zuber, M. S. (1963). Effects of soil and soil moisture levels on cold-test germination of corn. *Agron. J.* **55**, 453.

Hoshikawa, K. (1962). Studies on the ripening of wheat grain. 4. Influence of temperature upon the development of the endosperm. *Proc. Crop. Sci. Soc. Jap.* **30**, 228.

Hottes, C. F., and Huelsen, W. A. (1927). The determination of quality in sweet corn seed by means of the optical measurement of leached materials. *J. Agr. Res.* **35**, 147.

Howell, R. W., and Cartter, J. L. (1953). Physiological factors affecting composition of soybeans. I. Correlation of temperatures during certain portions of the pod filling stage with oil percentage in mature beans. *Agron. J.* **45**, 526.

Hsiao, A. I., and Vidaver, W. (1971a). Seed water content in relation to phytochrome-mediated germination of lettuce seeds (*Lactuca sativa* L. var Grand Rapids). *Can. J. Bot.* **49**, 111.

Hsiao, A. I., and Vidaver, W. (1971b). Water content and phytochrome-induced potential germination responses in lettuce seeds. *Plant Physiol.* **47**, 186.

Hulbert, H. W., and Whitney, G. M. (1934). Effect of seed injury upon the germination of *Pisum sativum*. *J. Amer. Soc. Agron.* **26**, 879.

Hunt, O. J., and Miller, D. G. (1965). Coleoptile length, seed size, and emergence in intermediate wheatgrass (*Agropyron intermedium* [Host] Beauv.). *Agron. J.* **57**, 192.

Hurd, A. M. (1921). Seed-coat injury and viability of seeds of wheat and barley as factors in susceptibility to molds and fungicides. *J. Agr. Res.* **21**, 99.

Ibañez, M. L. (1963). The point of irreversibility in cacao seed sensitivity to cold. *Turrialba* **13**, 127.

Isely, D. (1950). The cold test for corn. *Proc. Int. Seed. Test. Ass.* **16**, 299.

Isely, D. (1957). Vigor tests. *Proc. Ass. Off. Seed Anal.* **47**, 176.

Jensen, A., Stephansen, K., and Loeken, A. (1967). Seed ripening of Norwegian coniferous trees. II. Variation in the chemical content and germination of seeds of *Picea abies* and *Picea sitchensis*. *Medd. Vestlandets Forstl. Forsoekssta.* **13**, 191.

Kantor, D. J., and Webster, O. J. (1967). Effects of freezing and mechanical injury on viability of sorghum seed. *Crop Sci.* **7**, 196.

Kaufmann, M. R. (1969). Effects of water potential on germination of lettuce, sunflower, and citrus seeds. *Can. J. Bot.* **47**, 1761.

Kaufmann, M. R., and Ross, K. J. (1970). Water potential, temperature, and kinetin effects on seed germination in soil and solute systems. *Amer. J. Bot.* **57**, 413.

Kerr, E. A. (1961). The relation between seedling vigour and tree size in a sweet-cherry population. *Rep. Ont. Hort. Exp. Sta.,* p. 45.

Kerr, E. A. (1962). Germination of tomato seed as affected by fermentation time, variety, fruit maturity, plant maturity, and harvest date. *Rep. Ont., Hort. Exp. Sta.,* p. 79.

Kidd, F., and West, C. (1918). Physiological pre-determination: The influence of the physiological condition of the seed upon the course of subsequent growth and upon yield. I. The effects of soaking seeds in water. *Ann. Appl. Biol.* **5**, 1.

Kittock, D. L., and Law, A. G. (1968). Relation of seedling vigor to respiration and tetrazolium chloride reduction by germinating wheat seeds. *Agron. J.* **60**, 286.

Klein, L. M. (1969). A new seed combine. *OSU Seed Process. Conf. Pap.* (oral paper).

Klein, L. M., and Harmond, J. E. (1966). Effect of varying cylinder speed and clearance on threshing cylinders in combining crimson clover. *Trans. ASAE (Amer. Soc. Agr. Eng.)* **9**, 499.

Klein, S., and Pollock, B. M. (1968). Cell fine structure of developing lima bean seeds related to seed desiccation. *Amer. J. Bot.* **55**, 658.

Koller, D. (1962). Preconditioning of germination of lettuce at time of fruit ripening. *Amer. J. Bot.* **49**, 841.

Kollöffel, C. (1968). Activity of alcohol dehydrogenase in the cotyledons of peas germinated under different environmental conditions. *Acta Bot. Neer.* **17**, 70.

Kolp, B. J., Miller, D. G., Pratt, G. A., and Hwang, S. (1967). Relation of coleoptile structure to coleoptile strength and seedling emergence under compacted soil conditions in six varieties of winter wheat. *Crop. Sci.* **7**, 413.

Kotowski, F. (1926). Temperature relations to germination of vegetable seeds. *Proc. Amer. Soc. Hort. Sci.* **23**, 176.

Lang, A. (1965). Effects of some internal and external conditions on seed germination. *In* "Handbuch der Pflanzenphysiologie" (W. Ruhland, ed.), Vol. 15, Part 2, p. 848. Springer-Verlag, Berlin and New York.

Larson, M. M. (1961). Seed size, germination dates, and survival relationships of ponderosa pine in the Southwest. *Rocky Mnt. For. Range Exp. Sta. Res. Notes* **66**, 1.

Laude, H. M. (1956). The seedling emergence of grasses as affected by low temperature. *Agron. J.* **48**, 558.

Lawrence, T. (1967). Effect of age on the germination and emergence of Russian wild ryegrass (*Elymus junceus*) seed harvested by two methods at progressive stages of maturity. *Can. J. Plant Sci.* **47**, 181.

LeBaron, M. J. (1966). Zinc in Idaho bean production. *Idaho Univ. Ext. Idaho Curr. Inform. Ser.* **31**, 2 p.

Linko, P. (1961). Simple and rapid manometric method for determining glutamic acid decarboxylase activity as quality index of wheat. *J. Agr. Food Chem.* **9**, 310.

Linko, P., and Milner, M. (1959). Enzyme activation in wheat grains in relation to water content. Glutamic acid–alanine transaminase, and glutamic acid decarboxylase. *Plant Physiol.* **34**, 392.

McAlister, D. F. (1943). The effect of maturity on the viability and longevity of the seeds of western range and pasture grasses. *J. Amer. Soc. Agron.* **35**, 442.

McCollum, J. P. (1953). Factors affecting cotyledonal cracking during the germination of beans. *Plant Physiol.* **28**, 267.

McCoy, O. D. (1962–1963). Effect of age and germination temperature on the viability of great lakes 650 lettuce seed. *Annu. Vege. Crops Progr. Rep., 14th* (oral paper).

McFadden, A. D. (1963). Effect of seed source on comparative test results in barley. *Can. J. Plant Sci.* **43**, 295.

McIlrath, W. J., Abrol, Y. P., and Heiligman, F. (1963). Dehydration of seeds in intact tomato fruits. *Science* **142**, 1681.

McLemore, B. F. (1959). Cone maturity affects germination of longleaf pine (*Pinus palustris*) seed. *J. Forest.* **57**, 648.

McMaster, G. M., LeBaron, M. J., Corey, G. L., Hawthorn, L. R., and Toole, V. K. (1960). Irrigation of snap beans grown for seed in Idaho. *Idaho, Agr. Exp. Sta., Bull.* **336**, 19 p.

McMaster, G. M., LeBaron, M. J., Corey, G. L., Hawthorn, L. R., and Toole, V. K. (1965).

The influence of soil moisture on snap bean seed production. *Idaho, Agr. Exp. Sta., Bull.* **435**, 18 p.

Manohar, M. S., and Heydecker, W. (1964). Effects of water potential on germination of pea seeds. *Nature (London)* **202**, 22.

Manohar, M. S., and Mathur, M. K. (1967). Effects of alternating water potentials on germination of peas (*Pisum sativum* L.). *Naturwissenshaften* **54**, 48.

Marar, M. M. K., and Shambhu, K. (1961). Coconut nursery studies. III. Vigour of seedlings in relation to the floating position of seednuts in water. *Indian Coconut J.* **14**, 45.

Marshall, H. G. (1969). Effect of seed source and seedling age on the freezing resistance of winter oats. *Crop Sci.* **9**, 202.

Martens, M. (1960). L'importance de la qualité de la graine de betterave sucrière et du lit de germination en vue de la mécanisation des travaux de printemps. *Inst. Belge Amelior. Betterave. P. Tech.* **28**, 45.

Mart'yanova, K. L., Gubanova, Z. P., and Zhurikhin, V. K. (1961). Presowing drought hardening of tomatoes under condition of a production experiment. *Fiziol. Rast.* **8**, 638.

Matthews, S., and Bradnock, W. T. (1967). The detection of seed samples of wrinkle-seeded peas (*Pisum sativum* L.) of potentially low planting value. *Proc. Int. Seed Test. Ass.* **32**, 553.

Matthews, S., and Bradnock, W. T. (1968). Relationships between seed exudation and field emergence in peas and French beans. *Hort. Res.* **8**, 89.

Meyers, M. T. (1924). The influence of broken pericarp on the germination and yield of corn. *J. Amer. Soc. Agron.* **16**, 540.

Millier, W. F., and Sooter, C. (1967). Improving emergence of pelleted vegetable seed. *Trans. ASAE (Amer. Soc. Agr. Eng.)* **10**, 658.

Moore, R. P. (1963). Previous history of seed lots and differential maintenance of seed viability and vigor in storage. *Proc. Int. Seed Test. Ass.* **28**, 691.

Moore, R. P. (1965). Natural destruction of seed quality under field conditions as revealed by tetrazolium tests. *Proc. Int. Seed Test. Ass.* **30**, 995.

Moore, R. P. (1966). Tetrazolium tests for diagnosing causes for seed weaknesses and for predicting and understanding performance. *Proc. Ass. Off. Seed Anal.* **56**, 70.

Morton, C. T., and Buchele, W. F. (1960). Emergence energy of plant seedlings. *Agr. Eng.* **41**, 428 and 453.

Nagato, K., and Ebata, M. (1965). Effects of high temperature during ripening period on the development and the quality of rice kernels. *Proc. Crop Sci. Soc. Jap.* **34**, 59.

Nanda, K. K., Chinoy, J. J., and Gupta, S. M. (1959). A method for the determination of the rate of seedling growth and its application to the study of the effect of presowing hardening treatment on wheat grain. *Phyton* **12**, 153.

Nelson, S. O., Stetson, L. E., Sone, R. B., Webb, J. C., Pettibone, C. A., Works, D. W., Kehr, W. R., and Van Riper, G. E. (1964). Comparison of infrared, radiofrequency, and gas-plasma treatments of alfalfa seed for hard-seed reduction. *Trans. ASAE (Amer. Soc. Agr. Eng.)* **7**, 276.

Nichols, M. A., and Heydecker, W. (1968). Two approaches to the study of germination data. *Proc. Int. Seed Test. Ass.* **33**, 531.

Nobbe, F. (1872). Ueber die Wirkungen des Maschinendrusches auf die Keimfähigkeit des Getreides. *Landwirt. Vers.-Sta.* **15**, 252.

Obendorf, R. L., and Hobbs, P. R. (1970). Effect of seed moisture on temperature sensitivity during imbibition of soybean. *Crop Sci.* **10**, 563.

Oelke, E. A., Ball, R. B., Wick, C. M., and Miller, M. D. (1969). Influence of grain moisture at harvest on seed yield, quality, and seedling vigor of rice. *Crop Sci.* **9**, 144.

Ovcharov, K. E., Kolotova, S. S., Filippova, K. F., and Kharitonova, N. I. (1967). Effect of the presowing treatment of seeds with vitamins on the uptake of trace elements into the organs of broad beans. *Chem. Abstr.* **71**, 36458. (1969).

Oyer, E. B., and Koehler, D. E. (1966). A method of treating tomato seeds to hasten germination and emergence at suboptimal temperatures. *Proc. Int. Hort. Congr., 17th, 1965* Vol. 1, p. 626.

Paquet, J. (1968). Action d'une élévation brusque de température sur l'évalution de la teneur en protéines du grain de blé tendre. *Ann. Amelior. Plant.* **18**, 17.

Parmar, M. T., and Moore, R. P. (1968). Carbowax 6000, mannitol, and sodium chloride for simulating drought conditions in germination studies of corn (*Zea mays*) of strong and weak vigor. *Agron. J.* **60**, 192.

Pauli, A. W., and Harriott, B. L. (1968). Lettuce seed selection and treatment for precision planting. *Agr. Eng.* **49**, 18.

Pavlychenko, T. K., and Harrington, J. B. (1934). Competitive efficiency of weeds and cereal crops. *Can. J. Res.* **10**, 77.

Pearson, R., and Parkinson, D. (1961). The sites of excretion of ninhydrin-positive substances by broad bean seedlings. *Plant Soil* **13**, 391.

Perry, D. A. (1969a). A vigour test for peas based on seedling evaluation. *Proc. Int. Seed Test. Ass.* **34**, 265.

Perry, D. A. (1969b). Seed vigour in peas (*Pisum sativum* L.). *Proc. Int. Seed Test. Ass.* **34**, 221.

Phillips, J. C., and Youngman, V. E. (1971). Effect of initial seed moisture control on emergence and yield of grain sorghum. *Crop Sci.* **11**, 354.

Pittman, U. J., and Anstey, T. H. (1967). Magnetic treatment and seed orientation of single-harvest snap beans (*Phaseolus vulgaris* L.). *Proc. Amer. Soc. Hort. Sci.* **91**, 310.

Pollock, B. M. (1962). Temperature control of physiological dwarfing in peach seedlings. *Plant Physiol.* **37**, 190.

Pollock, B. M. (1962). Unpublished data.

Pollock, B. M. (1969). Imbibition temperature sensitivity of lima bean seeds controlled by initial seed moisture. *Plant Physiol.* **44**, 907.

Pollock, B. M., and Manalo, J. R. (1969). Controlling substrate moisture-oxygen levels during the imbibition stage of germination. *J. Amer. Soc. Hort. Sci.* **94**, 574.

Pollock, B. M., and Manalo, J. R. (1970). Simulated mechanical damage to garden beans during germination. *J. Amer. Soc. Hort. Sci.* **95**, 415.

Pollock, B. M., and Manalo, J. R. (1971). The influence of seed-lot history on sensitivity of lettuce seed to temperature and moisture stress. *HortSci.* **6**, 444.

Pollock, B. M., and Toole, V. K. (1966). Imbibition period as the critical temperature sensitive stage in germination of lima bean seeds. *Plant Physiol.* **41**, 221.

Pollock, B. M., Manalo, J. R., and Roos, E. E. (1970). Unpublished data.

Pollock, B. M., Roos, E. E., and Manalo, J. R. (1969). Vigor of garden bean seeds and seedlings influenced by initial seed moisture, substrate oxygen, and imbibition temperature. *J. Amer. Soc. Hort. Sci.* **94**, 577.

Presley, J. T. (1958). Relations of protoplast permeability to cotton seed viability and predisposition to seedling disease. *Plant Dis. Rep.* **42**, 852.

Rajagopalan, K., and Bhuvaneswari, K. (1964). Effect of germination of seeds and host exudations during germination on foot-rot disease of rice. *Phytopathol. Z.* **50**, 221.

Rangaswamy, N. S. (1963). Control of fertilization and embryo development. *In* "Recent Advances in the Embryology of Angiosperms" (P. Maheshwari, ed.), p. 327. Univ. of Delhi, Delhi, India.

Rasmussen, P. E., and Boawn, L. C. (1969). Zinc seed treatment as a source of zinc for beans. *Agron. J.* **61,** 674.

Rice, W. N. (1960). Development of the cold test for seed evaluation. *Proc. Ass. Off. Seed Anal.* **50,** 118.

Robertson, L. D., and Curtis, B. C. (1967). Germination of immature kernels of winter wheat. *Crop Sci.* **7,** 269.

Rodionenko, G. I. (1955). Connection between polycotyledonous seedlings and the structure of the mature plant. Translation (OTS 60-21912) 12039 from *Akad. Nauk SSSR. Bot. Inst.* im. V. L. Komarova. Trudy 6,310.

Rohmeder, E. (1962). Die Bedeutung der Keimschnelligkeit bei forstlichen Samenarten. *Proc. Int. Seed Test. Ass.* **27,** 657.

Roos, E. E., and Manalo, J. R. (1971). Testing vigor of beans following unfavorable storage conditions. *HortSci.* **6,** 347.

Roos, E. E., and Pollock, B. M. (1971). Soaking injury in lima beans.*Crop Sci.* **11,** 78.

Rosenow, D. T., Casady, A. J., and Heyne, E. G. (1962). Effects of freezing on germination of sorghum seed. *Crop Sci.* **2,** 99.

Rowe, J. S. (1964). Environmental preconditioning with special reference to forestry. *Ecology* **45,** 399.

Rowe, P. R., and Andrew, R. H. (1964). Phenotypic stability in corn as related to timing of moisture stress during early development. *Bot. Gaz.* **125,** 211.

Rush, G. E., and Neal, N. P. (1951). The effect of maturity and other factors on stands of corn at low temperatures. *Agron. J.* **43,** 112.

Scheer, D. F., and Ellison, J. H. (1960). Asparagus performance as related to seedling vigor. *Proc. Amer. Soc. Hort. Sci.* **76,** 370.

Scheibe, J., and Lang, A. (1967). Lettuce seed germination: A phytochrome-mediated increase in the growth rate of lettuce seed radicles. *Planta* **72,** 348.

Schroth, M. N., and Cook, J. R. (1964). Seed exudation and its influence on pre-emergence damping-off of bean. *Phytopathology* **54,** 670.

Schroth, M. N., and Hildebrand, D. C. (1964). Influence of plant exudates on root-infecting fungi. *Annu. Rev. Phytopathology* **2,** 101.

Schroth, M. N., and Synder, W. C. (1961). Effect of host exudates on chlamydospore germination of the bean rootrot fungus. *Fusarium solani* f. *phaseoli. Phytopathology* **51,** 389.

Schroth, M. N., Weinhold, A. R., and Hayman, D. S. (1966). The effect of temperature on quantitative differences in exudates from germinating seeds of bean, pea, and cotton. *Can. J. Bot.* **44,** 1429.

Schweizer, C. J., and Ries, S. K. (1969). Protein content of seeds: Increase improves growth and yield. *Science* **165,** 73.

Scott, D. H., and Waugh, J. C. (1941). Treatment of peach seed as affecting germination and growth of seedlings in the greenhouse. *Proc. Amer. Soc. Hort. Sci.* **38,** 291.

Sedgley, R. H. (1963). The importance of liquid-seed contact during the germination of *Medicago tribuloides* Desr. *Aust. J. Agr. Res.* **14,** 646.

Sherwin, T., and Simon, E. W. (1969). The appearance of lactic acid in *Phaseolus* seeds germinating under wet conditions. *J. Exp. Bot.* **20,** 776.

Simak, M. (1957). The x-ray contrast method for seed testing Scots pine *Pinus silvestris. Medd. Statens Skogsforskningsinst. (Sweden)* **47,** 1.

Simak, M., and Kamra, S. K. (1963). Comparative studies on Scots pine seed germinability with tetrazolium and x-ray contrast methods. *Proc. Int. Seed Test. Ass.* **28,** 3.

Sinha, R. N. (1969). Effect of presoaking seeds with plant-growth regulators and nutrient solution on dry matter production of rice. *Madras Agr. J.* **56,** 16.

Smith, N. A. (1961). An evaluation of methylcellulose and paper seed ribbons for the precision seeding of lettuce (*Lactuca sativa* L.) and other vegetables. *Diss. Abstr.* **22**, 383.

Smith, O., Yen, W. W. L., and Lyons, J. M. (1968). The effects of kinetin in overcoming high-temperature dormancy of lettuce seed. *J. Amer. Soc. Hort. Sci.* **93**, 444.

Solbrig, O. T. (1970). "Principles and Methods of Plant Biosystematics." Macmillan, New York.

Stearns, F. (1960). Effects of seed environment during maturation on seedling growth. *Ecology* **41**, 221.

Stermer, R. A. (1968). Environmental conditions and stress cracks in milled rice. *Cereal Chem.* **45**, 365.

Szukalski, H. (1961). The influence of a high phosphorus content of seeds on the development and yields of plants. II. Investigations on flax. *Rocz. Nauk Roln.* **84**, 789.

Takayanagi, K., and Harrington, J. F. (1971). Enhancement of germination rate of aged seeds by ethylene. *Plant Physiol.* **47**, 521.

Takayanagi, K., and Murakami, K. (1968). Rapid germinability test with exudates from seed. *Nature (London)* **218**, 493.

Tatum, L. A. (1954). Seed permeability and "cold-test" reaction in *Zea mays. Agron. J.* **46**, 8.

Tatum, L. A., and Zuber, M. S. (1943). Germination of maize under adverse conditions. *J. Amer. Soc. Agron.* **35**, 48.

Terman, G. L., Ramig, R. E., Dreier, A. F., and Olson, R. A. (1969). Yield-protein relationships in wheat grain, as affected by nitrogen and water. *Agron. J.* **61**, 755.

Thomas, C. A. (1960). Relations of variety, temperature, and seed immaturity to pre-emergence damping-off of castor bean. *Phytopathology* **50**, 473.

Thompson, R. C. (1936). Some factors associated with dormancy of lettuce seed. *Proc. Amer. Soc. Hort. Sci.* **33**, 610.

Tompkins, D. R. (1966). Broccoli maturity and production as influenced by seed size. *Proc. Amer. Soc. Hort. Sci.* **88**, 400.

Toole, E. H., and Toole, V. K. (1960). Viability of stored snap bean seed as affected by threshing and processing injury. *U.S., Dep. Agr., Tech. Bull.* **1213**.

Toole, E. H., Wester, R. E., and Toole, V. K. (1941). The effect of fruit rot of eggplant on seed germination. *Proc. Amer. Soc. Hort. Sci.* **38**, 496.

Toole, E. H., Toole, V. K., Lay, B. J., and Crowder, J. T. (1951). Injury to seed beans during threshing and processing. *U.S., Dep. Agr., Circ.* **874**.

Toole, E. H., Hendricks, S. B., Borthwick, H. A., and Toole, V. K. (1956). Physiology of seed germination. *Annu. Rev. Plant Physiol.* **7**, 299.

Tseng, S. T., and Lin, C. I. (1962). Studies on the physiological quality of pure rice seed. *Proc. Int. Seed Test. Ass.* **27**, 459.

Tucker, H., and Wright, L. N. (1965). Estimating rapidity of germination. *Crop Sci.* **5**, 398.

U.S. Department of Agriculture. (1948). Woody plant seed manual. *U.S., Dep. Agr., Misc. Publ.* **654**.

U.S. Department of Agriculture. (1968). Rules and regulations under the Federal Seed Act. (as amended) *U.S. Consumer Mkt. Serv.*

U.S. Department of Agriculture. (1969). A national program of research for vegetable crops.

Vegis, A. (1964). Dormancy in higher plants. *Annu. Rev. Plant Physiol.* **15**, 185.

Wanjura, D. F., Hudspeth, E. B., Jr., and Bilbro, J. D., Jr. (1969). Emergence time, seed quality, and planting depth effects on yield and survival of cotton. *Agron. J.* **61**, 63.

Waters, E. C., Jr., and Atkin, J. D. (1959). Performance of snap bean (*Phaseolus vulgaris*) seedlings having transversely broken cotyledons. *Proc. Amer. Soc. Hort. Sci.* **74**, 591.

Webster, L. V., and Dexter, S. T. (1961). Effects of physiological quality of seeds on total germination, rapidity of germination, and seedling vigor. *Agron. J.* **53**, 297.

Weibel, D. E. (1958). Vernalization of immature winter wheat embryos. *Agron. J.* **50**, 267.

Wellington, P. S. (1965). Germinability and its assessment. *Proc. Int. Seed Test. Ass.* **30**, 73.

Wester, R. E., and Magruder, R. (1938). Effect of size, condition, and production locality on germination and seedling vigor of baby fordhook bush lima bean seed. *Proc. Amer. Soc. Hort. Sci.* **36**, 614.

Wester, R. E., and Jorgensen, H. (1956). Relation of chlorophyll fading from cotyledons to germination and vigor of some green-seeded lima beans. *Seed World* **78**, 8.

Whalley, R. D. B., McKell, C. M., and Green, L. R. (1966). Seedling vigor and the early nonphotosynthetic stage of seedling growth in grasses. *Crop Sci.* **6**, 147.

Williams, F. J., Hollis, W. L., and Day, M. H. (1966). Incidence of hypocotyl collar rot of *Phaseolus vulgaris* in the field and in germination tests. *Phytopathology* **56**, 531.

Williams, W. A. (1956). Evaluation of the emergence force exerted by seedlings of small seeded legumes using probit analysis. *Agron. J.* **48**, 273.

Williams, W. A., Black, J. N., and Donald, C. M. (1968). Effect of seed weight on the vegetative growth of competing annual Trifoliums. *Crop Sci.* **8**, 660.

Wirth, M., Estabrook, G. F., and Rogers, D. J. (1966). A graph theory model for systematic biology, with an example for the Oncidiinae (Orchidaceae). *Syst. Zool.* **15**, 59.

Woodstock, L. W., and Combs, M. F. (1965). Effects of gamma-irradiation of corn seed on the respiration and growth of the seedling. *Amer. J. Bot.* **52**, 563.

Woodstock, L. W., and Grabe, D. F. (1967). Relationships between seed respiration during imbibition and subsequent seedling growth in *Zea mays* L. *Plant Physiol.* **42**, 1071.

Woodstock, L. W., and Pollock, B. M. (1965). Physiological predetermination: Imbibition, respiration, and growth of lima bean seeds. *Science* **150**, 1031.

Zagaja, S. W. (1961). The effect of stratification of embryos on the subsequent cherry (*Pr. avium*) seedlings growth. *Bull. Acad. Pol. Sci. Ser. B* **9**, 267.

Zagaja, S. W., Hough, L. F., and Bailey, C. H. (1960). The responses of immature peach embryos to low temperature treatments. *Proc. Amer. Soc. Hort. Sci.* **75**, 171.

Zink, F. W. (1967). Coated celery seed aids mechanization efforts. *Calif. Agr.* **21**, 4.

AUTHOR INDEX

Numbers in italics refer to the pages on which the complete references are listed.

A

Abbe, E. C., 234, *304*
Abbott, D. L., 322, 323, *376*
Abbott, H., 12, *18*
Abdul-Baki, A. A., 323, 338, 361, *376*
Abraham, P. G., 371, *376*
Abrol, Y. P., 351, *382*
Adams, M. W., 346, 347, 358, 360, *376, 379*
Afzelius, K., 127, *137*
Agarwal, S., 93, 94, 98, 108, 109, 112, *137, 138, 141*
Agrawal, J. S., 126, *137*
Ahluwalia, K., 125, 126, *142*
Ahuja, M. R., 127, *141*
Alam, Z., 371, *376*
Alberts, H. W., 353, 357, *376*
Alfert, M., 296, 300, *304, 309*
Alfieri, F. J., 263, *306*
Allen, G. S., 24, 35, 65, *69*
Allen, L. R., 324, *376*
Allen, R., 204, *217*
Alvarez, M. R., 98, *137*
Ambegaokar, K. B., 133, *145*
Amen, R. D., 16, *18*
Amici, G. B., 78, *137*
Anchori, H., 226, 228, *306*
Anderson, A. M., 361, *376*
Anderson, J. D., 361, *376*
Anderson, S. R., 349, *376*
Andrew, R. H., 341, *385*
Andrews, H. N., 22, 64, *69*
Andrews, H. N., Jr., 2, *18*
Anstey, T. H., 372, *384*
Archibald, E. E. A., 127, *137*
Arditti, J., 123, *137*
Arekal, G. D., 113, 114, 115, *137*
Arnold, C. A., 22, *69*
Arnott, H. J., 124, *138*
Arora, U., 83, *146*
Arrigoni, O., 289, *304*

Ashri, A., 161, *217*
Asnani, S., 119, 120, *137*
Atal, C. K., 208, *219*
Atkin, J. D., 357, 359, *376, 386*
Atsatt, R. R., 190, *217*
Austin, R. B., 318, 343, 371, 373, *376*
Avebury, J. L., 225, *304*
Avery, G. S., Jr., 229, 231, *304*
Ayyangar, G. S., 86, *137*

B

Bacchi, O., 127, *137*
Bacq, Z. M., 227, *309*
Baer, D. F., 234, *304*
Bailey, C. H., 365, *387*
Bailey, I. W., 10, *18, 278, 279, 304*
Bain, J. M., 164, *217*
Bainer, R., 353, 354, 355, 360, *376*
Baird, A. M., 24, 37, 44, 45, 47, *69*
Bajpai, P. D., 371, *376*
Baker, D., 337, *376*
Baker, D. B., 285, 288, *304, 310*
Baker, K. C., 169, *220*
Baldauf, D., 331, *378*
Baldovinos De La Pena, G., 248, *304*
Ball, R. B., 351, *383*
Banerjee, S. K., 40, *72*
Barkley, W. D., 136, *143*
Barner, H., 35, *69*
Baron, F. J., 251, *304*
Barriga, C., 358, *376*
Bartel, A. T., 351, *376*
Bashaw, E. C., 128, *139*
Baskin, C. C., 323, *378*
Basu, B., 108, 109, *138*
Battaglia, E., 39, *69, 126, 137*
Batygina, T. B., 104, 122, *137*
Baum, H., 101, *137, 160, 217*
Bay, C., 227, *311*
Beeridge, E. M., 41, *69*

389

SUBJECT INDEX

A

Abnormal ovule, 88, 89
Abnormal seedlings, 326–329
  loss of storage tissue, 328
  meristematic damage, 328, 329
  other abnormalities, 329
Abortion, 346
Abscission, 130, 151, 153–161, 163, 205, 217
Abscission layer, 155, 161
Abscission prevention, 161
Abscission tissue, 117, 153, 158, 160, 196–198
Abscission zone, 154–157, 159, 160, 173, 182, 202, 206
Absorption, 14, 96
  of water, 159, 224, 227, 232, 267, 288, 331, 351, 356, *see also* Hydration, Imbibition
Achene, 6, 88, 166, 167
Acid phosphatase, 298
Acid protease, 298
Adhesion, 209
Adsorption, 14
Adventitious roots, 250
Adventive embryo, 124, 126, 127, 129
Aeration, 15, 336, 341
Aerenchyma, 132, 133
After-ripening, 227, 323, 363, 365
Aggregate fruit, 166, 168
Aging, 323, 352, 361, 365, 372, 373, *see also* Senescence
Air space, 133, 191
Albedo, 168
Albino, 236
Alcohol dehydrogenase, 338
Aleurone grain, 229, 260, 268, 295–297, 303
Aleurone layer, 229
Allelopathy, 14–16
Alkaloids, 10, 15
Alternating temperature, 330, 344, *see also* Temperature

Alveolation, 66, 68
Alveolus, 48, 49
Amino acids, 37
Amitosis, 117
Ammonia lyase, 287
Amphitropous ovule, 80
Amyloplast, 40
Anacampylotropous ovule, 80, 81
Anatropous ovule, 80, 84, 88, 130
Anemochory, 151, 161, 182–192, 201, 202, 217
Angiosperm seeds, development of, 77–136
  embryo, 118–124
  endosperm, 106–118, *see also* Nuclear endosperm, Cellular endosperm, Helobial endosperm
  female gametophyte, 89–99
  mature seed, 133, 134
  ovule, 79–89
  pollination and fertilization, 99–106
  polyembryony, 124–129
  seed coat, 129–133
Angriffsfläche, 184
Anlage, 234, 256
Antheridial cell, 38–40
Anther, 32, 85, 99, 102, 341
Anthesis, 86, 131, 155, 164, 167, 340, 341
Anthocyanin, 86, 162, 163
Antibiotics, 362
Antipodal chamber, 98
Antipodals, 79, 90, 91, 95, 97–99, 107, 110, 124, 125, 129, 136
Antitelechory, 151, 161, 173, 195, 200–209, 213, 216
Ants, 175
Apical dome, 274
Apical initials, 26, 257, 268, 269, 296, 299
Apical meristem, 234, 256, 258, 260, 273, 275, 328
Apical pole, 118
Apocarpy, 168
Apomixis, 7
Apospory, 127